Lecture Notes on Numerical Methods in Engineering and Sciences

Aims and Scope of the Series

This series publishes text books on topics of general interest in the field of computational engineering sciences.

The books will focus on subjects in which numerical methods play a fundamental role for solving problems in engineering and applied sciences. Advances in finite element, finite volume, finite differences, discrete and particle methods and their applications are examples of the topics covered by the series.

The main intended audience is the first year graduate student. Some books define the current state of a field to a highly specialised readership; others are accessible to final year undergraduates, but essentially the emphasis is on accessibility and clarity.

The books will be also useful for practising engineers and scientists interested in state of the art information on the theory and application of numerical methods.

More information about this series at http://www.springer.com/series/8548

Michał Kleiber · Piotr Kowalczyk

Introduction to Nonlinear Thermomechanics of Solids

Springer

Michał Kleiber
Institute of Fundamental Technological Research
Polish Academy of Sciences
Warsaw
Poland

Piotr Kowalczyk
Institute of Fundamental Technological Research
Polish Academy of Sciences
Warsaw
Poland

ISSN 1877-7341 ISSN 1877-735X (electronic)
Lecture Notes on Numerical Methods in Engineering and Sciences
ISBN 978-3-319-81517-6 ISBN 978-3-319-33455-4 (eBook)
DOI 10.1007/978-3-319-33455-4

Printed on acid-free paper

This Springer imprint is published by Springer Nature
The registered company is Springer International Publishing AG Switzerland

Contents

Chapter 1
Introduction

1.1 General Remarks on the Book Content

The design of engineering structures and machines consists of several stages. Among them are the design and description of their geometry, the choice of appropriate materials, the design of the technological process of manufacturing and the frequently tedious mechanical computations aimed at ensuring that the designed object is going to be safe, sufficiently resilient and able to meet extensive and problem-specific requirements regarding functionality. The last stage includes in particular assessment of internal forces, displacements, deformations, accelerations, temperatures, etc. appearing in structural elements, both during the production and exploitation stage, in order to show that these quantities fit certain safety limits. The above-mentioned design stages are nowadays widely assisted by advanced computer methods and are frequently mutually interdependent. Thus, a user or designer of CAD/CAM systems, apart from a mathematical and information science background, should be equipped with the fundamentals of engineering mechanics and its computational aspects. Similarly, an expert specialized in mechanical computations cannot do without at least a basic knowledge of computer assisted design, analytical geometry, computer graphics and several related topics.

The key objective of this book is to acquaint readers with selected issues of contemporary thermomechanics in to the extent that would enable more conscious use of advanced CAD/CAM systems in the context of computational applications.

The authors had to take into account the following circumstances when selecting the material and the way of its presentation:

1. Contemporary thermomechanics of solids—even in its first, linear approximation—is an enormously vast field of knowledge. In fact, years of studies are necessary to learn about the most significant nonlinear effects in the mechanical behaviour of deformable bodies.

M. Kleiber and P. Kowalczyk, *Introduction to Nonlinear Thermomechanics of Solids*, Lecture Notes on Numerical Methods in Engineering and Sciences,
DOI 10.1007/978-3-319-33455-4_1

2. The way of the material presentation should not exclude from studying it readers whose background is limited to merely fundamentals of analytical mechanics and thus does not require any experience in dealing with continuum mechanics problems.

In view of the above, several important aspects traditionally included in advanced lectures on nonlinear thermomechanics of solids [1–14] had to be omitted. Among them are a classification of tensor functions with their symmetry groups, an extended presentation of constitutive equations with their physical background or a more thorough discussion of the thermodynamical fundamentals of the issues involved. On the other hand, all aspects of the presented theory that may have an important impact on computational applications are discussed in more detail than it is typically done in traditional textbooks. These aspects include the so-called updated Lagrangian description of nonlinear mechanics problems, frequent use of finite increments of time and its functions instead of derivatives and infinitesimal differentials, as well as a thorough discussion related to variational formulations of the problems. Besides, in the second part of the book, continuum mechanics considerations are supplemented by approximate discrete formulations that constitute the basis for numerical solution methods for differential equation systems typical of solid thermomechanics.

Until some 2–3 decades ago, it was considered natural that the latest achievements in nonlinear mechanics were not included in the standard curriculum of any faculty of engineering. This was surely impairing the students' education but it did not seem to have any practical consequences as the new theories were usually so complex that there were in fact no perspectives of their practical application in engineering design and computations. This has fundamentally changed during the last twenty years. Spectacular progress in computer hardware and development of new methods of computer modelling and simulation have made it possible to efficiently handle the analysis and synthesis of problems described by even most complex theoretical models. The researchers and designers were thus faced with the choice of whether to use in their practice only the traditional, widely accepted methods of analysis, or to employ available advanced models and methods offered by several commercial programs—this time without knowing their theoretical background (which, in practice, precludes any thorough critical analysis of the results obtained). Let us mention in this context advanced models of plastic hardening, models of fracture and damage of materials, inclusion of large deformation and strain effects, large scale optimization and assessment of structural reliability. All these are just examples of problems whose efficient computational handling was until quite recently unimaginable, while their solution methods are now widely available and implemented in several computer systems. This book makes an effort to present mathematical models of at least some of the mentioned problems in a "computer-oriented" way, in order to make students of engineering faculties familiar with computational methods of nonlinear mechanics, even in the case they have no advanced background in solid mechanics, tensor analysis or variational methods.

1.2 Nonlinear Continuum Thermomechanics as a Field of Research and Its Industrial Applications

Contemporary physics indicates that all material objects on the microscopic scale have a discontinuous structure and consist of molecules, elementary particles and atoms. On the other hand, in our everyday life we typically encounter pieces of matter having much greater size and our interest thus usually concentrates not on atomic properties but on a proper understanding of the behaviour of the whole material part. Ideally, we can think of trying to characterize the macroscopic system behaviour by an appropriate integration of its microscopic elements. It is still rare, however, that such an approach proves effective—which is why in the framework of continuum mechanics we typically base our studies on testing and analysis of macroscopic material elements and we neglect their discrete structure. In other words, we assume the material to be continuously distributed in a certain space region, thus constituting a material continuum. This allows us to identify individual particles with their momentary space positions and to define, at each point of such a medium, physical properties such as density, displacement or velocity by means of continuous, or sometimes piece-wise continuous functions. Even if in theory this assumption may not always be easy to convincingly justify, its potential in terms of virtually unlimited successful applications in engineering practice has determined its common acceptance.

Mechanics is a scientific discipline which investigates motion under the action of forces. Thermomechanics additionally takes into account thermal effects. As a result, on top of the typical mechanical variables of displacements, velocities, accelerations and stresses in thermomechanics we also have the variables of temperatures and heat fluxes. Formal description of the thermomechanical behaviour of deformable bodies requires as a rule rather advanced mathematics, tensor algebra and differential-integral equations. Studying problems of contemporary thermomechanics is thus typically based on the analysis of properties of specific sets of partial differential or differential-integral equations as well as the development of effective techniques for solving corresponding initial-boundary value problems. While analytical studies in some simple cases offer a way to better understand the nature of the problem on hand, the vast majority of situations of any practical significance require today the use of advanced numerical methods.

This book consists of three main parts. In the first part, we present the mathematical background needed later to precisely describe the basic problem of continuum thermomechanics. The book then concentrates on developing governing equations for the problem dealing in turn with the kinematics of material continuum, description of the stress state, discussion of the fundamental conservation laws of underlying physics, formulation of the initial-boundary value problems and presentation of the so-called weak or variational formulations. In the final part of the book, we address the crucial issue of developing techniques for solving specific problems of thermomechanics. To this aim, we present the discretized formulation of the governing equations, discuss the fundamentals of the finite element method and develop some basic algorithms for solving algebraic and ordinary differential equations typical of the problems on hand. Clearly, because of the very nature of the issue addressed in this book, students during the course must additionally be given the chance to experiment with any of the existing finite element codes dealing with applications of nonlinear mechanics—only then after completing the course they will be able to make effective use of the knowledge acquired.

The book is based on years of the authors' experience in lecturing on the subject of computational nonlinear thermomechanics for graduate students at the Department of Applied Mathematics and Information Science at the Warsaw University of Technology. The background of students made it appropriate on the one hand to frame the presentation in a thorough mathematical formalism, while on the other hand to introduce basic physical concepts of nonlinear solid thermomechanics using some elementary examples. Clearly, the book can also be employed as a basis for graduate students from any engineering department, perhaps after complementing the presentation with some additional explanatory material which would focus on the fundamentals of tensor analysis and variational methods in particular, while possibly omitting the elementary introductions to solid mechanics and the finite element method. In other words, a part of the book can be used for an introductory solid mechanics course at the undergraduate level, whereas the advanced material included in the main part of the book aims at supporting a graduate course on computational thermomechanics of solids.

References

1. Bhatia A.B., Singh R.N., 1985. *Mechanics of Deformable Media*. A. Hilger.
2. Boley B.A., Weiner J.H., 1960. *Theory of Thermal Stresses*. John Wiley & Sons.
3. Chadwick P., 1976. *Continuum Mechanics*. John Wiley & Sons.
4. Chen W.F., Saleeb A.F., 1982. *Constitutive Equations for Engineering Materials*. John Wiley & Sons.
5. Desai C.S., Siriwardane H.J., 1984. *Constitutive Laws for Engineering Materials*. Prentice-Hall.
6. Eringen A.C., 1967. *Mechanics of Continua*. John Wiley & Sons.
7. Fung Y.C., 1965. *Foundations of Solid Mechanics*. Prentice-Hall.
8. Gurtin M., 1981. *An Introduction to Continuum Mechanics*. Academic Press.
9. Jaunzemis W., 1967. *Continuum Mechanics*. Collier-Macmillan.
10. Khan A.S., Huang S., 1995. *Continuum Theory of Plasticity*. John Wiley & Sons.
11. Malvern L.E., 1969. *Introduction to the Mechanics of a Continuous Medium*. Prentice-Hall.
12. Maugin G.A., 1992. *The Thermomechanics of Plasticity and Fracture*. Cambridge University Press.
13. Spencer A.J.M., 1980. *Continuum Mechanics*. Longman.
14. Ziegler H., 1983. *An Introduction to Thermomechanics, 2nd edition*. North-Holland.

Chapter 2
Fundamental Concepts of Mechanics

Before passing to the core lecture on continuum mechanics in three-dimensional (3D) space, let us first consider a simple one-dimensional model example of a deformable solid. This will be useful to define a number of intuitive notions to be appropriately generalized in subsequent chapters. This example will also serve to highlight problems that appear when one attempts to reach beyond one dimension in an analysis of deformable continuum—how far such an attempt complicates the mathematics needed to describe the phenomena typical of a more general case.

2.1 Statics of a Bar

Consider the system presented in Fig. 2.1. This is a bar of length l and a constant cross-section area A, stretched[1] along its axis by the force P. Let us assume that the bar is elastic, i.e. its length extension Δl is proportional to the force P,

$$P = k\,\Delta l, \tag{2.1}$$

k being the proportionality factor (bar stiffness coefficient).

Let us discuss this system in a one-dimensional (1D) approach, treating each cross section of the bar as its material point characterized by a coordinate x and the geometric property assumed constant along the bar's length, $A(x) = \text{const}$.

Due to the extension, a point with the initial location x moves to another location $\bar{x}$. There is a transformation $\bar{x} = \bar{x}(x)$ that describes the location of all points after the deformation of the bar. One may also define the *displacement* field of the bar points $u(x) = \bar{x} - x$, $x \in [0, l]$, assuming values $u = 0$ at $x = 0$ and $u = \Delta l$ at $x = l$.

[1] The following discussion and conclusions are valid for the case of compressed bar too—in that case $P < 0$.

M. Kleiber and P. Kowalczyk, *Introduction to Nonlinear Thermomechanics of Solids*, Lecture Notes on Numerical Methods in Engineering and Sciences,
DOI 10.1007/978-3-319-33455-4_2

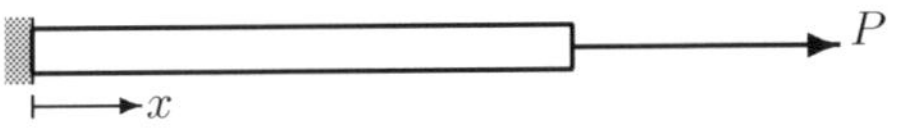

Fig. 2.1 Extended bar

In a continuum approach we postulate: (i) continuity of the field (function) $\bar{x}(x)$ (and, consequently, $u(x)$) and (ii) positive definition of the derivative $\frac{d\bar{x}}{dx}$. The two postulates are implied by the physical observation that neighbouring material particles are never separated and cannot occupy the same location in space.

The form of the function $u(x)$ within the interval $(0, l)$ is unknown and its determination generally requires solving the appropriate equations. It is, however, intuitively clear that in our case it can be assumed linear, $u(x) = x\Delta l/l$. Let us define the notions of *deformation* λ and *strain* ε, as

$$\lambda = \frac{d\bar{x}}{dx}, \qquad \varepsilon = \frac{du}{dx}. \tag{2.2}$$

According to the assumed postulate, $\lambda > 0$. In the considered case the two quantities are constant with respect to x and equal $\lambda(x) = 1 + \Delta l/l$ and $\varepsilon(x) = \Delta l/l$, respectively.

Deformation and strain are fundamental geometrical notions in continuum kinematics. In subsequent chapters, their definitions will be generalized to the 3D case. In an undeformed medium, strain (i.e. relative extension) ε assumes the zero value while deformation λ equals one. In addition, positive values of strain indicate increase in the distance between material points of a medium (extension), while negative—its decrease (shortening).

The next introduced notion is *stress*. This is a resultant force of inter-particle reactions in material, related to the unit cross-section area. To determine stress in the considered bar, let us first mention the next fundamental law of mechanics, the rule of force equilibrium which implies that stresses in a material part must remain in equilibrium with other stresses and with external forces acting on it. In particular, let us imagine that we cut off a fragment of the bar (see Fig. 2.2) and replace internal forces acting in the material with equivalent external forces acting on the cross-section surface. To assure equilibrium, the load force P acting on its right end must be equilibrated by the cross-section forces, i.e. their resultant F must equal P and be directed outwards with respect to the cross-section surface. Assuming the forces are uniformly distributed over the cross-section, one can compute stress as

$$\sigma = \frac{F}{A}. \tag{2.3}$$

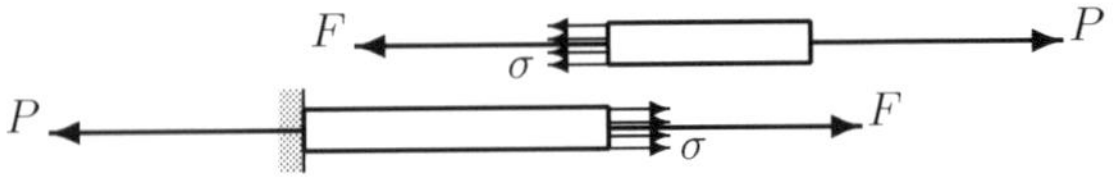

Fig. 2.2 Stress in the cross-section of the extended bar

An identical force F (although acting in the opposite direction) and corresponding stress appear on the adjacent cross-section surface of the remaining part of the bar. The force remains in equilibrium with the reaction force P acting at the support pointof the bar. Let us assume the convention that stress is positive if it is directed outwards of the cross-section surface (tension) and negative in the opposite case (compression). This way, stresses in the two adjacent cross-sections are the same, both in terms of their value and sign. In subsequent chapters, the notion of stress will be generalized to the 3D case.

As can be seen in Fig. 2.2, the stress distribution along the bar's length is constant, $\sigma(x) = P/A$.

Let us finally introduce the notion of *constitutive equation*. This is the relationship between stress and strain, specific to a given material. For elastic materials, this relationship has a linear form known as the *Hooke law*,

$$\sigma = E\varepsilon, \tag{2.4}$$

where E is a material constant called the Young modulus or the longitudinal stiffness modulus. Looking at the formulae (2.4) and (2.1) and at the definitions of stress and strain, one can easily conclude that the stiffness coefficient of the bar in Eq. (2.1) can be expressed as

$$k = \frac{EA}{l}, \tag{2.5}$$

i.e., it depends on both the bar's material properties and its geometry.

Distributions of displacement and stress can be similarly determined in a bar loaded in a more complex way, e.g. with forces distributed along its length (Fig. 2.3). Stress in an arbitrary cross section initially located at x can be computed from the equilibrium equation as an integral of forces acting on the bar on one side of the cross-section (signs should be treated with careful attention), divided by the cross section area A, e.g.

$$\sigma(x) = \frac{1}{A}\int_x^l p(\xi)\,\mathrm{d}\xi. \tag{2.6}$$

If the bar is additionally subjected to concentrated loads P_i acting at material points with initial locations x_i, respectively, these forces may be described as distributions of the Dirac delta function type, $\bar{p}(x) = \sum_i P_i\delta(x - x_i)$, which are added to the distribution $p(x)$ in the integrand in Eq. (2.6). Strains $\varepsilon(x)$ are determined by dividing σ by the stiffness modulus E (see Eq. (2.4)), while displacements determined by the

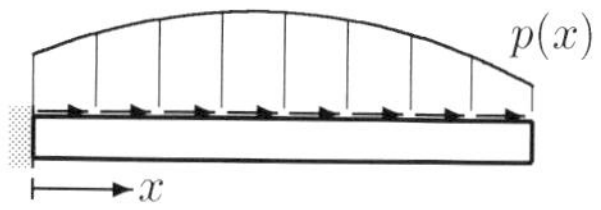

Fig. 2.3 Extended bar, general load case

integration of strains along x (see Eq. $(2.2)_2$) with the boundary conditions (here $u(0) = 0$) taken into account.

Note that Eq. (2.6) implies

$$\frac{\mathrm{d}\sigma}{\mathrm{d}x} = -\frac{p}{A}. \tag{2.7}$$

This is the differential *equilibrium equation* of the bar in the 1D case.

In summary, given the bar's geometry, material properties, loads and the way it is fixed, one can directly determine displacements, strains and stresses in particular material points (cross-sections) of the bar.

2.2 Trusses

In the previous section, probably the simplest of all solid mechanics problems has been described. Its solution scheme may now be a basis for solving a wide class of engineering structures named *trusses*. These are systems of several bars joined together with hinges located at the bars' ends and with loads and boundary conditions imposed only at the joints, called the *truss nodes*.

Consider a simple example of a planar truss consisting of four bars, fixed and loaded as presented in Fig. 2.4. The side length of the square composed by the bars is l. The cross-sections and mechanical properties of the bars are known. Let us try to determine displacements, strains and stresses in particular points of the system. Assuming linear displacement distributions along each bar's length which implies constant strain and stress in each bar, one may conclude that the task consists in determining the forces (stretching or compressing) acting in particular bars and the displacements of particular nodes. Additionally, it is necessary to determine unknown reaction forces in the fixed nodes. It will be shown now that the rules of force equilibrium and displacement continuity turn out to be sufficient to compute all these quantities.

Fig. 2.4 Planar four-bar truss

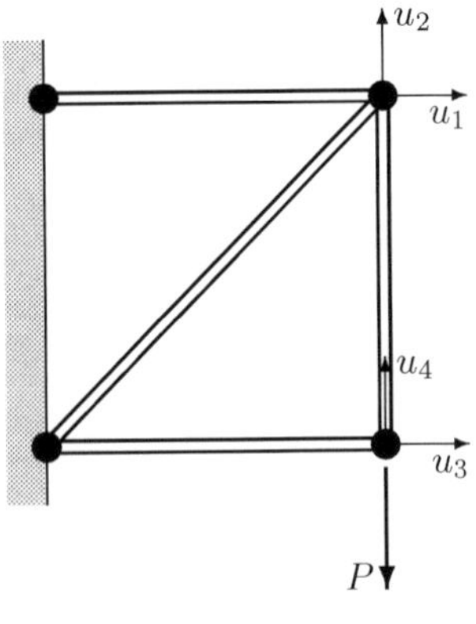

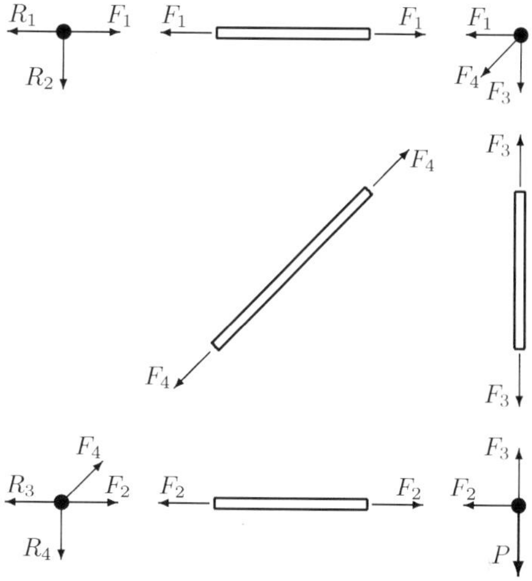

Fig. 2.5 Planar four-bar truss—decomposition into elements

Let us fictitiously decompose the truss into primary elements—bars and nodes—as shown in Fig. 2.5. It can be seen that there are 12 unknown quantities to be found:

$$u_1,\ u_2,\ u_3,\ u_4,\ F_1,\ F_2,\ F_3,\ F_4,\ R_1,\ R_2,\ R_3,\ R_4.$$

Equilibrium equations for each bar have been taken into account in the definition of unknowns, while the equilibrium equation of nodes written separately for the horizontal and vertical directions can be formulated as follows:

$$\begin{array}{lrr} \text{left upper node:} & F_1 - R_1 = 0, & -R_2 = 0, \\ \text{left lower node:} & F_2 + \frac{1}{\sqrt{2}} F_4 - R_3 = 0, & \frac{1}{\sqrt{2}} F_4 - R_4 = 0, \\ \text{right upper node:} & -F_1 - \frac{1}{\sqrt{2}} F_4 = 0, & -F_3 - \frac{1}{\sqrt{2}} F_4 = 0, \\ \text{right lower node:} & -F_2 = 0, & F_3 - P = 0 \end{array} \tag{2.8}$$

(it is assumed that the nodal displacements are very small and thus the shape and geometric proportions of the system remain the same). The above 8 equations are sufficient to determine the unknown forces:

$$\begin{array}{ll} R_1 = P, & F_1 = P, \\ R_2 = 0, & F_2 = 0, \\ R_3 = -P, & F_3 = P, \\ R_4 = -P, & F_4 = -P\sqrt{2} \end{array} \tag{2.9}$$

(negative value of F_4 obviously indicates compression of the particular bar while negative values of R_3 and R_4 mean that the forces act in the directions opposite to those shown in Fig. 2.5).

Nodal displacements can be determined from Eq. (2.1) in which the coefficients k for each bar can be determined from Eq. (2.5) while the length extension of each bar—from the geometry in Fig. 2.4. Four further equations take thus the form:

$$\begin{aligned} F_1 &= k_1 u_1, \\ F_2 &= k_2 u_3, \\ F_3 &= k_3 (u_2 - u_4), \\ F_4 &= k_4 \tfrac{1}{\sqrt{2}} (u_1 + u_2). \end{aligned} \tag{2.10}$$

Their solution (qualitatively illustrated in the sketch below) is:

$$\begin{aligned} u_1 &= \tfrac{P}{k_1}, \\ u_2 &= -P(\tfrac{1}{k_1} + \tfrac{2}{k_4}), \\ u_3 &= 0, \\ u_4 &= -P(\tfrac{1}{k_1} + \tfrac{2}{k_4} + \tfrac{1}{k_3}). \end{aligned} \tag{2.11}$$

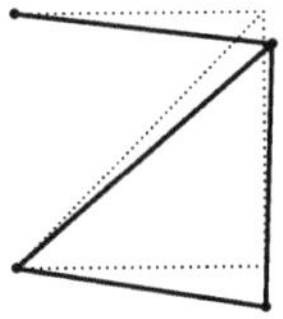

Now, the stresses in particular bars can be computed by dividing the forces F_i by appropriate cross-section areas A_i while strains by subsequently dividing stresses by E_i.

It is interesting to note that the lower horizontal bar does not participate in carrying the load—its force F_2 is null and its stiffness k_2 does not affect the solution. Note, however, that if the loading force P had a non-zero horizontal component, the bar would obviously become a significant load-carrying element of the system.

Remark. The system of equations for the 12 unknown quantities in the above discussed truss was quite easy to solve because equilibrium equations for bars (2.8) were actually independent of their elasticity equations (2.10). We took advantage of this feature by first determining the forces (8 unknowns) and only then displacements (4 unknowns). This is not always possible, however. Imagine a truss in which a fifth bar has been added as shown in Fig. 2.6. This makes another unknown force F_5 appear in the equation set (2.8)—terms containing this variable (the reader is encouraged to determine their form as an exercise) appear on the left-hand sides of equations for the upper left and lower right nodes. Determining the 9 unknown forces from the 8 equations is thus impossible. The problem of the truss statics has the solution, though, because the equations set (2.10) is extended by an additional equation, $F_5 = k_5 \frac{1}{\sqrt{2}}(u_3 - u_4)$. Hence, we have 13 equations with 13 unknowns altogether. It is easy to check in a computational exercise that in this case $F_2 \leq 0$ and $u_3 \leq 0$ and the values of the quantities depend on proportions between stiffness coefficients of particular bars.

Trusses for which the equations sets of equilibrium and elasticity cannot be decoupled are called *statically indeterminate*.

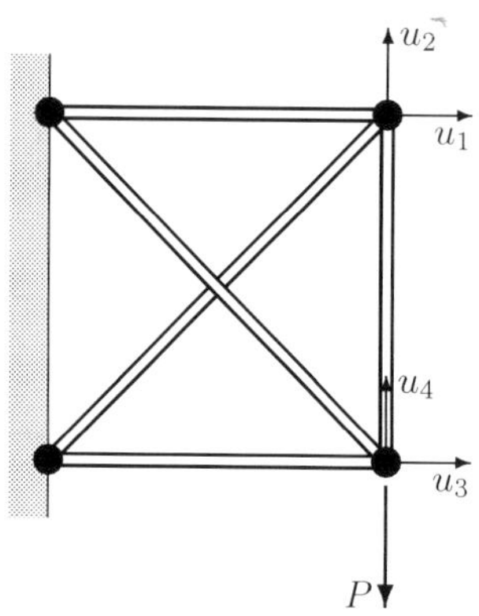

Fig. 2.6 Planar, statically indeterminate truss

As it can be seen in the examples above, the solution of the problem of linear elastic truss statics consists in formulating the system equations of displacement continuity, force equilibrium and elasticity in the truss nodes and bars and solving the system. Extending this approach to 3D trusses is straightforward—in fact, the only change is a significant increase in the number of equations and unknowns. For actual truss structures the size of such a system ranges from about 10^4 to 10^6 equations and its coefficient matrix is usually very sparsely populated by non-zero terms. This makes the solution task quite easy and efficient for well-known numerical solution procedures dedicated to such systems.

2.3 Two-Dimensional Continuum Generalization

Consider a structure of the same shape as that shown in Fig. 2.4 or 2.6, made up not of bars but rather of a plane sheet with a specified thickness h and with a square shape (Fig. 2.7). Let us try to answer the question of the distribution of displacements and internal forces in the material points of the sheet, fulfilling all the equations postulated in the previous section.

It can be easily seen that this task cannot be solved with the methods we have employed for bars and trusses. Firstly, this is because the system under consideration cannot be reasonably divided into simple primary elements in which the unknown quantities could be defined in a straightforward manner, as it was done in trusses. Secondly, the way we have defined certain fundamental notions in Sect. 2.1 is not applicable to the structure considered now. In fact, only the displacements u can be quite easily generalized to this case by defining

$$\boldsymbol{u} = \boldsymbol{x}^{\text{deformed}} - \boldsymbol{x}, \tag{2.12}$$

Fig. 2.7 Plane sheet

where $\boldsymbol{x}$ and $\boldsymbol{x}^{\text{deformed}}$ stand for the initial and deformed location of a given material point, respectively (bold face indicates quantities defined by a set of components related to the assumed Cartesian coordinate system axes). The function $\boldsymbol{u}(\boldsymbol{x})$ is then a vector field in the 2D space. Defining stress, for example is a much more difficult task here, however. Following our experience from the example of a 1D bar, we could obviously imagine cutting the sheet into two parts and consider stresses as internal forces "revealed" by such a cut on the adjacent faces of the cross-section line, locally related to the section area unit. Such internal forces are continuously distributed along the cutting line, see Fig. 2.8a. A definition of stress that naturally comes to mind in this case would be

$$\boldsymbol{\sigma} = \frac{1}{h}\frac{\mathrm{d}\boldsymbol{f}}{\mathrm{d}s} = \frac{\mathrm{d}\boldsymbol{f}}{\mathrm{d}A},$$

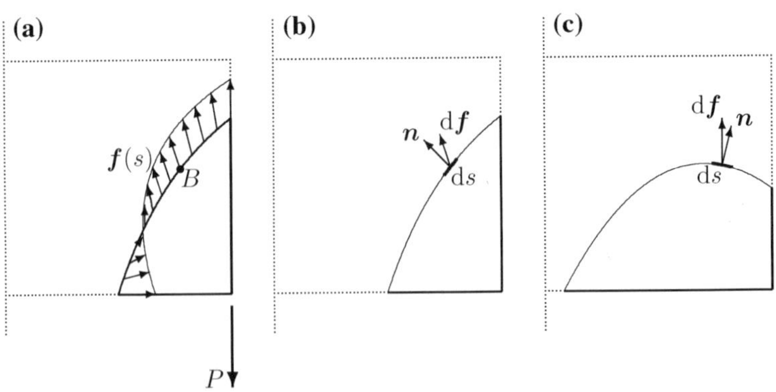

Fig. 2.8 Stress in plane sheet

where s is a coordinate that parametrizes the section line and $\mathrm{d}A$ is an infinitesimal element of the cross-section area. This definition is ambiguous, though, which can be seen in Fig. 2.8b, c: depending on the direction of the section line at a given point B, the so-defined stress vector $\boldsymbol{\sigma}$ has different values. A thorough investigation of this issue leads to the conclusion that stress in a material point of a continuum is described by a second order tensor. The rigorous definition of this tensor will be given in Chap. 5. Analogous discussion allows us to conclude that strain and deformation are also second order tensors whose relations to the displacement field will be discussed in Chap. 4. Equilibrium equations written down in terms of so-defined quantities assume the form of partial differential equations with respect to scalar, vector and tensor fields.

To conclude this introductory chapter, let us finally mention that in the presented simple examples and fundamental definitions we have actually neglected the fact that the considered media (bars, sheets) are deformable in the sense of significantly changing their geometric shape. We have defined the displacements of material points and introduced the notion of strain but implicitly assumed that the deformation is small enough to neglect any substantial changes in geometry of the system. Hence, changes in geometry were not considered in formulating equilibrium equations, definition of stress, etc. However, the variable geometry of the system must be taken into account when dealing with a general case of a deformed continuum. These issues are going to be rigorously discussed in subsequent chapters of the book.

Chapter 3
Fundamentals of Tensor Algebra and Analysis

3.1 Introduction

The introductory discussion presented in the previous chapter leads to the conclusion that the general formulation of solid mechanics equations requires mathematical background in tensor algebra and analysis. The current chapter is an introduction to this area of knowledge. It contains not only fundamental definitions and descriptions of the properties of defined objects. Its another objective is to introduce specific notation used by the authors in subsequent chapters of the book. Thus, studying this chapter is recommended to all readers, including those who already feel familiar with the basic mathematical issues mentioned in its title.

The objects called tensors may be defined in several equivalent ways. Here, the definition referring to a (right-handed) Cartesian coordinate system will be exploited. As it is going to be shown later, a tensor is an object that exists apart from any coordinate system—the system is only necessary to quantitatively express its components in relation to this system.

Let us start our considerations with the definitions of fundamental notions: the Euclidean space, coordinate systems and their transformations, and then move on to simple objects in the space, like scalars and vectors.

3.1.1 Euclidean Space and Coordinate Systems

Let us assume that the model of the physical space we are interested in is the three-dimensional Euclidean space of points E^3. Consider a certain Cartesian coordinate

M. Kleiber and P. Kowalczyk, *Introduction to Nonlinear Thermomechanics of Solids*, Lecture Notes on Numerical Methods in Engineering and Sciences,
DOI 10.1007/978-3-319-33455-4_3

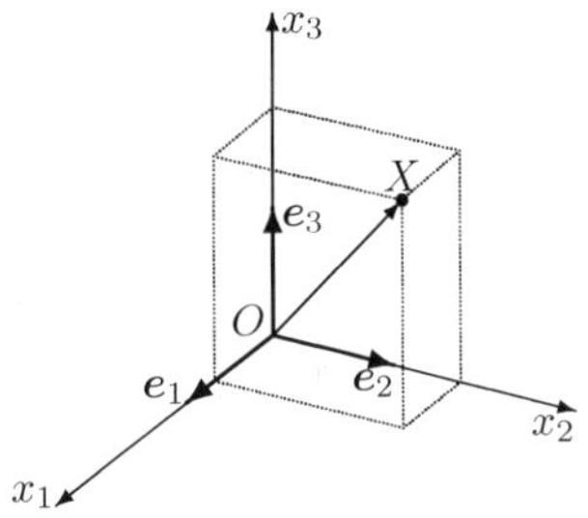

Fig. 3.1 Cartesian coordinate system

system $\{x_1, x_2, x_3\}$ in this space, defined by its origin O and three versors $\boldsymbol{e}_1, \boldsymbol{e}_2, \boldsymbol{e}_3$, i.e. three orthogonal vectors of equal unit length.[1]

An arbitrary point X in the space E^3 is described by a set of three numbers x_1, x_2, x_3—its coordinates in the selected coordinate system (Fig. 3.1), i.e.

$$X = (x_1, x_2, x_3) = (x_i), \qquad i = 1, 2, 3. \tag{3.1}$$

The point's *radius vector*[2] connecting the origin of the coordinate system with the point is defined as

$$\boldsymbol{x} = x_1\boldsymbol{e}_1 + x_2\boldsymbol{e}_2 + x_3\boldsymbol{e}_3 = \sum_{i=1}^{3} x_i\boldsymbol{e}_i \,. \tag{3.2}$$

The coordinates x_i are thus lengths of the radius vector projections onto coordinate system axes, i.e. dimensions of the cuboid shown in Fig. 3.1. Due to an obvious unique relation between the point's coordinates and its radius vector, the use of the same notation for both should not lead to any misunderstandings.

The coordinate system in the Euclidean space may be obviously selected in several ways. It can be shown that the transformation between the coordinates of points in the space under transition from a selected coordinate set (x_1, x_2, x_3) to another arbitrary set (x'_1, x'_2, x'_3) is a combination of two operations: *translation*

$$x'_i = x_i + \Delta x_i \tag{3.3}$$

and *orthogonal transformation*

$$x'_i = \sum_{j=1}^{3} Q_{ij}x_j \tag{3.4}$$

[1]Except for in Sect. 3.5, where tensor representations in curvilinear coordinate systems are discussed, no other coordinate systems than Cartesian will be used throughout this book. Hence, in further derivations, the notion "coordinate system" should be understood as Cartesian coordinate system.

[2]The radius vector is not a vector according to the strict definition given later in this chapter; a commonly accepted "colloquial" name is used here.

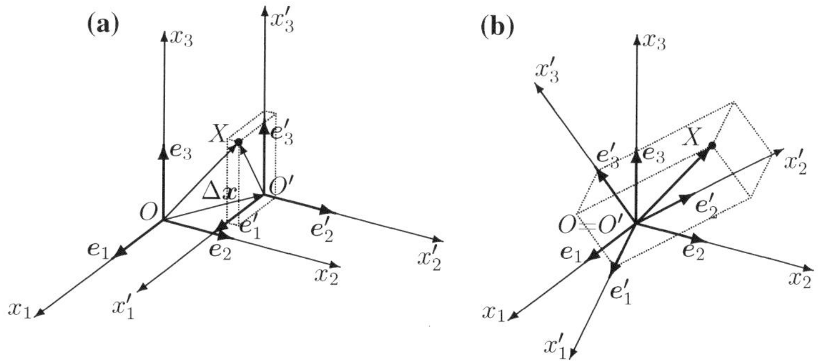

Fig. 3.2 Transformations of Cartesian coordinates: **a** translation, **b** orthogonal transformation

(see Fig. 3.2). The *orthogonal matrix* $[Q_{ij}]$ consists of directional cosines of the new coordinate system axes:

$$Q_{ij} = \cos(x'_i, x_j), \tag{3.5}$$

which implies the following property:

$$\sum_{k=1}^{3} Q_{ik} Q_{jk} = \delta_{ij}, \qquad \delta_{ij} = \begin{cases} 0 & \text{if } i \neq j, \\ 1 & \text{if } i = j, \end{cases} \tag{3.6}$$

where δ_{ij} is a very useful quantity known as the *Kronecker symbol*. Before we justify this property, let us make the following remark about the notation used in this book.

Remark. Equations (3.2), (3.4), (3.6), as well as several other relationships containing sums over indices 1, 2, 3, may be written down much more concisely according to the so-called summation convention. According to this convention, a doubly repeated index in a monomial indicates the summation operation. Consequently, the formulae (3.2), with respect to this index within the range 1, 2, 3. (3.4) and (3.6) may be written as

$$\boldsymbol{x} = x_i \boldsymbol{e}_i \,, \tag{3.7}$$

$$x'_i = Q_{ij} x_j \,, \tag{3.8}$$

$$Q_{ik} Q_{jk} = \delta_{ij} \,. \tag{3.9}$$

The repeated index is called the dummy index—it can be used in the monomial with an arbitrary letter (as long as there is no collision with other indices in the monomial) and $Q_{ij} x_j$ can be replaced by $Q_{ik} x_k$ in Eq. (3.8), for instance.

If there are two or more pairs of dummy indices in one monomial, the summation is done over the pairs of repeated indices independently, generating a sum consisting of 3^n components.

Contrary to dummy indices, indices appearing only once are called free indices—as the indices i, j in the expressions on both sides of Eqs. (3.8) and (3.9). Change of the free index notation letter is only allowed if it is consequently done in all other monomials containing this index, e.g. $Q_{ik}Q_{lk} = \delta_{il}$ in Eq. (3.9). The new notation letter must not collide with other indices in the monomials.

It must be stressed that the summation convention is not valid if an index is repeated more than twice in a monomial.

In further parts of the book the above convention will be consequently used to express operations on quantities with indices running over the range 1, 2, 3.

The property (3.9) results from the formula for the inverse transformation,

$$x_i = Q_{ji}x'_j \tag{3.10}$$

(the property $\cos(x'_i, x_j) \equiv \cos(x_j, x'_i)$ has been used), which, when substituted into Eq. (3.8), yields

$$x'_i = Q_{ik}x_k = Q_{ik}Q_{jk}x'_j \,.$$

The matrix $[Q_{ik}Q_{jk}]$ (with free indices i, j) multiplied by the components x'_j yields the same components x'_i, i.e. it must be the unit matrix $[\delta_{ij}]$. Similarly, substituting the formulae in the opposite order, one arrives at the alternative definition of the orthogonal matrix,

$$Q_{ki}Q_{kj} = \delta_{ij} \,. \tag{3.11}$$

Orthogonality of the matrix $[Q_{ij}]$ implies the property

$$\det[Q_{ij}] = \pm 1. \tag{3.12}$$

It is noteworthy that the determinant's sign indicates mutual orientation of the old and new coordinate systems. If the determinant equals $+1$ then the system x'_i has the same orientation as the system x_i, i.e. it is right-handed as well. The orthogonal transformation is then just a rotation of the coordinate system around a certain axis in space by a certain angle, $Q_{ij} = R_{ij}$. Otherwise, we have to do with a combination of the rotation R_{ij} with the planar symmetric flip S_{ij}, the latter of which changes the system's orientation to left-handed.

Exercise 3.1
The transformation matrix Q_{ij} of a coordinate system has the form

$$[Q_{ij}] = \begin{bmatrix} Q_{11} & Q_{12} & Q_{12} \\ Q_{21} & Q_{22} & Q_{22} \\ Q_{31} & Q_{32} & Q_{32} \end{bmatrix} = \begin{bmatrix} \frac{12}{25} & -\frac{9}{25} & \frac{4}{5} \\ \frac{3}{5} & \frac{4}{5} & 0 \\ -\frac{16}{25} & \frac{12}{25} & \frac{3}{5} \end{bmatrix}.$$

Show that the point $(0, 1, -1)$ has in the transformed coordinate system coordinates $(-\frac{29}{25}, \frac{4}{5}, -\frac{3}{25})$ and that the transformation is a rotation (i.e. it fulfills Eqs. (3.9) and (3.12) with the positive determinant's sign).

3.1.2 *Scalars and Vectors*

Let us now define certain classes of objects in the space E^3. The first of them are *scalars*. A scalar is a quantity described by a real number whose value is independent of any selected coordinate system.

Quantities like mass density, temperature, pressure and distance between points, may be listed as examples of scalars. Examples of quantities that are described by real numbers but are not scalars are coordinates x_i of points in space or components of vectors, as we shall discuss these later.

The next class of objects, featuring a somewhat more complex structure than scalars, are *vectors*. A vector $\boldsymbol{v}$ in the space E^3 with a selected Cartesian coordinate system is a physical quantity described by an ordered set of three real numbers $\{v_1, v_2, v_3\}$ which

- remain unchanged under translation of the coordinate system,
- change values under an orthogonal transformation of the coordinate system defined by the matrix $[Q_{ij}]$ according to

$$v'_i = Q_{ij} v_j \tag{3.13}$$

The above definition immediately yields the previously mentioned conclusion that the "radius vector" does not meet the definition of a vector as its components x_i get changed under translation of the coordinate system.

Displacement of a point, velocity, force and heat flux are examples of vectors. Probably the simplest example of this kind of object is an ordered pair of points in the space E^3. Let us arbitrarily select two different points X and Y with coordinates $\boldsymbol{x} = (x_1, x_2, x_3)$ and $\boldsymbol{y} = (y_1, y_2, y_3)$, respectively. Let us then define

$$\boldsymbol{v} = \boldsymbol{x} - \boldsymbol{y} \tag{3.14}$$

with the components

$$v_i = x_i - y_i, \qquad i = 1, 2, 3. \tag{3.15}$$

Equations (3.3) and (3.8) written down for both x'_i and y'_i imply the fulfillment of both the conditions in the vector definition.

A set of all so-defined vectors will be called the vector space V^3. It is a linear space which may be seen by examining that—in view of the given definition—both the sum of two vectors $v_i + w_i$ and the product of a vector and a scalar av_i are vectors. The neutral element required by the definition of a linear space is the *zero vector* $\boldsymbol{0}$ whose components (in any coordinate system) are zeros. Generalization of the above notions in E^3 and V^3 to cases of larger or lower space dimensions is trivial.

Note that the Euclidean space E^3 is related to the vector space V^3 by a projection

$$E^3 \times E^3 \longrightarrow V^3.$$

In other words, an ordered pair of points $(\boldsymbol{x} \in E^3, \boldsymbol{y} \in E^3)$ defines a vector $\boldsymbol{v} = \boldsymbol{y} - \boldsymbol{x} \in V^3$.

A *scalar product* of two vectors, being a projection $V^3 \times V^3 \longrightarrow R$, is defined as

$$\boldsymbol{v} \cdot \boldsymbol{w} = v_i w_i\,. \tag{3.16}$$

A norm (length) of a vector $\boldsymbol{v}$ is the number

$$|\boldsymbol{v}| = \sqrt{\boldsymbol{v} \cdot \boldsymbol{v}} = \sqrt{v_1^2 + v_2^2 + v_3^2}\,. \tag{3.17}$$

In particular, vectors of the length 1 are called *unit vectors*. The angle φ between vectors $\boldsymbol{v}$ and $\boldsymbol{w}$ can be expressed as

$$\cos\varphi = \frac{\boldsymbol{v} \cdot \boldsymbol{w}}{|\boldsymbol{v}||\boldsymbol{w}|}, \qquad 0 \le \varphi \le \pi. \tag{3.18}$$

The above definitions imply that the scalar product of two orthogonal vectors is zero and that the projection of a vector $\boldsymbol{v}$ onto the direction of the vector $\boldsymbol{w}$ has length

$$l = |\boldsymbol{v}| \cos\varphi = \frac{\boldsymbol{v} \cdot \boldsymbol{w}}{|\boldsymbol{w}|} = \boldsymbol{v} \cdot \boldsymbol{n}, \tag{3.19}$$

where $\boldsymbol{n}$ is the unit length vector (called versor) of the direction $\boldsymbol{w}$.

Exercise 3.2
Let us see whether the scalar product is a scalar, according to the definition given above. In order to do this, let us denote for two arbitrary vectors $\boldsymbol{v}$, $\boldsymbol{w}$ the quantity $a = v_i w_i$ and calculate its value $a' = v'_i w'_i$ after a transformation of the coordinate system. From the vector definition, under translation we have $v'_i = v_i$, $w'_i = w_i$ and thus $a' = a$. Under an orthogonal transformation we have

$$a' = (Q_{ik} v_k)\,(Q_{il} w_l) = (Q_{ik} Q_{il})\, v_k w_l = \delta_{kl}\, v_k w_l = v_k w_k = a,$$

and thus the scalar product value does not change in this case, either. The scalar product is thus a scalar, similarly to the vector length and the angle between two vectors defined above on the basis of the scalar product.

Remark. Recalling the summation convention, let us note that the expression $(Q_{ik} v_k)\,(Q_{il} w_l)$ appearing in the above formula contains 3 pairs of repeating indices and thus it should be interpreted as the sum of $3^3 = 27$ monomials in the form

$$\sum_{i=1}^{3} \sum_{k=1}^{3} \sum_{l=1}^{3} (Q_{ik} v_k)\,(Q_{il} w_l).$$

In addition, the multiplication of an expression by the Kronecker symbol is—in the summation convention—equivalent to the elimination of one of the Kronecker symbol indices from the expression and its replacement by the other one. In other words,

$$a_i\, b_j\, \delta_{ij} = a_i b_i \equiv a_j b_j \qquad \text{and} \qquad a_i\, \delta_{ij} = a_j, \tag{3.20}$$

which also illustrates the rule that any change in the dummy index notation does not change the expression's value.

Exercise 3.3
Show that the area of a parallelogram spanned on two vectors $\boldsymbol{v}$ and $\boldsymbol{w}$ equals

$$\begin{aligned} A &= \sqrt{(\boldsymbol{v} \cdot \boldsymbol{v})(\boldsymbol{w} \cdot \boldsymbol{w}) - (\boldsymbol{v} \cdot \boldsymbol{w})^2} \\ &= \sqrt{(v_1^2 + v_2^2 + v_3^2)(w_1^2 + w_2^2 + w_3^2) - (v_1 w_1 + v_2 w_2 + v_3 w_3)^2}\,. \end{aligned}$$

Solution. The parallelogram's area is $A = lh$ where l is its base length, i.e. length of one of the two vectors, e.g. $l = |\boldsymbol{w}| = \sqrt{\boldsymbol{w} \cdot \boldsymbol{w}}$, and h is its height, i.e.

the length of the projection of the other vector onto the direction orthogonal to the base. From Eq. (3.19) we can first determine the projection of $\boldsymbol{v}$ onto the base direction,

$$t = \frac{\boldsymbol{v} \cdot \boldsymbol{w}}{|\boldsymbol{w}|} .$$

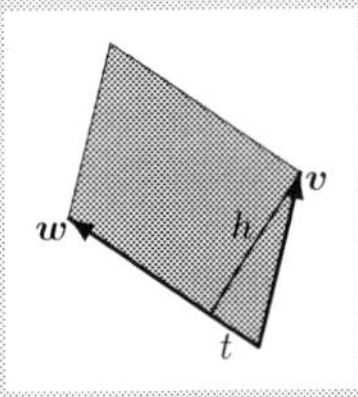

From Eq. (3.17) and the Pythagoras theorem, the other projection's length (i.e. height) can be now computed as

$$h = \sqrt{|\boldsymbol{v}|^2 - t^2} = \sqrt{\boldsymbol{v} \cdot \boldsymbol{v} - \frac{(\boldsymbol{v} \cdot \boldsymbol{w})^2}{\boldsymbol{w} \cdot \boldsymbol{w}}} .$$

Thus, the area equals

$$A = lh = \sqrt{\boldsymbol{w} \cdot \boldsymbol{w}} \sqrt{\boldsymbol{v} \cdot \boldsymbol{v} - \frac{(\boldsymbol{v} \cdot \boldsymbol{w})^2}{\boldsymbol{w} \cdot \boldsymbol{w}}} = \sqrt{(\boldsymbol{v} \cdot \boldsymbol{v})(\boldsymbol{w} \cdot \boldsymbol{w}) - (\boldsymbol{v} \cdot \boldsymbol{w})^2} .$$

The *vector product* of two vectors, denoted as $\boldsymbol{u} = \boldsymbol{v} \times \boldsymbol{w}$, $\boldsymbol{u} = \{u_1, u_2, u_3\}$, is defined as

$$u_k = v_i w_j \epsilon_{ijk} , \tag{3.21}$$

where ϵ_{ijk} stands for the *permutation symbol*, defined as

$$\epsilon_{ijk} = \begin{cases} 1 & ijk = 123 \vee 231 \vee 312, \\ -1 & ijk = 321 \vee 213 \vee 132, \\ 0 & \text{otherwise.} \end{cases} \tag{3.22}$$

Having summed up the components, one can express u_i in a closed form as

$$\{u_1, u_2, u_3\} = \{\, v_2 w_3 - v_3 w_2, v_3 w_1 - v_1 w_3, v_1 w_2 - v_2 w_1 \} . \tag{3.23}$$

The first conclusion we can draw from this definition is that the sign of the vector product's result depends on the order of its factors, i.e. $\boldsymbol{v} \times \boldsymbol{w} = -\boldsymbol{w} \times \boldsymbol{v}$. In particular, the equality $\boldsymbol{v} \times \boldsymbol{v} = \boldsymbol{0}$ holds truei, which directly results from Eq. (3.23) for $v_i = w_i$.

It is also easy to show that $\boldsymbol{v} \cdot (\boldsymbol{v} \times \boldsymbol{w}) = \boldsymbol{0}$ and $(\boldsymbol{v} \times \boldsymbol{w}) \cdot \boldsymbol{w} = \boldsymbol{0}$. Let us, for example, examine the latter relationship: the expression $v_i w_j w_k \epsilon_{ijk}$ is the sum of 27 components of which, according to the definition (3.22), only 6 are non-zero:

$$v_i w_j w_k \epsilon_{ijk} = v_1 w_2 w_3 + v_2 w_3 w_1 + v_3 w_1 w_2 - v_1 w_3 w_2 - v_2 w_1 w_3 - v_3 w_2 w_1 .$$

Note that the three components with the sign "+" are pair-wise equal to appropriate components with the sign "–", hence the sum is zero. This leads to the conclusion that the vector being the result of the vector product of two other vectors is orthogonal to each of them and, consequently, to the plane defined by the two vectors.[3] Regarding the product's length, i.e. the norm $\sqrt{u_1^2 + u_2^2 + u_3^2}$ that can be computed from Eq. (3.23), one can realize after simple transformations that it equals the area A of the parallelogram spanned on the vectors $\boldsymbol{v}$ and $\boldsymbol{w}$—cf. the result of Exercise 3.3. With $\boldsymbol{n}$ denoting the unit vector orthogonal to the parallelogram's plane, one can write

$$\boldsymbol{v} \times \boldsymbol{w} = A\boldsymbol{n} . \tag{3.24}$$

Remark. The property (3.24), which is going to be very useful later in this book, requires a comment. Since each plane has two sides, it has also two normal versors $\boldsymbol{n}$ in two opposite directions. The formula (3.24) is correct for only one of them—the one for which the ordered triplet of vectors $\boldsymbol{v}, \boldsymbol{w}, \boldsymbol{n}$ has the right-handed orientation. Once we change the order of vectors $\boldsymbol{v}, \boldsymbol{w}$ in the product, the area A remains the same but the triplet's orientation changes and so does the sign of the versor $\boldsymbol{n}$ in Eq. (3.24).

The following properties result from the definition of the permutation symbol:

- insensibility to the so-called even permutations of indices,

$$\epsilon_{ijk} = \epsilon_{jki} = \epsilon_{kij} ,$$

- antisymmetry with respect to all pairs of indices,

$$\epsilon_{ijk} = -\epsilon_{jik} = -\epsilon_{ikj} = -\epsilon_{kji} .$$

Among other properties we may list the following:

$$\epsilon_{ijk}\epsilon_{ijk} = 6, \tag{3.25}$$

$$\epsilon_{ijk}\epsilon_{ijl} = 2\delta_{kl} , \tag{3.26}$$

$$\epsilon_{ijk}\epsilon_{ilm} = \delta_{jl}\delta_{km} - \delta_{jm}\delta_{kl} , \tag{3.27}$$

$$\epsilon_{ijk}A_{il}A_{jm}A_{kn} = \epsilon_{lmn}\det[A_{pq}] \quad \text{(for any matrix } [A_{pq}]). \tag{3.28}$$

The latter property will help us to examine whether the vector product result is a vector. The invariance of components u_k at the translation of the coordinate system is obvious. At the orthogonal transformation one can write

$$u'_k = v'_i w'_j \epsilon_{ijk} = (Q_{il}v_l)(Q_{jm}w_m)\,\epsilon_{ijk} = v_l w_m\, Q_{il}Q_{jm}\epsilon_{ijk} .$$

[3] Unless they are co-linear—but in that case the result is zero.

Note that in view of the properties of the Kronecker symbol and the orthogonality condition of the matrix $[Q_{ij}]$ one can write

$$\epsilon_{ijk} = \delta_{kn}\epsilon_{ijn} = Q_{kp}Q_{np}\epsilon_{ijn} .$$

Substituting the above expression into the previous equation and making use of the property (3.28) we obtain

$$\begin{aligned} u'_k = \cdots &= v_l w_m \, Q_{il} Q_{jm} Q_{kp} Q_{np} \epsilon_{ijn} = Q_{kp} v_l w_m \, \epsilon_{lmp} \det[Q_{rs}] \\ &= Q_{kp} u_p \, \det[Q_{rs}]. \end{aligned}$$

What we can learn from this result is that the definition of vector (3.13) is only partially fulfilled here. It is only if the orthogonal transformation is a rotation (i.e. $\det[Q_{rs}] = 1$) that the vector product $\boldsymbol{u}$ appears to be a vector; otherwise, if the orthogonal transformation switches the orientation of coordinates, $\boldsymbol{u}$ no longer satisfies the vector definition. However, since orientation-changing transformations will not be used in any further derivations, we assume now for simplicity that the vector product is a projection $V^3 \times V^3 \longrightarrow V^3$, keeping in mind the above warning.

Remark. The readers have surely noticed that two kinds of notation are alternatively used in this book to denote vectors: the absolute and the index notation. The first one, characterized by bold face symbols, allows to express simple operations on vectors in a concise way, independently of the selected coordinate system. In the index notation, however, one has to operate on vector components in a given coordinate system which makes it possible to write down even the most complex algebraic transformations on vectors. In Sect. 3.6 which closes this chapter, yet another notation will be discussed in the form of the so-called matrix notation, widely used in several fields of mathematics.

The notion of vector has been defined in the Euclidean space of points E^3. We observe that vectors are in no way "tied" to any particular points in the space. They define certain directions in the space, but have no predefined localization. However, in many applications, it is necessary to somehow associate certain types of vectors (e.g. force or momentum) with points in the space. This association will become more understandable upon introduction of the vector field notion in Sect. 3.4. We can then say that a vector is "applied" at a point $\boldsymbol{x} \in E^3$.

In such a case, if a vector $\boldsymbol{v}$ is applied at a point $\boldsymbol{x}$, we can define its *moment* with respect to another point $\boldsymbol{y}$ as a vector $\boldsymbol{m}$ given by the following formula:

$$\boldsymbol{m} = (\boldsymbol{x} - \boldsymbol{y}) \times \boldsymbol{v}. \tag{3.29}$$

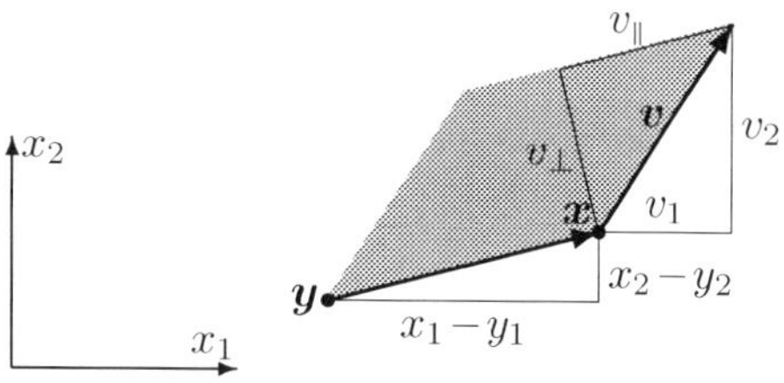

Fig. 3.3 Moment of a vector; the component m_3 equals the area of the marked parallelogram. It is clear from the graph that any change of the component's $v_\parallel$ length has no influence on the value of m_3

Its components given as

$$
\begin{aligned}
m_1 &= (x_2 - y_2)\, v_3 - (x_3 - y_3)\, v_2\,,\\
m_2 &= (x_3 - y_3)\, v_1 - (x_1 - y_1)\, v_3\,,\\
m_3 &= (x_1 - y_1)\, v_2 - (x_2 - y_2)\, v_1
\end{aligned}
\tag{3.30}
$$

(cf. Eq. (3.23)) are called moments of the vector $\boldsymbol{v}$ with respect to three orthogonal straight lines (moment axes) containing the point $\boldsymbol{y}$ and parallel to the coordinate system axes, respectively. The moment's value depends on the lengths and the angle between the vectors $\boldsymbol{v}$ and $\boldsymbol{x} - \boldsymbol{y}$ (the latter called the *moment's arm*). The moment is zero if $\boldsymbol{y} = \boldsymbol{x}$ (zero arm) or if $\boldsymbol{v} = \alpha(\boldsymbol{x} - \boldsymbol{y})$ (vector is parallel to the arm). In fact, the moment's value is only affected by the component of the vector $\boldsymbol{v}$ orthogonal to the moment's arm; it is easy to prove that the other, parallel component is of no importance.

For example, if $\{v_i\} = \{v_1, v_2, 0\}$ and $\{x_i - y_i\} = \{x_1 - y_1, x_2 - y_2, 0\}$ (both the vector and the arm lie in the plane orthogonal to x_3), then the moment $\boldsymbol{m}$ is parallel to x_3 ($m_1 = m_2 = 0$); the value of m_3 is graphically illustrated in Fig. 3.3.

In particular, a moment of a vector $\boldsymbol{v}$ with respect to the origin of the coordinate system equals $\boldsymbol{m} = \boldsymbol{x} \times \boldsymbol{v}$.

Exercise 3.4
Prove the properties (3.25) and (3.26).

Solution. The expression $\epsilon_{ijk}\epsilon_{ijk}$ is the sum of $3^3 = 27$ components of which only 6 are non-zero, cf. Eq. (3.22):

$$\epsilon_{ijk}\epsilon_{ijk} = \epsilon_{123}^2 + \epsilon_{231}^2 + \epsilon_{312}^2 + \epsilon_{213}^2 + \epsilon_{321}^2 + \epsilon_{132}^2 = 3 \cdot 1^2 + 3 \cdot (-1)^2 = 6.$$

The expression $\epsilon_{ijk}\epsilon_{ijl}$ is for each pair of indices k, l the sum of 9 components. If $k \neq l$ then all the components are zero—the indices may only assume 3 values, hence, independently of the values of i, j, there must be at least one pair of repeated indices in at least one of the terms $\epsilon_{ijk}, \epsilon_{ijl}$, which makes the entire monomial vanish by the definition (3.22). Otherwise if, say, $k = l = 1$ then the sum $\epsilon_{ij1}\epsilon_{ij1}$ has 2 non-zero components: $\epsilon_{231}\epsilon_{231} = 1^2 = 1$ and

$\epsilon_{321}\epsilon_{231} = (-1)^2 = 1$, which sum up to 2; similarly for $k = l = 2$ and 3. Recalling the definition of the Kronecker symbol, we can generally express the entire sum as $2\delta_{ij}$.

Exercise 3.5
Show that for any three vectors $\boldsymbol{u}$, $\boldsymbol{v}$, $\boldsymbol{w}$ the following equalities hold true:

$$(\boldsymbol{u} \times \boldsymbol{v}) \cdot \boldsymbol{w} = (\boldsymbol{v} \times \boldsymbol{w}) \cdot \boldsymbol{u} = (\boldsymbol{w} \times \boldsymbol{u}) \cdot \boldsymbol{v}$$

and the geometrical meaning of this quantity is the volume of the parallelepiped spanned on the three vectors.

Solution. Let us write down in the index notation

$$(\boldsymbol{u} \times \boldsymbol{v}) \cdot \boldsymbol{w} = (\boldsymbol{u} \times \boldsymbol{v})_k w_k = \epsilon_{ijk}\, u_i v_j w_k \,.$$

In view of properties of the permutation symbol it is clear that the above expression is invariant with respect to even permutations of the indices ijk. Hence, it can as well be written down in the forms

$$\epsilon_{ijk}\, u_i v_j w_k = \epsilon_{ijk}\, v_i w_j u_k = \epsilon_{ijk}\, w_i u_j v_k \,.$$

which are equivalent to the expressions $(\boldsymbol{v} \times \boldsymbol{w}) \cdot \boldsymbol{u}$ and $(\boldsymbol{w} \times \boldsymbol{u}) \cdot \boldsymbol{v}$.

The interpretation of such a double product can be given as follows. Since the length of the vector $\boldsymbol{u} \times \boldsymbol{v}$ equals the area of the parallelogram spanned on the vectors $\boldsymbol{u}$ and $\boldsymbol{v}$, the volume of the parallelepiped can be computed by multiplying this vector's length by the parallelogram's height h,

$$V = |\boldsymbol{u} \times \boldsymbol{v}|\, h.$$

The height is the projection's length of the vector $\boldsymbol{w}$ onto the direction of the vector $\boldsymbol{u} \times \boldsymbol{v}$ (orthogonal to the plane of $\boldsymbol{u}$ and $\boldsymbol{v}$), i.e., according to Eq. (3.19),

$$h = \frac{\boldsymbol{w} \cdot (\boldsymbol{u} \times \boldsymbol{v})}{|\boldsymbol{u} \times \boldsymbol{v}|} \,.$$

Hence, in view of the commutative property of the scalar product one can write

$$V = (\boldsymbol{u} \times \boldsymbol{v}) \cdot \boldsymbol{w} \,.$$

Note: In the above derivation, we have tacitly assumed that the vectors $\boldsymbol{u}$, $\boldsymbol{v}$, $\boldsymbol{w}$ have right-handed orientation. Reversing the order of any two of the vectors makes the result's sign change—the computed volume appears to be negative. Explain the reason.

Exercise 3.6
Show that for any four vectors $\boldsymbol{t}$, $\boldsymbol{u}$, $\boldsymbol{v}$, $\boldsymbol{w}$ the following equation holds true:

$$(\boldsymbol{t} \times \boldsymbol{u}) \cdot (\boldsymbol{v} \times \boldsymbol{w}) + (\boldsymbol{u} \times \boldsymbol{v}) \cdot (\boldsymbol{t} \times \boldsymbol{w}) + (\boldsymbol{v} \times \boldsymbol{t}) \cdot (\boldsymbol{u} \times \boldsymbol{w}) = 0.$$

Solution. Let us write down in the index notation the first term in the sum, introducing auxiliary vectors $\boldsymbol{a} = \boldsymbol{t} \times \boldsymbol{u}$ and $\boldsymbol{b} = \boldsymbol{v} \times \boldsymbol{w}$ whose components equal $a_k = t_i u_j \epsilon_{ijk}$ and $b_n = v_l w_m \epsilon_{lmn}$, respectively:

$$(\boldsymbol{t} \times \boldsymbol{u}) \cdot (\boldsymbol{v} \times \boldsymbol{w}) = \boldsymbol{a} \cdot \boldsymbol{b} = a_k b_k = t_i u_j \epsilon_{ijk} v_l w_m \epsilon_{lmk} \,.$$

From the properties of the permutation symbol and from Eq. (3.27) we obtain

$$\begin{aligned} t_i u_j \epsilon_{ijk} v_l w_m \epsilon_{lmk} &= t_i u_j v_l w_m \left(\delta_{il}\delta_{jm} - \delta_{im}\delta_{jl}\right) = t_i u_j v_i w_j - t_i u_j v_j w_i \\ &= (\boldsymbol{t}\boldsymbol{v})(\boldsymbol{u}\boldsymbol{w}) - (\boldsymbol{t}\boldsymbol{w})(\boldsymbol{u}\boldsymbol{v}). \end{aligned}$$

Repeating the above derivation for the second and the third term (by simple permutations of the symbols $\boldsymbol{t}$, $\boldsymbol{u}$, $\boldsymbol{v}$), we transform them to the forms

$$(\boldsymbol{u}\boldsymbol{t})(\boldsymbol{v}\boldsymbol{w}) - (\boldsymbol{u}\boldsymbol{w})(\boldsymbol{v}\boldsymbol{t}), \qquad (\boldsymbol{v}\boldsymbol{u})(\boldsymbol{t}\boldsymbol{w}) - (\boldsymbol{v}\boldsymbol{w})(\boldsymbol{t}\boldsymbol{u}).$$

Summing up the three results and grouping the terms we arrive at

$$(\boldsymbol{v}\boldsymbol{u} - \boldsymbol{u}\boldsymbol{v})(\boldsymbol{t}\boldsymbol{w}) + (\boldsymbol{t}\boldsymbol{v} - \boldsymbol{v}\boldsymbol{t})(\boldsymbol{u}\boldsymbol{w}) + (\boldsymbol{u}\boldsymbol{t} - \boldsymbol{t}\boldsymbol{u})(\boldsymbol{v}\boldsymbol{w}) = 0.$$

3.1.3 Basis of Vector Space

Coordinates x_i of a point X are projections of its radius vector onto the coordinate system axes. Making use of Eq. (3.19) one can express x_i in terms of the axes' versors as $x_i = \boldsymbol{x} \cdot \boldsymbol{e}_i$, which in view of Eq. (3.2) leads to the relationship

$$\boldsymbol{x} = (\boldsymbol{x} \cdot \boldsymbol{e}_i)\, \boldsymbol{e}_i \,. \tag{3.31}$$

By analogy, any ordered pair of points $\boldsymbol{v} = \boldsymbol{x} - \boldsymbol{y}$ can be expressed as

$$v_i = \boldsymbol{v} \cdot \boldsymbol{e}_i \,, \qquad \boldsymbol{v} = v_i \boldsymbol{e}_i \,, \tag{3.32}$$

i.e. its components are identified with the vector projections onto the coordinate system axes. This conclusion may be easily generalized to any vector $\boldsymbol{v}$ fulfilling the definition. The set of three unit vectors $(\boldsymbol{e}_1, \boldsymbol{e}_2, \boldsymbol{e}_3)$, whose components are $(\{1, 0, 0\}, \{0, 1, 0\}, \{0, 0, 1\})$, respectively, constitutes the *basis of the vector space* V^3, and each element of the space is a linear combination of the basis vectors.

Orthogonality of the coordinate system axes implies that

$$\boldsymbol{e}_i \cdot \boldsymbol{e}_j = \begin{cases} 0 & \text{if } i \neq j, \\ 1 & \text{if } i = j, \end{cases} \qquad \text{i.e.} \quad \boldsymbol{e}_i \cdot \boldsymbol{e}_j = \delta_{ij} \,. \tag{3.33}$$

Note also that

$$\boldsymbol{v} \cdot \boldsymbol{w} = (v_i \, \boldsymbol{e}_i) \cdot (w_j \, \boldsymbol{e}_j) = v_i w_j \, \boldsymbol{e}_i \cdot \boldsymbol{e}_j = v_i w_j \, \delta_{ij} = v_i w_i \,, \tag{3.34}$$

which conforms to the definition of the scalar product of two vectors (3.16).

Applying the vector product formula to the basis vectors one obtains

$$\begin{aligned} \boldsymbol{e}_1 \times \boldsymbol{e}_1 &= \boldsymbol{0}, & \boldsymbol{e}_1 \times \boldsymbol{e}_2 &= \boldsymbol{e}_3, & \boldsymbol{e}_1 \times \boldsymbol{e}_3 &= -\boldsymbol{e}_2, \\ \boldsymbol{e}_2 \times \boldsymbol{e}_1 &= -\boldsymbol{e}_3, & \boldsymbol{e}_2 \times \boldsymbol{e}_2 &= \boldsymbol{0}, & \boldsymbol{e}_2 \times \boldsymbol{e}_3 &= \boldsymbol{e}_1, \\ \boldsymbol{e}_3 \times \boldsymbol{e}_1 &= \boldsymbol{e}_2, & \boldsymbol{e}_3 \times \boldsymbol{e}_2 &= -\boldsymbol{e}_1, & \boldsymbol{e}_3 \times \boldsymbol{e}_3 &= \boldsymbol{0}, \end{aligned} \tag{3.35}$$

which can be concisely written as

$$\boldsymbol{e}_i \times \boldsymbol{e}_j = \epsilon_{ijk} \, \boldsymbol{e}_k \,. \tag{3.36}$$

The vector product of two vectors can thus be expressed as

$$\boldsymbol{u} = \boldsymbol{v} \times \boldsymbol{w} = (v_i \boldsymbol{e}_i) \times (w_j \boldsymbol{e}_j) = v_i w_j \epsilon_{ijk} \, \boldsymbol{e}_k = u_k \boldsymbol{e}_k \,, \tag{3.37}$$

which fully corresponds to the definition (3.21).

Orthogonal transformation of the coordinate system also changes the system basis. The new axes' versors $(\boldsymbol{e}'_1, \boldsymbol{e}'_2, \boldsymbol{e}'_3)$ become the basis vectors while the radius vector of an arbitrary space point may be written as

$$\boldsymbol{x} = x_i \boldsymbol{e}_i = x'_i \boldsymbol{e}'_i \,. \tag{3.38}$$

Similarly, an arbitrary vector $\boldsymbol{v}$ can be expressed in the new basis as

$$\boldsymbol{v} = v'_i \boldsymbol{e}'_i = (\boldsymbol{v} \cdot \boldsymbol{e}'_i) \, \boldsymbol{e}'_i \,. \tag{3.39}$$

Note that the transformation matrix (3.5) can be defined as

$$Q_{ij} = \cos(\boldsymbol{e}'_i, \boldsymbol{e}_j) = \boldsymbol{e}'_i \cdot \boldsymbol{e}_j \,. \tag{3.40}$$

Also note that the rule of vector decomposition in the basis vectors (3.32) applies in particular to the transformed system's versors,

$$\boldsymbol{e}'_i = (e'_i)_j \boldsymbol{e}_j = (\boldsymbol{e}'_i \cdot \boldsymbol{e}_j)\boldsymbol{e}_j = Q_{ij}\boldsymbol{e}_j \tag{3.41}$$

and

$$\boldsymbol{e}_i = (\boldsymbol{e}_i \cdot \boldsymbol{e}'_j)\boldsymbol{e}'_j = Q_{ji}\boldsymbol{e}'_j \,. \tag{3.42}$$

Let us now substitute the above formula and Eq. (3.13) into Eq. (3.39). The result is

$$\boldsymbol{v} = v'_i \boldsymbol{e}'_i = (Q_{ij} v_j)(Q_{ik} \boldsymbol{e}_k) = \delta_{jk} v_j \boldsymbol{e}_k = v_k \boldsymbol{e}_k \,, \tag{3.43}$$

which proves that decomposition of a vector in different orthogonal bases is consistent with the vector definition (3.13).

Let us conclude the above discussion by stating that Eq. $(3.32)_2$, expressing the decomposition of a vector $\boldsymbol{v}$ in the orthogonal basis of a Cartesian coordinate system versors, is in fact an alternative definition of vector. The relationships between the vector components in different coordinate systems result naturally from transformations of the basis versors between the coordinate systems, e.g. (3.41).

Exercise 3.7
Show that for arbitrary vectors $\boldsymbol{u}$, $\boldsymbol{v}$, $\boldsymbol{w}$ the following relation holds true:

$$\boldsymbol{u} \times (\boldsymbol{v} \times \boldsymbol{w}) = (\boldsymbol{u} \cdot \boldsymbol{w})\,\boldsymbol{v} - (\boldsymbol{u} \cdot \boldsymbol{v})\,\boldsymbol{w}.$$

Solution. Making use of Eqs. (3.32) and (3.36) let us write

$$\begin{aligned}\boldsymbol{u} \times (\boldsymbol{v} \times \boldsymbol{w}) &= u_j \boldsymbol{e}_j \times (v_k \boldsymbol{e}_k \times w_l \boldsymbol{e}_l) = u_j v_k w_l\, \boldsymbol{e}_j \times (\boldsymbol{e}_k \times \boldsymbol{e}_l) \\ &= u_j v_k w_l\, \boldsymbol{e}_j \times (\epsilon_{kli} \boldsymbol{e}_i) = u_j v_k w_l \epsilon_{kli} \epsilon_{jin}\, \boldsymbol{e}_n \,.\end{aligned}$$

In view of the relations $\epsilon_{kli} = \epsilon_{ikl}$, $\epsilon_{jin} = \epsilon_{inj}$ and Eq. (3.27) we have

$$\begin{aligned}u_j v_k w_l \epsilon_{kli} \epsilon_{jin} \boldsymbol{e}_n &= u_j v_k w_l \,(\delta_{kn}\delta_{lj} - \delta_{kj}\delta_{ln})\,\boldsymbol{e}_n = u_l v_k w_l\, \boldsymbol{e}_k - u_k v_k w_l\, \boldsymbol{e}_l \\ &= (\boldsymbol{u} \cdot \boldsymbol{w})\,\boldsymbol{v} - (\boldsymbol{u} \cdot \boldsymbol{v})\,\boldsymbol{w}.\end{aligned}$$

Note that this task could obviously be solved without reference to the versors $\boldsymbol{e}_i$, i.e. by operating only on the vector components (as we have done in previous exercises). Let us stress, however, that the notation employed here is concise and seems much more immune to mistakes in indices' notation.

3.2 Tensors

In the previous section, definitions of scalars and vectors in the three-dimensional Euclidean space have been given. Scalars and vectors are special cases of objects called *tensors*; a scalar is a tensor of the zero order[4] while a vector is a tensor of the first order. This terminology will become clear in this section, once tensors of the second and higher orders are defined.

3.2.1 *Definitions*

Second Order Tensors

A second order tensor in the space E^3 with a Cartesian coordinate system $\{x_1, x_2, x_3\}$ is a physical quantity $\boldsymbol{A}$ described by $3^2 = 9$ real numbers (components) A_{ij} which

- remain unchanged under translation of the coordinate system,
- are subject to the following transformation under an orthogonal transformation of the coordinate system defined by the matrix $[Q_{ij}]$:

$$A'_{ij} = Q_{ik} Q_{jl} A_{kl} , \qquad A_{ij} = Q_{ki} Q_{lj} A'_{kl} , \tag{3.44}$$

where A'_{ij} are the components of the tensor $\boldsymbol{A}$ in the system (x'_1, x'_2, x'_3). It is convenient to present components of a second order tensor in the form of a 3×3 matrix $[A_{ij}]$.

In view of the above definition, both a um of two tensors $\boldsymbol{A} + \boldsymbol{B}$ with components $A_{ij} + B_{ij}$ and a product of a tensor and a scalar $a\boldsymbol{A}$ with components aA_{ij} are second order tensors. Introducing a zero tensor $\boldsymbol{0}$ whose components are all zeros, we may define a space of second order tensors which will be denoted by $\mathcal{T}_2$.

As we shall soon see, quantities such as stress, strain or thermal conductivity may be represented by second order tensors.

An interesting example of a tensor is a *dyad* which is a result of the *tensor product* of two vectors whose components have the form

$$A_{ij} = v_i w_j , \qquad \{v_i\}, \{w_i\} \in V^3 . \tag{3.45}$$

In the absolute notation, this operation reads as

$$\boldsymbol{A} = \boldsymbol{v} \otimes \boldsymbol{w} . \tag{3.46}$$

Fulfillment of the tensor definition by the dyad $\boldsymbol{A}$ is directly implied by the fact that $\boldsymbol{v}, \boldsymbol{w}$ are vectors. Under orthogonal transformation of the coordinate system we have

[4] The order of a tensor is also called its *degree* or *rank*, however, definitions of the terms used in the literature are ambiguous.

$$A'_{ij} = v'_i w'_j = (Q_{ik} v_k)(Q_{jl} w_l) = Q_{ik} Q_{jl} A_{kl} \,. \tag{3.47}$$

The tensor product is thus a projection $V^3 \times V^3 \longrightarrow \mathcal{T}_2$.

Not every second order tensor is a dyad. However, note that application of the tensor product operation to versors of the coordinate system axes $\boldsymbol{e}_i$ leads to the following property:

$$\boldsymbol{A} = A_{ij}\, \boldsymbol{e}_i \otimes \boldsymbol{e}_j \,, \tag{3.48}$$

for an arbitrary second order tensor $\boldsymbol{A}$. Indeed, the component matrix of the dyad $\boldsymbol{e}_i \otimes \boldsymbol{e}_j$ consists of 8 zeros and 1 unit located at the intersection of ith row and jth column. Hence, each second order tensor can be presented as a linear combination of such 9 dyads, which thus constitute the *basis of the tensor space* $\mathcal{T}_2$. For example, the tensor defined in Eq. (3.46) can be expressed in view of Eq. (3.32)$_2$ as $\boldsymbol{A} = (v_i \boldsymbol{e}_i) \otimes (w_j \boldsymbol{e}_j) = v_i w_j\, \boldsymbol{e}_i \otimes \boldsymbol{e}_j$. The space $\mathcal{T}_2$ has therefore a structure of nine-dimensional vector space

$$\mathcal{T}_2 = V^3 \otimes V^3 \,. \tag{3.49}$$

The formula (3.48) can be considered as an alternative definition of a second order tensor, similarly as the formula (3.32)$_2$ in the previous section constituted an alternative definition of a vector.

The set of 9 numbers collected in the matrix $[A_{ij}]$ is called the representation of the tensor $\boldsymbol{A}$ in the basis $\boldsymbol{e}_i \otimes \boldsymbol{e}_j$. Once the basis is selected, the tensor may be identified with its representation $[A_{ij}]$. This fact is frequently a reason for misunderstandings resulting from unjustified identifying tensors with their component matrices in general. Similarly as a vector (a first order tensor) in the space V^3 is not just a triplet of numbers (as the triplet must observe the transformation rule under any change of coordinate system), a tensor is not just a matrix of its components—we still have to keep in mind the transformation (3.44) that must be fulfilled by the components.

Exercise 3.8

In a given coordinate system $\{x_i\}$, a tensor $\boldsymbol{A}$ has the following matrix representation:

$$[A_{ij}] = \begin{bmatrix} 1 & -1 & 0 \\ 0 & 0 & 0 \\ 0 & 2 & 0 \end{bmatrix} .$$

Show that in the system defined by transformation $x'_i = Q_{ij} x_i$, where

$$[Q_{ij}] = \begin{bmatrix} 1/\sqrt{2} & 1/\sqrt{2} & 0 \\ -1/\sqrt{2} & 1/\sqrt{2} & 0 \\ 0 & 0 & 1 \end{bmatrix} ,$$

the tensor is represented as

$$[A'_{ij}] = \begin{bmatrix} 0 & -1 & 0 \\ 0 & 1 & 0 \\ 1/\sqrt{2} & 1/\sqrt{2} & 0 \end{bmatrix} .$$

Higher Order Tensors

The definition of a second order tensor given at the beginning of this section can easily be generalized. The physical quantity $\boldsymbol{A}$, described in a selected Cartesian coordinate system by 3^N real numbers (components) $A_{ij\ldots k}$ is called an Nth order tensor if the components remain unchanged under translation of the coordinate system while under an orthogonal transformation they are subject to the transformation

$$\underbrace{A'_{ij\ldots k}}_{N \text{ indices}} = \underbrace{Q_{ip} Q_{jr} \cdots Q_{ks}}_{N \text{ matrices}} \underbrace{A_{pr\ldots s}}_{N \text{ indices}} . \tag{3.50}$$

For $N = 3$ we can write in particular

$$A'_{ijk} = Q_{ip} Q_{jr} Q_{ks} A_{prs} , \tag{3.51}$$

while for $N = 4$

$$A'_{ijkl} = Q_{ip} Q_{jr} Q_{ks} Q_{lt} A_{prst} . \tag{3.52}$$

In analogy to the second order tensors, one can show that Nth order tensors constitute a space, being the result of a multiple tensor product of vector spaces

$$\mathcal{T}_N = \underbrace{V^3 \otimes V^3 \otimes \cdots \otimes V^3}_{N \text{ times}} , \tag{3.53}$$

whose elements can be expressed as linear combinations of the basis tensors[5]

$$\boldsymbol{A} = \underbrace{A_{ij\ldots k}}_{N \text{ indices}} \boldsymbol{e}_i \otimes \boldsymbol{e}_j \otimes \cdots \otimes \boldsymbol{e}_k . \tag{3.54}$$

The set of 3^N numbers $A_{ij\ldots k}$ is called the representation of the tensor $\boldsymbol{A}$ in the basis $\boldsymbol{e}_i \otimes \boldsymbol{e}_j \otimes \cdots \otimes \boldsymbol{e}_k$.

[5]In our considerations we assume the same basis $\boldsymbol{e}_i$ of Cartesian coordinate axes' versors in all N vector spaces.

Exercise 3.9
Check whether the permutation symbol ϵ_{ijk} (3.22) is a representation of a third order tensor.

Solution. Consider an orthogonal transformation of the coordinate system defined by a matrix $[Q_{ij}]$. In the transformed coordinate system we have

$$\epsilon'_{ijk} = Q_{il} Q_{jm} Q_{kn} \epsilon_{lmn} .$$

The property (3.28) allows us to further transform it as follows:

$$\epsilon'_{ijk} = \epsilon_{ijk} \det[Q^{\mathrm{T}}_{pq}] = \pm\epsilon_{ijk} .$$

Since the permutation symbol is expected to fit its definition in each coordinate system, the above result means that the hypothesis formulated above is not generally true. Only if we limit ourselves to transformations preserving the coordinate system's orientation (for which $\det[Q_{pq}] = \det[Q^{\mathrm{T}}_{pq}] = 1$), the permutation symbol ϵ_{ijk} meets the requirements of the third order tensor definition.

3.2.2 *Operations on Tensors*

Tensor Contraction

If $\boldsymbol{A}$ is a tensor of the order $M \geq 2$ with the components $A_{i\ldots jklm\ldots n}$ in a certain coordinate system, then the object with components $A_{i\ldots jkkm\ldots n}$ (being, according to our summation convention, the sums of 3 components in which $k = 1, 2, 3$, respectively) is called the *contraction* of $\boldsymbol{A}$ on the selected pair of indices kl. This is a tensor of the order $M-2$. Let us write down its components in the case of an orthogonal transformation of the coordinate system. Note that

$$A'_{i\ldots jklm\ldots n} = Q_{ip} \ldots Q_{jr} Q_{ks} Q_{lt} Q_{mu} \ldots Q_{nv} A_{p\ldots rstu\ldots v} . \tag{3.55}$$

Hence, cf. Eq. (3.11),

$$\begin{aligned} A'_{i\ldots jkkm\ldots n} &= Q_{ip} \ldots Q_{jr} \underbrace{Q_{ks} Q_{kt}}_{\delta_{st}} Q_{mu} \ldots Q_{nv} A_{p\ldots rstu\ldots v} \\ &= Q_{ip} \ldots Q_{jr} Q_{mu} \ldots Q_{nv} A_{p\ldots rssu\ldots v} . \end{aligned} \tag{3.56}$$

Using the tensor decomposition in the tensor space basis, tensor $\boldsymbol{A}$ may be represented as

$$\boldsymbol{A} = A_{i\ldots jklm\ldots n}\, \boldsymbol{e}_i \otimes \cdots \otimes \boldsymbol{e}_j \otimes \boldsymbol{e}_k \otimes \boldsymbol{e}_l \otimes \boldsymbol{e}_m \otimes \cdots \otimes \boldsymbol{e}_n\,. \tag{3.57}$$

Its contraction on the indices kl is defined as

$$A_{i\ldots jklm\ldots n}\, \boldsymbol{e}_i \otimes \cdots \otimes \boldsymbol{e}_j \underbrace{(\boldsymbol{e}_k \cdot \boldsymbol{e}_l)}_{\delta_{kl}} \otimes \boldsymbol{e}_m \otimes \cdots \otimes \boldsymbol{e}_n\,, \tag{3.58}$$

i.e.

$$A_{i\ldots jkkm\ldots n}\, \boldsymbol{e}_i \otimes \cdots \otimes \boldsymbol{e}_j \otimes \boldsymbol{e}_m \otimes \cdots \otimes \boldsymbol{e}_n\,. \tag{3.59}$$

Contraction can be performed on any other pair of indices, not necessarily neighbouring, which yields different tensors as results.

In particular, if $\boldsymbol{A}$ is a second order tensor, there is only one pair of indices to choose for the contraction operation, i.e.

$$\operatorname{tr} \boldsymbol{A} = A_{ii} = A_{11} + A_{22} + A_{33}. \tag{3.60}$$

The result is a scalar (zero-th order tensor) fulfilling the transformation rule $A'_{ii} = A_{ii}$, called the *trace* of the tensor $\boldsymbol{A}$. If $\boldsymbol{A}$ is a dyad then its trace $\operatorname{tr}(\boldsymbol{v} \otimes \boldsymbol{w}) = \boldsymbol{v} \cdot \boldsymbol{w}$.

Tensor Transpose

If $\boldsymbol{A}$ is a tensor of the order $M \geq 2$ with the components $A_{i\ldots jklm\ldots n}$ in a certain coordinate system,

$$\boldsymbol{A} = A_{i\ldots jklm\ldots n}\, \boldsymbol{e}_i \otimes \cdots \otimes \boldsymbol{e}_j \otimes \boldsymbol{e}_k \otimes \boldsymbol{e}_l \otimes \boldsymbol{e}_m \otimes \cdots \otimes \boldsymbol{e}_n\,,$$

then a tensor of the same order

$$\boldsymbol{A}^{\mathrm{T}} = A_{i\ldots jlkm\ldots n}\, \boldsymbol{e}_i \otimes \cdots \otimes \boldsymbol{e}_j \otimes \boldsymbol{e}_k \otimes \boldsymbol{e}_l \otimes \boldsymbol{e}_m \otimes \cdots \otimes \boldsymbol{e}_n\,,$$

(i.e. a tensor with the changed ordering of indices kl) is called the *transpose* of $\boldsymbol{A}$ with respect to this pair of indices. In particular, if a second order tensor is represented by the component matrix $[A_{ij}]$, then its transpose $\boldsymbol{A}^{\mathrm{T}}$ is represented by the matrix transpose $[A_{ji}]$.

Note that the operation of transposition is insensitive to transformations of the coordinate system. In other words, under the orthogonal transformation the new components of the transpose $[A'_{ji}]$ are still transpose of the transformed component matrix $[A'_{ij}]$.

If for a given pair of indices of the tensor $\boldsymbol{A}$ the equality $\boldsymbol{A}^{\mathrm{T}} = \boldsymbol{A}$ holds true, then the tensor is called *symmetric* with respect to this pair of indices. In the case of $\boldsymbol{A}^{\mathrm{T}} = -\boldsymbol{A}$ the tensor is called *antisymmetric* with respect to this pair of indices.

Tensor (outer) Product of Two Tensors

If $\boldsymbol{A} \in \mathcal{T}_M$ and $\boldsymbol{B} \in \mathcal{T}_N$, then their *tensor product* $\boldsymbol{C} = \boldsymbol{A} \otimes \boldsymbol{B}$ is a tensor of the order $M+N$, i.e. $\boldsymbol{C} \in \mathcal{T}_{M+N}$. In the index notation

$$\boldsymbol{A} = A_{ij\ldots k}\, \underbrace{\boldsymbol{e}_i \otimes \boldsymbol{e}_j \otimes \cdots \otimes \boldsymbol{e}_k}_{M}\,,$$

$$\boldsymbol{B} = B_{pr\ldots s}\, \underbrace{\boldsymbol{e}_p \otimes \boldsymbol{e}_r \otimes \cdots \otimes \boldsymbol{e}_s}_{N}\,,$$

$$\boldsymbol{C} = \boldsymbol{A} \otimes \boldsymbol{B} = A_{ij\ldots k}\, B_{pr\ldots s}\, \underbrace{\boldsymbol{e}_i \otimes \boldsymbol{e}_j \otimes \cdots \otimes \boldsymbol{e}_k \otimes \boldsymbol{e}_p \otimes \boldsymbol{e}_r \otimes \cdots \otimes \boldsymbol{e}_s}_{M+N}\,.$$

Note that the tensor product is generally not commutative, i.e. $\boldsymbol{A} \otimes \boldsymbol{B} \neq \boldsymbol{B} \otimes \boldsymbol{A}$.

Inner Product of Two Tensors

If $\boldsymbol{A} \in \mathcal{T}_M$ and $\boldsymbol{B} \in \mathcal{T}_N$, then their *inner product* $\boldsymbol{C} = \boldsymbol{A}\boldsymbol{B}$, is a tensor of the order $M+N-2$ defined by the following operations performed on $\boldsymbol{A}$ and $\boldsymbol{B}$:

- compute the tensor product $\boldsymbol{A} \otimes \boldsymbol{B}$,
- contract the result on the last index of $\boldsymbol{A}$ and the first index of $\boldsymbol{B}$.

In more detail, we have

$$\boldsymbol{A} = A_{ij\ldots kl}\, \underbrace{\boldsymbol{e}_i \otimes \boldsymbol{e}_j \otimes \cdots \otimes \boldsymbol{e}_k \otimes \boldsymbol{e}_l}_{M}\,,$$

$$\boldsymbol{B} = B_{pr\ldots st}\, \underbrace{\boldsymbol{e}_p \otimes \boldsymbol{e}_r \otimes \cdots \otimes \boldsymbol{e}_s \otimes \boldsymbol{e}_t}_{N}\,,$$

$$\boldsymbol{C} = \boldsymbol{A}\boldsymbol{B} = A_{ij\ldots kl}\, B_{lr\ldots st}\, \underbrace{\boldsymbol{e}_i \otimes \boldsymbol{e}_j \otimes \cdots \otimes \boldsymbol{e}_k \otimes \boldsymbol{e}_r \otimes \cdots \otimes \boldsymbol{e}_s \otimes \boldsymbol{e}_t}_{M+N-2}\,.$$

Similarly to the outer product, the inner product is not commutative, i.e. $\boldsymbol{A}\boldsymbol{B} \neq \boldsymbol{B}\boldsymbol{A}$ in general.

Full Inner Product of Two Tensors

If $\boldsymbol{A} \in \mathcal{T}_M$, $\boldsymbol{B} \in \mathcal{T}_N$, then their *full inner product* denoted as $\boldsymbol{C} = \boldsymbol{A} \cdot \boldsymbol{B}$, is a tensor $\boldsymbol{C} \in \mathcal{T}_{|M-N|}$ defined by the following operations performed on $\boldsymbol{A}$ and $\boldsymbol{B}$:

- compute the tensor product $\boldsymbol{A} \otimes \boldsymbol{B}$,
- if $N \geq M$, then contract the result M times on all indices of $\boldsymbol{A}$ and the initial M indices of $\boldsymbol{B}$, otherwise, contract the result N times on the last N indices of $\boldsymbol{A}$ and all indices of $\boldsymbol{B}$.

For $M > N$ we have, for instance,

$$
\begin{aligned}
\boldsymbol{A} &= A_{i\ldots jk\ldots l}\ \underbrace{\boldsymbol{e}_i \otimes \cdots \otimes \boldsymbol{e}_j}_{M-N} \otimes \underbrace{\boldsymbol{e}_k \otimes \cdots \otimes \boldsymbol{e}_l}_{N}\,, \\
\boldsymbol{B} &= B_{k\ldots l}\ \underbrace{\boldsymbol{e}_k \otimes \cdots \otimes \boldsymbol{e}_l}_{N}\,, \\
\boldsymbol{C} &= \boldsymbol{A} \cdot \boldsymbol{B} = A_{i\ldots jk\ldots l} B_{k\ldots l}\ \underbrace{\boldsymbol{e}_i \otimes \cdots \otimes \boldsymbol{e}_j}_{M-N}\ .
\end{aligned}
$$

In particular, the full inner product of two tensors of the same order is a scalar. If these are vectors then the result is their scalar product $\boldsymbol{v} \cdot \boldsymbol{w}$ (the use of the same symbol $\cdot$ is thus not confusing) while for two second order tensors the result is $\boldsymbol{A} \cdot \boldsymbol{B} = A_{ij} B_{ij}$. The full inner product is commutative for the same order tensors only.

Let us also note that the full inner product of an arbitrary tensor and vector is equivalent to their inner product, $\boldsymbol{A} \cdot \boldsymbol{v} \equiv \boldsymbol{A}\boldsymbol{v}$. Hence, the scalar product of two vectors $\boldsymbol{v} \cdot \boldsymbol{w}$ can in our notation be alternatively written as $\boldsymbol{v}\boldsymbol{w}$.

Exercise 3.10
Show that if $\boldsymbol{A} \in \mathcal{T}_M$, $\boldsymbol{B} \in \mathcal{T}_N$, then their products $\boldsymbol{A} \otimes \boldsymbol{B}$, $\boldsymbol{A}\boldsymbol{B}$ and $\boldsymbol{A} \cdot \boldsymbol{B}$ are tensors, too.

A particular case of a tensor of an arbitrary order is the *zero tensor* $\boldsymbol{0}$, i.e. a tensor whose components (in any coordinate system) are all zero, $A_{ij\ldots kl} \equiv 0$. All products involving this tensor are zero, too.

3.2.3 Isotropic Tensors

Tensor components are always defined with respect to a given coordinate system. An orthogonal transformation of the system leads in general to a change of the component values, according to the formula (3.50).

An *isotropic tensor* is a special kind of tensor (of any order) whose components remain the same under all orthogonal transformations of the coordinate system. In other words, for an isotropic tensor the following equality holds true for any orthogonal matrix Q_{ij}:

$$A_{ij\ldots k} = A'_{ij\ldots k} = Q_{ip} Q_{jr} \ldots Q_{ks} A_{pr\ldots s}\ . \tag{3.61}$$

Every zero tensor $A_{ij\ldots k} = 0$ is obviously isotropic. Besides, every zero-th order tensor (scalar) is isotropic by definition. On the other hand, there is no non-zero tensor of the first order (vector) that meets the requirements of isotropy.

Consider now tensors of the orders 2, 3 and 4. In the first case, the isotropic property is only featured by tensors of the form $a\boldsymbol{I}$ where a is a scalar and $\boldsymbol{I}$ is the unit tensor represented by the component matrix

$$[\delta_{ij}] = \begin{bmatrix} 1 & 0 & 0 \\ 0 & 1 & 0 \\ 0 & 0 & 1 \end{bmatrix} .$$

Indeed,

$$a\delta'_{ij} = a Q_{ik} Q_{jl} \delta_{kl} = a Q_{ik} Q_{jk} = a\delta_{ij} ,$$

cf. definition (3.6). Such tensors are called the *spherical tensors* and will be discussed in more detail in Sect. 3.3.1.

The permutation symbol which has been defined in Exercise 3.9 (with the reservation regarding coordinate system orientation) is the third order isotropic tensor, for instance.

Among fourth order tensors, the most general form of an isotropic tensor appears to be the following:

$$A_{ijkl} = \alpha\, \delta_{ij} \delta_{kl} + \beta\, \delta_{ik} \delta_{jl} + \gamma\, \delta_{il} \delta_{jk} , \tag{3.62}$$

where α, β, γ are scalars.

3.3 Second Order Tensors

Most of the tensors we deal with in solid mechanics are second order tensors, i.e. tensors represented in a Cartesian coordinate system by 3×3 matrices. In this section, selected properties of this class of tensors, important from the point of view of further considerations in this book, are presented.

3.3.1 *Definitions and Properties*

Operations on Tensors

An inner product of two second order tensors is another second order tensor:

$$\boldsymbol{C} = \boldsymbol{A}\boldsymbol{B} , \qquad C_{ij} = A_{ik} B_{kj} .$$

Inner products of a tensor and a vector, $\boldsymbol{A}\boldsymbol{v}$ and $\boldsymbol{v}\boldsymbol{A}$, are vectors. Inner product operations are not commutative, i.e. their result generally depends on the ordering of the product factors.

Remark. Readers familiar with the matrix calculus have surely noticed that the inner product operation on two second order tensors corresponds to operation of multiplication of their two component matrices—components of the tensor $\boldsymbol{C} = \boldsymbol{A}\boldsymbol{B}$ are expressed by a matrix being a result of the following multiplication:

$$\begin{bmatrix} C_{11} & C_{12} & C_{13} \\ C_{21} & C_{22} & C_{23} \\ C_{31} & C_{32} & C_{33} \end{bmatrix} = \begin{bmatrix} A_{11} & A_{12} & A_{13} \\ A_{21} & A_{22} & A_{23} \\ A_{31} & A_{32} & A_{33} \end{bmatrix} \begin{bmatrix} B_{11} & B_{12} & B_{13} \\ B_{21} & B_{22} & B_{23} \\ B_{31} & B_{32} & B_{33} \end{bmatrix}.$$

Similarly, an inner product of a second order tensor and a vector, $\boldsymbol{w} = \boldsymbol{A}\boldsymbol{v}$ is a vector whose components $w_i = A_{ij} v_j$ are results of the following multiplication of a matrix by a vector:

$$\begin{Bmatrix} w_1 \\ w_2 \\ w_3 \end{Bmatrix} = \begin{bmatrix} A_{11} & A_{12} & A_{13} \\ A_{21} & A_{22} & A_{23} \\ A_{31} & A_{32} & A_{33} \end{bmatrix} \begin{Bmatrix} v_1 \\ v_2 \\ v_3 \end{Bmatrix}.$$

For this reason, the inner product of two tensors of the order ≤ 2 is frequently referred to as their "multiplication" (in addition, we can talk about left- or right-multiplying of a tensor by another tensor, etc.).

Note, however, that the analogy between our notation and the matrix notation is not complete. An example is the inner product $\boldsymbol{w} = \boldsymbol{v}\boldsymbol{A}$ (a left-multiplication of $\boldsymbol{v}$ by $\boldsymbol{A}$) which in that notation can only be expressed in terms of the row (transpose) vectors that do not exist in the notation used here. Section 3.6 lists details of notations employed in this book for representing vectors and tensors.

A full inner product of two tensors is a scalar,

$$\boldsymbol{A} \cdot \boldsymbol{B} = A_{ij} B_{ij} .$$

This operation is commutative.

Unit Tensor

As it has already been mentioned in Sect. 3.2.3, the unit tensor, traditionally denoted as $\boldsymbol{I}$, is a tensor represented by the unit matrix $[\delta_{ij}]$ (the Kronecker symbol). Among its properties we can mention the fact that its inner product with another arbitrary tensor $\boldsymbol{A}$ of any order $n > 0$ (no matter how the product factors are ordered) results in the same tensor: $\boldsymbol{I}\boldsymbol{A} = \boldsymbol{A}\boldsymbol{I} = \boldsymbol{A}$ ($\delta_{ip} A_{pj...kl} = A_{ij...kp}\delta_{pi} = A_{ij...kl}$). In particular, for vectors we have $\boldsymbol{I}\boldsymbol{v} = \boldsymbol{v}\boldsymbol{I} = \boldsymbol{v}$ ($\delta_{ik} v_k = v_i$).

Symmetric and Antisymmetric Tensors

Tensor transposition and properties of symmetry and antisymmetry have been discussed in the previous section for tensors of an arbitrary order. Regarding second order tensors in particular, we recall that the transpose of a second order tensor $\boldsymbol{A} = A_{ij}\boldsymbol{e}_i \otimes \boldsymbol{e}_j$ is a tensor $\boldsymbol{A}^{\mathrm{T}} = A_{ji}\boldsymbol{e}_i \otimes \boldsymbol{e}_j$, i.e. a tensor with the reverse order of indices in its matrix representation. It is easy to show that the transpose of an inner product can be expressed as $(\boldsymbol{AB})^{\mathrm{T}} = \boldsymbol{B}^{\mathrm{T}}\boldsymbol{A}^{\mathrm{T}}$.

A tensor $\boldsymbol{A}$ is called *symmetric* if $\boldsymbol{A} = \boldsymbol{A}^{\mathrm{T}}$, i.e. in each coordinate system $A_{ij} = A_{ji}$. A tensor $\boldsymbol{A}$ is called *antisymmetric* if $\boldsymbol{A} = -\boldsymbol{A}^{\mathrm{T}}$, i.e. $A_{ij} = -A_{ji}$.

Each tensor $\boldsymbol{A}$ can be uniquely expressed as a sum of a symmetric and an antisymmetric tensor,

$$\boldsymbol{A} = \operatorname{sym} \boldsymbol{A} + \operatorname{asym} \boldsymbol{A}, \tag{3.63}$$

where

$$\operatorname{sym} \boldsymbol{A} = \frac{1}{2}\left(\boldsymbol{A} + \boldsymbol{A}^{\mathrm{T}}\right), \qquad \operatorname{asym} \boldsymbol{A} = \frac{1}{2}\left(\boldsymbol{A} - \boldsymbol{A}^{\mathrm{T}}\right) \tag{3.64}$$

are called symmetric and antisymmetric part of the tensor $\boldsymbol{A}$, respectively.

If $\boldsymbol{A}$ is symmetric then its inner products with vectors are commutative, i.e. $\boldsymbol{A}\boldsymbol{v} = \boldsymbol{v}\boldsymbol{A}$. Besides, for two symmetric tensors we have $\boldsymbol{AB} = \boldsymbol{BA}$.

If $\boldsymbol{A}$ is a symmetric tensor and $\boldsymbol{B}$ is an arbitrary tensor then their full inner product has the same result as the full inner product of $\boldsymbol{A}$ and the symmetric part of $\boldsymbol{B}$:

$$\boldsymbol{A} \cdot \boldsymbol{B} = \boldsymbol{A} \cdot \operatorname{sym} \boldsymbol{B}. \tag{3.65}$$

Indeed, since the tensor $\boldsymbol{A}$ equals its symmetric part, we can write

$$\boldsymbol{A} \cdot \boldsymbol{B} = \frac{1}{2}\left(\boldsymbol{A} + \boldsymbol{A}^{\mathrm{T}}\right) \cdot \boldsymbol{B} = \frac{1}{2}\left(\boldsymbol{A} \cdot \boldsymbol{B} + \boldsymbol{A}^{\mathrm{T}} \cdot \boldsymbol{B}\right) = \frac{1}{2}(A_{ij}B_{ij} + A_{ji}B_{ij}).$$

Note, however, that the second term in the parentheses contains two pairs of dummy indices that could as well be denoted by other letters, e.g.

$$\frac{1}{2}(A_{ij}B_{ij} + A_{ji}B_{ij}) = \frac{1}{2}(A_{ij}B_{ij} + A_{ij}B_{ji}) = A_{ij}\,\frac{1}{2}(B_{ij} + B_{ji}) = \boldsymbol{A} \cdot \operatorname{sym} \boldsymbol{B}.$$

This result means in particular that the full inner product of a symmetric and antisymmetric tensor is zero.

Exercise 3.11
Show that the full inner product of a symmetric tensor $\boldsymbol{A} = \boldsymbol{A}^{\mathrm{T}}$ and the permutation symbol yields the zero vector:

$$A_{ij}\epsilon_{ijk} = 0. \tag{3.66}$$

Solution. This results directly from the above discussion about the full inner product of a symmetric and antisymmetric tensor which can be easily generalized to tensors of higher order. The permutation symbol is—as we remember—antisymmetric with respect to each pair of its indices, in particular $\epsilon_{ijk} = -\epsilon_{jik}$. We can verify the result by summing up expressions for particular components of the product (3.66) for $k = 1, 2, 3$ and skipping zero terms of ϵ_{ijk}, cf. (3.22):

$$\begin{aligned} k=1: &\quad A_{ij}\epsilon_{ij1} = A_{23} - A_{32} = 0,\\ k=2: &\quad A_{ij}\epsilon_{ij2} = A_{31} - A_{13} = 0,\\ k=3: &\quad A_{ij}\epsilon_{ij3} = A_{12} - A_{21} = 0. \end{aligned}$$

Spherical and Deviatoric Tensors

A *spherical tensor* is a tensor of the form $a\boldsymbol{I}$ where a is a scalar. A *deviatoric tensor* (or shorter: a *deviator*) is a tensor of zero trace, $\operatorname{tr}\boldsymbol{A} = A_{ii} = 0$. The two types of tensors are introduced together in view of another decomposition of an arbitrary tensor $\boldsymbol{A}$. Each tensor can be expressed as a sum of its spherical and deviatoric part:

$$\boldsymbol{A} = \boldsymbol{A}^{\mathrm{S}} + \boldsymbol{A}^{\mathrm{D}}, \tag{3.67}$$

where

$$\boldsymbol{A}^{\mathrm{S}} = \frac{1}{3}(\operatorname{tr}\boldsymbol{A})\boldsymbol{I}, \qquad \boldsymbol{A}^{\mathrm{D}} = \boldsymbol{A} - \frac{1}{3}(\operatorname{tr}\boldsymbol{A})\boldsymbol{I}. \tag{3.68}$$

Exercise 3.12
Show that the tensor $\boldsymbol{A}^{\mathrm{D}}$ defined as above is a deviator. Decompose the tensor represented by the following component matrix to its spherical and deviatoric part,

$$\begin{bmatrix} 6 & 2 & 7 \\ 5 & 1 & 3 \\ 9 & 4 & 8 \end{bmatrix}.$$

Tensor Determinant

The *determinant* of a tensor $\boldsymbol{A}$ is a number expressed by the formula

$$\det\boldsymbol{A} = \frac{1}{6}\epsilon_{ijk}\epsilon_{lmn}\,A_{il}A_{jm}A_{kn}\,, \tag{3.69}$$

where ϵ_{ijk} is the permutation symbol, cf. Eq. (3.22). The definition is thus identical to the definition of the tensor's component matrix determinant, $\det[A_{ij}]$. The determinant is a scalar. One can check this by expressing it in an orthogonally transformed coordinate system as

$$(\det \boldsymbol{A})' = \frac{1}{6}\epsilon_{ijk}\epsilon_{lmn} A'_{il} A'_{jm} A'_{kn}$$
$$= \frac{1}{6}\epsilon_{ijk}\epsilon_{lmn} (Q_{ip} Q_{lq} A_{pq}) (Q_{jr} Q_{ms} A_{rs}) (Q_{kt} Q_{nu} A_{tu}).$$

Recalling the property (3.28), we can write

$$Q_{ip} Q_{jr} Q_{kt} \epsilon_{ijk} = \epsilon_{prt} \det[Q_{ab}], \qquad Q_{lq} Q_{ms} Q_{nu} \epsilon_{lmn} = \epsilon_{qsu} \det[Q_{ab}]$$

and substitute it into the previous equation. Taking additionally into account the orthogonal matrix property $\det[Q_{ab}] = \pm 1$ and the definition (3.69) we finally arrive at the result

$$(\det \boldsymbol{A})' = \frac{1}{6} \epsilon_{prt} \epsilon_{qsu} (\det[Q_{ab}])^2 A_{pq} A_{rs} A_{tu} = \det \boldsymbol{A}.$$

The unit tensor $\boldsymbol{I}$ has the determinant equal to 1 while for a spherical tensor we have $\det(a\boldsymbol{I}) = a^3$.

A tensor whose determinant is zero is called *singular*. For each singular tensor $\boldsymbol{A}$ one can find such non-zero vectors $\boldsymbol{v}$ that $\boldsymbol{A}\boldsymbol{v} = \boldsymbol{0}$ (otherwise, if $\det \boldsymbol{A} \neq 0$, then the only vector satisfying such equality is a zero vector $\boldsymbol{v} = \boldsymbol{0}$). This property is widely used in the theory of linear equation systems, for instance.

Exercise 3.13
Based on the definition (3.69), show that $\det \boldsymbol{A} = \det \boldsymbol{A}^{\mathrm{T}}$.

An important and frequently useful property of the tensor determinant is the equality

$$\det(\boldsymbol{A}\boldsymbol{B}) = (\det \boldsymbol{A})(\det \boldsymbol{B}) \tag{3.70}$$

that holds true for arbitrary tensors $\boldsymbol{A}$ and $\boldsymbol{B}$. From the definition (3.69) we have

$$\det(\boldsymbol{A}\boldsymbol{B}) = \frac{1}{6}\epsilon_{ijk}\epsilon_{lmn} A_{ip} B_{pl} A_{jr} B_{rm} A_{ks} B_{sn} .$$

Recalling the property (3.28) and the result of Exercise 3.13, we can write

$$\epsilon_{ijk} A_{ip} A_{jr} A_{ks} = \epsilon_{prs} \det \boldsymbol{A}, \qquad \epsilon_{lmn} B_{pl} B_{rm} B_{sn} = \epsilon_{prs} \det \boldsymbol{B},$$

which, after substitution into the previous equation and making use of Eq. (3.25) yields

$$\det(\boldsymbol{AB}) = \frac{1}{6}\epsilon_{prs}\epsilon_{prs}\,(\det \boldsymbol{A})(\det \boldsymbol{B}) = (\det \boldsymbol{A})(\det \boldsymbol{B}).$$

Positive Definite Tensors

A tensor $\boldsymbol{A}$ is called *positive definite* if, for any non-zero vector $\boldsymbol{v} \neq \boldsymbol{0}$, the following inequality holds true:

$$\boldsymbol{vAv} = v_i A_{ij} v_j > 0. \tag{3.71}$$

Inverse Tensors

An *inverse tensor* to a given tensor $\boldsymbol{A}$ is a tensor $\boldsymbol{A}^{-1}$ that fulfills the condition

$$\boldsymbol{A}^{-1}\boldsymbol{A} = \boldsymbol{I}. \tag{3.72}$$

Left-multiplying both sides of Eq. (3.72) by the tensor $\boldsymbol{A}$, i.e. $\boldsymbol{AA}^{-1}\boldsymbol{A} = \boldsymbol{A}$, one can also conclude that

$$\boldsymbol{AA}^{-1} = \boldsymbol{I} \tag{3.73}$$

(no other tensor but $\boldsymbol{I}$, when multiplied by $\boldsymbol{A}$, may yield the same tensor $\boldsymbol{A}$ as the result). Moreover, taking into account Eq. (3.70), one can also deduce that $\det \boldsymbol{A}^{-1} = 1/\det \boldsymbol{A}$ which means that a singular tensor does not have the inverse tensor.

Remark. Note that if the tensor $\boldsymbol{B}$ is an inverse of $\boldsymbol{A}$ then its components B_{ij} in a given coordinate system must meet the relations:

$$B_{ij}A_{jk} = \delta_{ik}\,,$$

cf. Eq. (3.72) written in the index notation. This is a system of 9 equations with 9 unknowns, which has a unique solution only if the determinant of the matrix $[A_{ij}]$ is non-zero. This is another way to prove that a singular tensor cannot have the inverse tensor.

It is easy to show that the inverse of an inner product of two tensors can be expressed as $(\boldsymbol{AB})^{-1} = \boldsymbol{B}^{-1}\boldsymbol{A}^{-1}$.

Orthogonal Tensors

An *orthogonal tensor*, usually denoted by the symbol $\boldsymbol{Q}$, is a tensor featuring the following property:

$$\boldsymbol{Q}^{\mathrm{T}}\boldsymbol{Q} = \boldsymbol{Q}\boldsymbol{Q}^{\mathrm{T}} = \boldsymbol{I} \tag{3.74}$$

(in other words, $\boldsymbol{Q}^{-1} = \boldsymbol{Q}^{\mathrm{T}}$). The orthogonal transformation matrix of a coordinate system, introduced in Eq. (3.5), is a representation of a corresponding orthogonal tensor in a given (non-transformed) coordinate system. Combining Eqs. (3.74) and (3.70) one can immediately derive the property (3.12), given earlier without a proof, i.e. $\det \boldsymbol{Q} = \pm 1$. If $\det \boldsymbol{Q} = 1$ then the orthogonal tensor is called a rotation tensor and is usually denoted as $\boldsymbol{R}$.

Exercise 3.14
Show that an orthogonal rotation matrix $[R_{ij}]$ is the representation of the rotation tensor $\boldsymbol{R}$ in the specified coordinate system.

Solution. Let $[Q_{ij}]$ be an orthogonal transformation matrix of the coordinate system. If $\boldsymbol{R}$ is a tensor then its transformed components should fulfill the condition (3.44)

$$R'_{kl} = Q_{ki} Q_{lj} R_{ij} \,.$$

Let us check if $[R'_{ij}]$ is a rotation matrix, i.e. it is orthogonal and its determinant is positive. The above formula (double product of matrices $[Q_{ij}]$ and $[R_{ij}]$) implies that $\det[R'_{ij}] = (\det[Q_{ij}])^2 \cdot \det[R_{ij}] = \det[R_{ij}]$, i.e. the determinant remains unchanged under such a transformation of coordinates. Let us now substitute the components R'_{ij} into Eq. (3.9):

$$\begin{aligned} R'_{il} R'_{ml} &= (Q_{ij} Q_{lk} R_{jk})(Q_{mp} Q_{lr} R_{pr}) \\ &= Q_{ij} R_{jk} Q_{mp} R_{pr} \underbrace{Q_{lk} Q_{lr}}_{\delta_{kr}} = Q_{ij} Q_{mp} \underbrace{R_{jk} R_{pk}}_{\delta_{jp}} = Q_{ij} Q_{mj} = \delta_{im} \,. \end{aligned}$$

Hence, $\boldsymbol{R}$ is an orthogonal rotation tensor.
Remark: For a general orthogonal tensor $\boldsymbol{Q}$ a similar derivation applies.

Exercise 3.15
Find values of angles α for which the rotation tensor $\boldsymbol{R}$ represented by the matrix

$$[R_{ij}] = \begin{bmatrix} \cos\alpha & -\sin\alpha & 0 \\ \sin\alpha & \cos\alpha & 0 \\ 0 & 0 & 1 \end{bmatrix}$$

is positive definite.

Tensor Exponentiation

The tensor $\boldsymbol{A}^2$, being an inner product of two equal tensors $\boldsymbol{A}$,

$$\boldsymbol{A}^2 = \boldsymbol{A}\boldsymbol{A},$$

is a second order tensor, $\boldsymbol{A}^2 \in \mathcal{T}_2$, called the square of $\boldsymbol{A}$. In the index notation

$$\boldsymbol{A} = A_{ij}\boldsymbol{e}_i \otimes \boldsymbol{e}_j, \qquad \boldsymbol{A}^2 = A_{ik}A_{kj}\boldsymbol{e}_i \otimes \boldsymbol{e}_j .$$

More generally, the tensor $\boldsymbol{A}^n \in \mathcal{T}_2$, being an $(n-1)$-tuple product of a tensor $\boldsymbol{A}$,

$$\boldsymbol{A}^n = \underbrace{\boldsymbol{A}\boldsymbol{A}\ldots\boldsymbol{A}}_{n},$$

is called the nth power of $\boldsymbol{A}$. In the index notation

$$\boldsymbol{A}^n = \underbrace{A_{ik}A_{kl}\ldots A_{mj}}_{n}\,\boldsymbol{e}_i \otimes \boldsymbol{e}_j .$$

Remark. Properties and definitions of tensors presented above in this section may seem difficult in perception due to their very formal appearance. This can be partly explained by the fact that—while vectors can be quite easily imagined as e.g. arrows of certain length and direction in space, whose components are simply their projections on the coordinate system axes—second (and higher) order tensors and their components do not lend themselves to such appealing interpretations. In order to at least partly overcome this difficulty, let us mention one of interpretations of a second order tensor, i.e. an operator of the vector space transformation $V^3 \longrightarrow V^3$:

$$\boldsymbol{w}(\boldsymbol{v}) = \boldsymbol{A}\boldsymbol{v}.$$

Let us examine in more detail what properties of this transformation correspond to selected properties and definitions regarding the tensor $\boldsymbol{A}$.

If the tensor is spherical, $\boldsymbol{A} = a\boldsymbol{I}$, then $\boldsymbol{w} = a\boldsymbol{v}$, i.e. all vectors from the transformation domain are proportionally extended or shortened with the scale coefficient a. If $a < 0$ the transformation also switches its direction. No rotation of the vector direction occurs.

If $\boldsymbol{A}$ is a rotation tensor, then it can be shown that $\boldsymbol{w}$ is the result of rotation of $\boldsymbol{v}$ around a certain line in space by a certain angle. If the tensor is orthogonal with negative determinant, then the transformation is defined as rotation combined with symmetric flip with respect to a certain plane.

If $\boldsymbol{A}$ is positive definite then the transformation features the property $\boldsymbol{v}\boldsymbol{w} > 0$ (unless $\boldsymbol{v}$ is the zero vector—in this case the result is zero, too). This means that the angle between the given vector and its transformation image is always acute ($\cos(\boldsymbol{v}, \boldsymbol{w}) > 0$).

The inner product of two or more tensors, e.g. $\boldsymbol{ABC}$, corresponds to a sequential combination of transformations described by the particular tensors (starting from the rightmost one) $\boldsymbol{C}$, $\boldsymbol{B}$, $\boldsymbol{A}$. Tensor exponentiation corresponds to multiple repetition of the same transformation on a vector $\boldsymbol{v}$ and its subsequent images.

The inverse tensor $\boldsymbol{A}^{-1}$ corresponds to the inverse transformation $\boldsymbol{v} = \boldsymbol{A}^{-1}\boldsymbol{w}$.

Other tensor properties in view of its interpretation as a transformation of a vector space will be discussed later in this section.

3.3.2 *Tensor Eigenproblem*

The product of a tensor $\boldsymbol{A} \in \mathcal{T}_2$ and a vector $\boldsymbol{v}$ is another vector $\boldsymbol{w}$

$$\boldsymbol{w} = \boldsymbol{A}\boldsymbol{v}. \tag{3.75}$$

Generally, the vectors $\boldsymbol{v}$ and $\boldsymbol{w}$ have different lengths and directions.

The *eigenproblem* for a tensor $\boldsymbol{A}$ consists in finding such vectors $\bar{\boldsymbol{v}}$ that preserve their direction upon the transformation (3.75). A vector $\bar{\boldsymbol{v}}$ for which

$$\boldsymbol{A}\bar{\boldsymbol{v}} = \lambda\bar{\boldsymbol{v}} \tag{3.76}$$

(where λ is a scalar) is called the *eigenvector* of the tensor $\boldsymbol{A}$ and defines the *principal direction* of this tensor. The parameter λ is called the *eigenvalue* corresponding to this particular *eigenvector*.

Since any vector $a\bar{\boldsymbol{v}}$ proportional to the eigenvector also fulfills the condition (3.76), we assume the convention that the term "eigenvector" and the notation $\bar{\boldsymbol{v}}$ will be referred only to a normalized (unit) vector of the principal direction.

Equation (3.76) may be presented in the index notation as

$$(A_{ij} - \lambda\delta_{ij})\bar{v}_j = 0, \tag{3.77}$$

and in the matrix notation as

$$\begin{bmatrix} A_{11} - \lambda & A_{12} & A_{13} \\ A_{21} & A_{22} - \lambda & A_{23} \\ A_{31} & A_{32} & A_{33} - \lambda \end{bmatrix} \begin{bmatrix} \bar{v}_1 \\ \bar{v}_2 \\ \bar{v}_3 \end{bmatrix} = \begin{bmatrix} 0 \\ 0 \\ 0 \end{bmatrix}, \tag{3.78}$$

i.e. as a homogeneous system of 3 equations with respect to 3 unknown components of the vector $\bar{\boldsymbol{v}} = \{\bar{v}_i\}$. The system has non-zero solutions for only such values of λ that

$$\det[A_{ij} - \lambda\delta_{ij}] = 0. \tag{3.79}$$

Trying to compute the above determinant one can see that transformations lead to the following equation for λ, called the *characteristic equation* of the tensor $\boldsymbol{A}$,

$$\lambda^3 - I_1\lambda^2 - I_2\lambda - I_3 = 0. \tag{3.80}$$

The coefficients I_1, I_2 and I_3 are called basic invariants of the tensor $\boldsymbol{A}$ and are defined as

$$\begin{aligned} I_1(\boldsymbol{A}) &= \operatorname{tr}\boldsymbol{A} = A_{ii}, \\ I_2(\boldsymbol{A}) &= \frac{1}{2}\left[\operatorname{tr}\boldsymbol{A}^2 - (\operatorname{tr}\boldsymbol{A})^2\right] = \frac{1}{2}(A_{ij}A_{ji} - A_{ii}A_{jj}), \\ I_3(\boldsymbol{A}) &= \det\boldsymbol{A}. \end{aligned} \tag{3.81}$$

Values of the invariants are independent of each other. They are also independent of the selected coordinate system, i.e. they are scalars (invariants of orthogonal transformations of the coordinate system for the tensor $\boldsymbol{A}$).

The characteristic equation (3.80) is a polynomial equation of the third order with respect to λ. Such an equation has at least one real root. If $\boldsymbol{A}$ is a symmetric tensor, one can prove that the equation has three real roots λ_1, λ_2, λ_3.

Properties of symmetric tensors in the context of their eigenproblem solution appear important. The following three cases are possible:

1. If $\lambda_1 \neq \lambda_2 \neq \lambda_3$ (i.e. the tensor $\boldsymbol{A}$ has 3 single roots), then each of the eigenvalues λ_i uniquely corresponds to a principal direction defined by the eigenvector $\boldsymbol{v}_i$ ($|\boldsymbol{v}_i| - 1$) such that

$$\boldsymbol{A}\bar{\boldsymbol{v}}_1 = \lambda_1\bar{\boldsymbol{v}}_1\,, \qquad \boldsymbol{A}\bar{\boldsymbol{v}}_2 = \lambda_2\bar{\boldsymbol{v}}_2\,, \qquad \boldsymbol{A}\bar{\boldsymbol{v}}_3 = \lambda_3\bar{\boldsymbol{v}}_3\,. \tag{3.82}$$

It can be shown that the vectors $\bar{\boldsymbol{v}}_1, \bar{\boldsymbol{v}}_2, \bar{\boldsymbol{v}}_3$ are mutually orthogonal. Let us examine this on the example of the vectors $\bar{\boldsymbol{v}}_1, \bar{\boldsymbol{v}}_2$. Left-multiplying them by the second and the first of Eqs. (3.82), respectively, we obtain

$$\bar{\boldsymbol{v}}_2\boldsymbol{A}\bar{\boldsymbol{v}}_1 = \lambda_1\bar{\boldsymbol{v}}_1\bar{\boldsymbol{v}}_2\,, \qquad \bar{\boldsymbol{v}}_1\boldsymbol{A}\bar{\boldsymbol{v}}_2 = \lambda_2\bar{\boldsymbol{v}}_1\bar{\boldsymbol{v}}_2\,. \tag{3.83}$$

Since the tensor $\boldsymbol{A}$ is symmetric, the left-hand sides of Eqs. (3.83) are identical by the definition. Hence, the right-hand sides are identical, too, which implies $\bar{\boldsymbol{v}}_1\bar{\boldsymbol{v}}_2 = 0$ (because $\lambda_1 \neq \lambda_2$).

Since the eigenvectors are normalized, we can write for $i, j = 1, 2, 3$

$$\bar{\boldsymbol{v}}_i \bar{\boldsymbol{v}}_j = \delta_{ij} \qquad \text{or} \qquad \bar{\boldsymbol{v}}_i \cdot \bar{\boldsymbol{v}}_j = \delta_{ij}\,, \tag{3.84}$$

and call them mutually *orthonormal*.

2. If there is one double root of the characteristic equation (3.80), e.g. $\lambda_1 \neq \lambda_2 = \lambda_3 = \lambda$, then the principal directions are not defined uniquely. While the eigenvalue λ_1 does uniquely correspond to the eigenvector $\bar{\boldsymbol{v}}_1$, the other two eigenvalues correspond to an infinite set of eigenvectors. It can be shown that each vector $\hat{\boldsymbol{v}}$ that is orthogonal to $\bar{\boldsymbol{v}}_1$ is the tensor's eigenvector related to $\lambda_2 = \lambda_3 = \lambda$, i.e.

$$\boldsymbol{A}\bar{\boldsymbol{v}}_1 = \lambda_1 \bar{\boldsymbol{v}}_1\,, \qquad \boldsymbol{A}\hat{\boldsymbol{v}} = \lambda \hat{\boldsymbol{v}}\,, \quad \boldsymbol{v}_1 \cdot \hat{\boldsymbol{v}} = 0. \tag{3.85}$$

3. If the characteristic equation has a triple root, i.e. $\lambda_1 = \lambda_2 = \lambda_3 = \lambda$, then every arbitrary normalized vector $\bar{\boldsymbol{v}}$ fulfills the condition (3.76) in this case. It is easy to show that this can only occur if the tensor $\boldsymbol{A}$ is spherical, i.e. $\boldsymbol{A} = \lambda \boldsymbol{I}$.

Exercise 3.16
Find eigenvalues and eigenvectors of a symmetric tensor $\boldsymbol{A} \in \mathcal{T}_2$ whose representation in the selected coordinate system (x_i) has the form

$$[A_{ij}] = \begin{bmatrix} 3 & -1 & 0 \\ -1 & 3 & 0 \\ 0 & 0 & 1 \end{bmatrix}.$$

Solution. The eigenvalues can be determined from Eq. (3.80) whose coefficient, cf. Eq. (3.81), can be computed as

$$I_1 = 7, \qquad I_2 = (3^2 + 3^2 + 1^2 + 1^2 + 1^2 - 7^2) = -14, \qquad I_3 = 9 - 1 = 8.$$

Thus, we have

$$\lambda^3 - 7\lambda^2 + 14\lambda - 8 = 0,$$

which can be subsequently transformed as follows[6]

$$\begin{aligned} &\lambda^3 - \lambda^2 - 6\lambda^2 + 6\lambda + 8\lambda - 8 = 0, \\ &\lambda^2(\lambda - 1) - 6\lambda(\lambda - 1) + 8(\lambda - 1) = 0, \\ &(\lambda^2 - 6\lambda + 8)(\lambda - 1) = 0, \\ &(\lambda - 4)(\lambda - 2)(\lambda - 1) = 0. \end{aligned}$$

[6] We apply here a simplified method of "guessing the solution"; in a general case, a well-known strict algorithm (Cardan's method) for solving 3rd order polynomial equations should be applied, see e.g. [1].

The roots of this equation are the eigenvalues sought:

$$\lambda_1 = 4, \quad \lambda_2 = 2, \quad \lambda_3 = 1.$$

To find eigenvectors corresponding to these eigenvalues, Eq. (3.78) is used. For $\lambda = \lambda_1 = 4$ we have

$$\begin{bmatrix} -1 & -1 & 0 \\ -1 & -1 & 0 \\ 0 & 0 & -3 \end{bmatrix} \begin{bmatrix} \bar{v}_1 \\ \bar{v}_2 \\ \bar{v}_3 \end{bmatrix} = \begin{bmatrix} 0 \\ 0 \\ 0 \end{bmatrix}.$$

From the first two equivalent equations one can deduce that $\bar{v}_1 = -\bar{v}_2$; the third equation implies $\bar{v}_3 = 0$. To strictly determine the components $\bar{v}_1$, $\bar{v}_2$ we additionally use the normalizing condition,

$$\bar{v}_1^2 + \bar{v}_2^2 + \bar{v}_3^2 = 1,$$

from which we obtain

$$\bar{v}_1 = \pm\frac{1}{\sqrt{2}}, \qquad \bar{v}_2 = \mp\frac{1}{\sqrt{2}}.$$

Of the two possible choices regarding the vector's sign, let us arbitrarily select $\bar{v}_1 = \frac{1}{\sqrt{2}}$, $\bar{v}_2 = -\frac{1}{\sqrt{2}}$, and finally write the components of the considered eigenvector as

$$\bar{\boldsymbol{v}}_1 = \left\{\frac{1}{\sqrt{2}}, -\frac{1}{\sqrt{2}}, 0\right\}.$$

Repeating the same algorithm for λ_2 and λ_3 one can determine the other two eigenvectors as

$$\bar{\boldsymbol{v}}_2 = \left\{\frac{1}{\sqrt{2}}, \frac{1}{\sqrt{2}}, 0\right\}, \qquad \bar{\boldsymbol{v}}_3 = \{0, 0, 1\}.$$

3.3.3 *Spectral Decomposition of Symmetric Tensor*

From the discussion above one can conclude that for each symmetric tensor $\boldsymbol{A}$ there exists a system of three mutually orthonormal eigenvectors $\bar{\boldsymbol{v}}_1$, $\bar{\boldsymbol{v}}_2$, $\bar{\boldsymbol{v}}_3$ corresponding to three eigenvalues λ_1, λ_2, λ_3, respectively, that fulfill Eq. (3.76). In some cases the choice of the orthonormal system is non-unique (repeated eigenvalues) but nevertheless always possible. Since the numbering of the eigenvalues is arbitrary, and so is the selection of the eigenvectors' signs, one can in each case define them so as to ensure the right orientation of the system $\bar{\boldsymbol{v}}_1$, $\bar{\boldsymbol{v}}_2$, $\bar{\boldsymbol{v}}_3$.

Let us note that the system may be considered a basis of a certain Cartesian coordinate system $\{x'_i\}$. Denoting by Q_{ij} the jth component of $\bar{\boldsymbol{v}}_i$ (i.e. the directional cosine that relates the coordinates x'_i and x_j) we can immediately conclude that the orthogonal matrix

$$[Q_{ij}] = \begin{bmatrix} \{\bar{\boldsymbol{v}}_1\}_1 & \{\bar{\boldsymbol{v}}_1\}_2 & \{\bar{\boldsymbol{v}}_1\}_3 \\ \{\bar{\boldsymbol{v}}_2\}_1 & \{\bar{\boldsymbol{v}}_2\}_2 & \{\bar{\boldsymbol{v}}_2\}_3 \\ \{\bar{\boldsymbol{v}}_3\}_1 & \{\bar{\boldsymbol{v}}_3\}_2 & \{\bar{\boldsymbol{v}}_3\}_3 \end{bmatrix} \tag{3.86}$$

is the transformation matrix of an arbitrarily selected coordinate system (x_i) to the specific system related to the principal directions of the tensor $\boldsymbol{A}$. Let us compute the tensor's components in this coordinate system. From Eq. (3.44) we have

$$A'_{ij} = Q_{ik} Q_{jl} A_{kl} ,$$

which, after substitution of Eqs. (3.76) and (3.86) leads to the following result:

$$[A'_{ij}] = [\text{diag}\,\{\lambda_1, \lambda_2, \lambda_3\}] = \begin{bmatrix} \lambda_1 & 0 & 0 \\ 0 & \lambda_2 & 0 \\ 0 & 0 & \lambda_3 \end{bmatrix} . \tag{3.87}$$

The coordinate system (x'_i) is thus a very special system for the tensor $\boldsymbol{A}$—in this system $\boldsymbol{A}$ has a diagonal representation.

It is interesting to compute the basic invariants of $\boldsymbol{A}$ in this coordinate system and express them as functions of the eigenvalues. Substituting the components (3.87) into Eq. (3.81) we arrive at

$$\begin{aligned} I_1 &= \text{tr}\,\boldsymbol{A} = \lambda_1 + \lambda_2 + \lambda_3 , \\ I_2 &= \frac{1}{2}[\text{tr}\,\boldsymbol{A}^2 - (\text{tr}\,\boldsymbol{A})^2] = -(\lambda_1\lambda_2 + \lambda_2\lambda_3 + \lambda_3\lambda_1) \\ I_3 &= \det \boldsymbol{A} = \lambda_1\lambda_2\lambda_3 . \end{aligned} \tag{3.88}$$

Exercise 3.17

Recalling the result of Exercise 3.16, we can write down the matrix of orthogonal transformation of the coordinate system to the system associated with principal directions as

$$Q_{ij} = \begin{bmatrix} \frac{1}{\sqrt{2}} & -\frac{1}{\sqrt{2}} & 0 \\ \frac{1}{\sqrt{2}} & \frac{1}{\sqrt{2}} & 0 \\ 0 & 0 & 1 \end{bmatrix} .$$

The representation of the considered tensor $\boldsymbol{A}$ in the so-defined coordinate system can be computed from Eq. (3.44) as

$$A'_{ij} = Q_{ik}Q_{jl}A_{kl} = 3Q_{i1}Q_{j1} - Q_{i1}Q_{j2} - Q_{i2}Q_{j1} + 3Q_{i2}Q_{j2} + Q_{i3}Q_{j3}\,.$$

Summing up the above terms for different values if $i, j = 1, 2, 3$ we arrive at, cf. Eq. (3.87),

$$[A'_{ij}] = \begin{bmatrix} 4 & 0 & 0 \\ 0 & 2 & 0 \\ 0 & 0 & 1 \end{bmatrix} = [\mathrm{diag}\,\{4, 2, 1\}].$$

Recalling our earlier considerations regarding decomposition of second order tensors in the basis associated with a selected coordinate system, cf. Eq. (3.48), let us express an arbitrary symmetric tensor $\boldsymbol{A}$ in such a form in the basis associated with its principal directions. The basis consists of tensor products of the eigenvectors $\bar{\boldsymbol{v}}_i \otimes \bar{\boldsymbol{v}}_j$ $(i, j = 1, 2, 3)$ and the coefficients in the decomposition formula are the tensor components in the coordinate system (x'_i), cf. Eq. (3.87). The tensor can thus be expressed in the form[7]

$$\boldsymbol{A} = \sum_{i=1}^{3} \lambda_i\, \bar{\boldsymbol{v}}_i \otimes \bar{\boldsymbol{v}}_i\,. \tag{3.89}$$

The above form is called the *spectral decomposition* of the tensor $\boldsymbol{A}$. This is an important feature of all symmetric second order tensors. Let us remind here that the spectral decomposition is non-unique if the tensor's eigenvalues are repeated.

Remark. Since the coordinate system associated with the principal directions is so special for each symmetric tensor, we may conclude that such tensors are in some way "oriented" in space. Similarly as each vector has its characteristic direction in space, each symmetric tensor has three orthogonal characteristic directions.

In the remark made at the end of Sect. 3.3.1 (p. 44), second order tensors have been interpreted as operators of linear vector transformations, $\boldsymbol{w} = \boldsymbol{A}\boldsymbol{v}$. Let us now see what special features of such a transformation may result from the symmetry of the operator $\boldsymbol{A}$ and what interpretation its eigenvalues and principal directions may have in this context.

[7] Note that in this particular formula the summation convention, successfully employed throughout this book, fails—in Eq. (3.89) the index i is repeated 3 times. Hence, a traditional notation with the summation operator must be used.

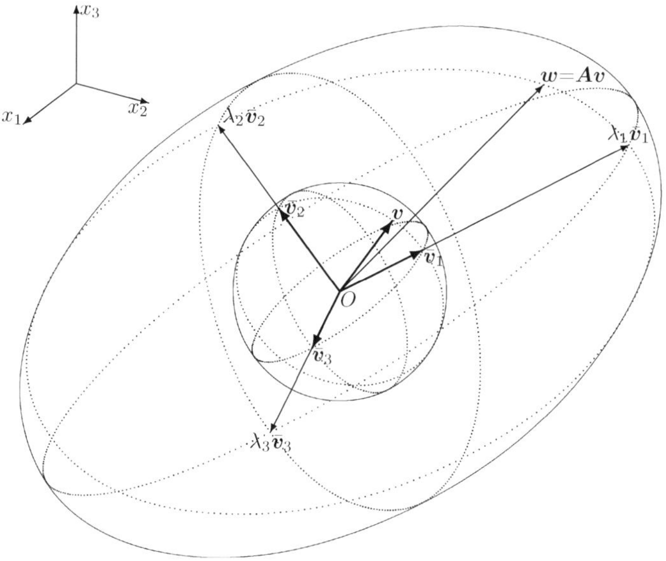

Fig. 3.4 Symmetric tensor as a vector transformation

Let V_n be a class of vectors featuring arbitrary directions and unit length. This class can be geometrically represented as a sphere of the unit radius and a centre located at a selected space point O—each point on the sphere corresponds to a vector $\boldsymbol{v} \in V_n$ applied at O (Fig. 3.4).

The image of such a class of vectors in the transformation $\boldsymbol{w} = \boldsymbol{A}\boldsymbol{v}$ is a class W_n. It appears that, if all vectors of this class are also applied at the point O, the locus of the class vectors' end points is an ellipsoid with semi-axes of lengths $\lambda_1, \lambda_2, \lambda_3$, parallel to the corresponding principal directions, respectively (Fig. 3.4). The vectors of V_n parallel to the directions (i.e. the tensor's eigenvectors) are simply elongated by the factors λ_i while all other vectors generally change their direction. In other words, the transformation "expands" (or "compresses") the unit sphere of vectors in each of the three principal directions, forming an ellipsoid.

The case of double eigenvalue corresponds to a transformation in which the image is an ellipsoid of revolution (it has two equal semi-axes and thus one of its main cross-sections is a circle) while a triple eigenvalue makes the ellipsoid-image a sphere.

If one of the eigenvalues is negative then the ellipsoid-image is symmetrically reflected about the plane orthogonal to the corresponding eigenvector. A case of zero eigenvalue (singular tensor) corresponds to a degenerated ellipsoid, i.e. "flattened" to a zero length of a corresponding semi-axis.

The fact that each symmetric tensor can be expressed in a diagonal form has interesting consequences. For example, it allows to perform operations on such tensors that are typically associated with real numbers only. To illustrate this, let us try to find a square root of a symmetric tensor $\boldsymbol{A}$, i.e. a tensor $\boldsymbol{B} = \sqrt{\boldsymbol{A}}$ such that $\boldsymbol{B}\boldsymbol{B} = \boldsymbol{A}$. To do this, we have to solve the tensor's eigenproblem, i.e. find its eigenvalues and eigenvectors, define the orthogonal transformation matrix $[Q_{ij}]$, cf. Eq. (3.86), and express the tensor in the diagonal form in the coordinate system associated with the eigenvectors,

$$[A'_{ij}] = [\text{diag}\,\{\lambda_1, \lambda_2, \lambda_3\}].$$

If, in this coordinate system,

$$[B'_{ij}] = \left[\text{diag}\,\left\{\sqrt{\lambda_1}, \sqrt{\lambda_2}, \sqrt{\lambda_3}\right\}\right]$$

then the square of it yields the diagonal representation of $\boldsymbol{A}$:

$$[B'_{ik}B'_{kj}] = [\text{diag}\,\{\lambda_1, \lambda_2, \lambda_3\}] = [A'_{ij}].$$

Components of the tensor $\boldsymbol{B}$ in the original, arbitrary coordinate system can be computed from the corresponding transformation formula in which the transpose of the matrix $[Q_{ij}]$ has to be used. Clearly, the necessary condition for existence of the tensor's square root $\sqrt{\boldsymbol{A}}$ is non-negative value of each of the tensor's eigenvalues, $\lambda_i \geq 0$.

Other operations on a symmetric tensors, such as $\ln \boldsymbol{A}, \exp \boldsymbol{A}$ or even trigonometric functions, can be defined in a similar way. One has only to keep in mind that in each case the tensor's eigenvalues must belong to the domain on which the given function is defined. The results are symmetric tensors, too, whose principal directions coincide with those of the operand tensor $\boldsymbol{A}$. In particular, the tensor $\boldsymbol{A}^{-1}$ inverse to a symmetric tensor $\boldsymbol{A}$ may be expressed in the coordinate system associated with its principal directions as

$$[A'^{-1}_{ij}] = [\text{diag}\,\{1/\lambda_1, 1/\lambda_2, 1/\lambda_3\}],$$

which implies that it only exists if all eigenvalues are non-zero.

Remark. Operations like $\ln \boldsymbol{A}, \exp \boldsymbol{A}, \sin \boldsymbol{A}$, etc. may also be defined as sums of infinite series, analogous to those well known form the mathematical analysis for functions of real arguments, e.g.

$$\exp \boldsymbol{A} = \boldsymbol{I} + \boldsymbol{A} + \frac{\boldsymbol{A}^2}{2} + \cdots + \frac{\boldsymbol{A}^n}{n!} + \cdots ,$$

$$\sin \boldsymbol{A} = \boldsymbol{A} - \frac{1}{3!}\boldsymbol{A}^3 + \cdots + (-1)^n \frac{\boldsymbol{A}^{2n+1}}{(2n+1)!} + \cdots .$$

These definitions do not require symmetry of $\boldsymbol{A}$; for symmetric tensors they are equivalent to the computational schemes presented above.

Exercise 3.18
Show that if a tensor $\boldsymbol{A}$ is symmetric then the tensor $\boldsymbol{A}^n$ has the same principal directions as $\boldsymbol{A}$. Find its eigenvalues as functions of the eigenvalues of $\boldsymbol{A}$.

Exercise 3.19
Show that a symmetric tensor $\boldsymbol{A}$ is positive definite if and only if all its eigenvalues are positive.

Solution. Compute the product $\boldsymbol{v}\boldsymbol{A}\boldsymbol{v}$ in the coordinate system associated with the principal directions of $\boldsymbol{A}$. The vector components are v_i' and the tensor components form the diagonal matrix diag $\{\lambda_i\}$, $i = 1, 2, 3$. It is easy to show that $\boldsymbol{v}\boldsymbol{A}\boldsymbol{v} = \lambda_1(v_1')^2 + \lambda_2(v_2')^2 + \lambda_3(v_3')^2$. This quantity is positive if and only if $\lambda_i > 0$, $i = 1, 2, 3$.

3.3.4 Polar Decomposition of Tensor

An arbitrary non-singular tensor $\boldsymbol{A} \in \mathcal{T}_2$ (i.e. det $\boldsymbol{A} \neq 0$) can be uniquely expressed as a product of a symmetric tensor ($\boldsymbol{U}$ or $\boldsymbol{V}$, respectively) right- or left-multiplied by an orthogonal tensor $\boldsymbol{Q}$:

$$\boldsymbol{A} = \boldsymbol{Q}\boldsymbol{U} = \boldsymbol{V}\boldsymbol{Q}. \tag{3.90}$$

To prove the first equality (3.90) let us define the tensor

$$\boldsymbol{C} = \boldsymbol{A}^{\mathrm{T}}\boldsymbol{A}, \qquad C_{ij} = A_{ki}A_{kj} .$$

This tensor is symmetric and positive definite. Symmetry results directly from the tensor's definition—reversing the indices i, j on the right-hand side does not change the result's value. To show the second property, let us recall the earlier mentioned property of non-singular tensors: if a vector $\boldsymbol{v} \neq \boldsymbol{0}$ then $\boldsymbol{w} = \boldsymbol{A}\boldsymbol{v} \neq \boldsymbol{0}$. Hence,

$$\boldsymbol{v}\boldsymbol{C}\boldsymbol{v} = \boldsymbol{v}\boldsymbol{A}^{\mathrm{T}}\boldsymbol{A}\boldsymbol{v} = (\boldsymbol{A}\boldsymbol{v})(\boldsymbol{A}\boldsymbol{v}) = \boldsymbol{w}\boldsymbol{w},$$

which for a non-zero vector $\boldsymbol{w}$ is always positive.

Let us denote $\boldsymbol{U} = \sqrt{\boldsymbol{C}}$, i.e. $\boldsymbol{U}\boldsymbol{U} = \boldsymbol{C} = \boldsymbol{A}^{\mathrm{T}}\boldsymbol{A}$. Such a tensor exists because—in view of the property shown in Exercise 3.19—all eigenvalues of $\boldsymbol{C}$ are positive. Besides, $\boldsymbol{U}$ is a symmetric tensor with a positive determinant (because it has positive eigenvalues, too), hence its inverse $\boldsymbol{U}^{-1}$ exists. Now, what remains to complete our proof is to show that the tensor $\boldsymbol{Q}$ defined as $\boldsymbol{A}\boldsymbol{U}^{-1}$ is orthogonal.

Symmetry of $\boldsymbol{U}$ implies $\boldsymbol{U}^{-1} = (\boldsymbol{U}^{-1})^{\mathrm{T}}$. The orthogonality condition (3.74) thus becomes

$$\boldsymbol{Q}^{\mathrm{T}}\boldsymbol{Q} = (\boldsymbol{A}\boldsymbol{U}^{-1})^{\mathrm{T}}(\boldsymbol{A}\boldsymbol{U}^{-1}) = \boldsymbol{U}^{-1}\boldsymbol{A}^{\mathrm{T}}\boldsymbol{A}\boldsymbol{U}^{-1} = \boldsymbol{U}^{-1}\boldsymbol{U}\boldsymbol{U}\boldsymbol{U}^{-1} = \boldsymbol{I},$$

which completes the proof.

The second equality of Eq. (3.90) can be proven in a similar manner, starting from the postulate $\boldsymbol{V} = \sqrt{\boldsymbol{B}}$, $\boldsymbol{B} = \boldsymbol{A}\boldsymbol{A}^{\mathrm{T}}$. Since determinants of both $\boldsymbol{U}$ and $\boldsymbol{V}$ are positive, one can conclude that the sign of the determinant of the orthogonal tensor $\boldsymbol{Q}$ depends on the sign of determinant of $\boldsymbol{A}$ (if $\det \boldsymbol{A} > 0$ then $\det \boldsymbol{Q} = 1$ and $\boldsymbol{Q} = \boldsymbol{R}$ is a rotation tensor; otherwise $\det \boldsymbol{Q} = -1$).

Remark. Continuing our discussion regarding interpretation of a tensor as a vector space transformation, we may now extend it to the case of an arbitrary non-singular tensor. As we already know, if a class of unit vectors V_n is considered and represented geometrically by a unit radius sphere, a symmetric tensor acting as a transformation operator on such a sphere "extends" it to an ellipsoid whose semi-axes correspond to eigenvectors and eigenvalues of this tensor. Now, the polar decomposition theorem implies that acting by an arbitrary non-singular tensor on this unit sphere will result in its extension combined with an orthogonal transformation (rotation or rotation with a symmetric flip). Vectors corresponding to semi-axes of the ellipsoid-image are thus generally not parallel to their counter-images.

Exercise 3.20

A tensor $\boldsymbol{A}$ is represented in a certain coordinate system by the following component matrix,

$$[A_{ij}] = \begin{bmatrix} 1 & -\frac{5}{6} & 0 \\ 0 & 1 & 0 \\ 0 & 0 & 1 \end{bmatrix}$$

Find components of the stretching tensor $\boldsymbol{U}$ and the rotation tensor $\boldsymbol{R}$ in the polar decomposition $\boldsymbol{A} = \boldsymbol{R}\boldsymbol{U}$.

Solution. Let us start from computing the determinant $\det \boldsymbol{A}$—it is easy to see that it equals 1 which ensures that the polar decomposition is possible in this case and that the corresponding orthogonal tensor is a rotation tensor. Furthermore, let us compute the components of $\boldsymbol{C} = \boldsymbol{A}^{\mathrm{T}}\boldsymbol{A}$ as

$$[C_{ij}] = [A_{ki}A_{kj}] = \begin{bmatrix} 1 & -\frac{5}{6} & 0 \\ -\frac{5}{6} & 1+\frac{25}{36} & 0 \\ 0 & 0 & 1 \end{bmatrix}.$$

Next, having written its characteristic equation

$$\det[C_{ij} - \lambda\delta_{ij}] = \det \begin{bmatrix} 1-\lambda & -\frac{5}{6} & 0 \\ -\frac{5}{6} & \frac{61}{36}-\lambda & 0 \\ 0 & 0 & 1-\lambda \end{bmatrix} = (1-\lambda)\left[1 - \tfrac{97}{36}\lambda + \lambda^2\right] = 0,$$

we can solve the eigenproblem with the method demonstrated in Exercise 3.16, to arrive at the following solution:

$$\begin{aligned} \lambda_1 &= \tfrac{4}{9}, \quad \{\bar{\boldsymbol{v}}_1\} = \tfrac{1}{\sqrt{13}}\{3, 2, 0\}, \\ \lambda_2 &= \tfrac{9}{4}, \quad \{\bar{\boldsymbol{v}}_2\} = \tfrac{1}{\sqrt{13}}\{-2, 3, 0\}, \\ \lambda_3 &= 1, \quad \{\bar{\boldsymbol{v}}_3\} = \{0, 0, 1\}. \end{aligned}$$

The transformation matrix (3.86) between the given coordinate system and the one associated with the principal directions of $\boldsymbol{C}$ is expressed as

$$[Q_{ij}] = \begin{bmatrix} \frac{3}{\sqrt{13}} & \frac{2}{\sqrt{13}} & 0 \\ -\frac{2}{\sqrt{13}} & \frac{3}{\sqrt{13}} & 0 \\ 0 & 0 & 1 \end{bmatrix},$$

while the tensors $\boldsymbol{U} = \sqrt{\boldsymbol{C}}$ and $\boldsymbol{U}^{-1}$ have in the latter system the diagonal representations in the form

$$\begin{aligned} [U'_{ij}] &= \left[\operatorname{diag}\left\{\sqrt{\lambda_1}, \sqrt{\lambda_2}, \sqrt{\lambda_3}\right\}\right] = \left[\operatorname{diag}\left\{\tfrac{2}{3}, \tfrac{3}{2}, 1\right\}\right], \\ [U'^{-1}_{ij}] &= \left[\operatorname{diag}\left\{1/\sqrt{\lambda_1}, 1/\sqrt{\lambda_2}, 1/\sqrt{\lambda_3}\right\}\right] = \left[\operatorname{diag}\left\{\tfrac{3}{2}, \tfrac{2}{3}, 1\right\}\right]. \end{aligned}$$

Transformation of the tensors' components to the original coordinate system leads to the following representations:

$$[U_{ij}] = [Q_{ki}Q_{lj}U'_{kl}] = \frac{1}{13}\begin{bmatrix} 12 & -5 & 0 \\ -5 & \frac{97}{6} & 0 \\ 0 & 0 & 13 \end{bmatrix},$$

$$[U_{ij}^{-1}] = [Q_{ki}Q_{lj}U'^{-1}_{kl}] = \frac{1}{13}\begin{bmatrix} \frac{97}{6} & 5 & 0 \\ 5 & 12 & 0 \\ 0 & 0 & 13 \end{bmatrix}.$$

Finally, the components of the rotation tensor $\boldsymbol{R} = \boldsymbol{A}\boldsymbol{U}^{-1}$ can be computed as

$$[R_{ij}] = [A_{ik}U_{kj}^{-1}] = \frac{1}{13}\begin{bmatrix} 12 & -5 & 0 \\ 5 & 12 & 0 \\ 0 & 0 & 13 \end{bmatrix}.$$

Figure 3.5 illustrates the transformation $\boldsymbol{w} = \boldsymbol{A}\boldsymbol{v} = \boldsymbol{R}\boldsymbol{U}\boldsymbol{v}$ for all unit vectors $\boldsymbol{v}$ for the above numerical data (the third dimension is not shown; v_3 is assumed 0).

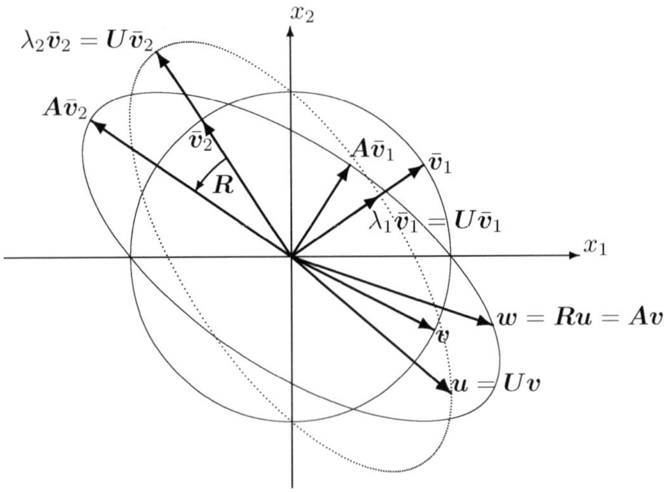

Fig. 3.5 Illustration to Exercise 3.20 for $|\boldsymbol{v}| = 1$, $v_3 = 0$: transformation $\boldsymbol{w} = \boldsymbol{A}\boldsymbol{v}$ as a combination of stretching $\boldsymbol{u} = \boldsymbol{U}\boldsymbol{v}$ and rotation $\boldsymbol{w} = \boldsymbol{R}\boldsymbol{u}$

3.4 Tensor Functions and Fields

The notion of *tensor function* is a natural generalization of the common mathematical function: if tensors of the order n, $\boldsymbol{X} \in D \subset \mathcal{T}_N$, are uniquely mapped on tensors of the order m, $\boldsymbol{Y} \in \mathcal{T}_M$, one can say that a tensor function of a tensor argument $f : \mathcal{T}_M \to \mathcal{T}_N$ of the form

$$\boldsymbol{Y} = f(\boldsymbol{X}). \tag{3.91}$$

is defined on the domain D. A natural generalization of this definition is a tensor function of more arguments,

$$\boldsymbol{Y} = f(\boldsymbol{X}, \boldsymbol{Z}, \boldsymbol{a}, \alpha, \ldots). \tag{3.92}$$

In a selected coordinate system, the function $\boldsymbol{Y} = f(\boldsymbol{X})$ can be expressed as

$$Y_{\underbrace{ij\ldots k}_{M}} = f_{ij\ldots k}(X_{\underbrace{pr\ldots s}_{N}}). \tag{3.93}$$

In other words, this is a set of 3^M real functions of 3^N real arguments. The following terminology can be adopted in a natural way:

- a tensor function of a scalar argument:

$$\boldsymbol{Y} = f(x), \qquad x \in R,\ \boldsymbol{Y} \in \mathcal{T}_M\,, \tag{3.94}$$

- a scalar function of a tensor argument:

$$y = f(\boldsymbol{X}), \qquad y \in R,\ \boldsymbol{X} \in \mathcal{T}_N\,. \tag{3.95}$$

As examples of scalar functions of a tensor argument, invariants of second order tensors defined in Sect. 3.3 may be quoted: $y = \operatorname{tr} \boldsymbol{X}$, $y = \det \boldsymbol{X}$. Among other tensor functions we can mention

$$\begin{array}{lll}
\mathcal{T}_1 \to \mathcal{T}_0 : & y(\boldsymbol{u}) = |\boldsymbol{u}|, & \\
\mathcal{T}_1 \to \mathcal{T}_0 : & y(\boldsymbol{u}) = \boldsymbol{u}\boldsymbol{A}\boldsymbol{u} & (\boldsymbol{A} \in \mathcal{T}_2), \\
\mathcal{T}_2 \to \mathcal{T}_0 : & y(\boldsymbol{X}) = \boldsymbol{v}\boldsymbol{X}\boldsymbol{w} & (\boldsymbol{v}, \boldsymbol{w} \in \mathcal{T}_1), \\
\mathcal{T}_2 \to \mathcal{T}_2 : & \boldsymbol{Y}(\boldsymbol{X}) = \boldsymbol{H} \cdot \boldsymbol{X} & (\boldsymbol{H} \in \mathcal{T}_4), \\
\mathcal{T}_2 \to \mathcal{T}_2 : & \boldsymbol{Y}(\boldsymbol{X}) = \boldsymbol{X}^{\mathrm{T}}\boldsymbol{X}, & \\
\mathcal{T}_2 \to \mathcal{T}_2 : & \boldsymbol{Y}(\boldsymbol{X}) = \exp \boldsymbol{X}, & \\
\mathcal{T}_2 \to \mathcal{T}_2 : & \boldsymbol{Y}(\boldsymbol{X}) = \dfrac{\partial y(\boldsymbol{X})}{\partial \boldsymbol{X}}\,, & Y_{ij} = \dfrac{\partial y(X_{kl})}{\partial X_{ij}}\,.
\end{array}$$

Tensor functions defined on the Euclidean space of points E^3 (or its subdomains) are called *tensor fields*. For obvious reasons (assigning tensor values to physical locations), tensor fields play an important role in modelling of solid mechanics problems. Thus, if a point $\boldsymbol{x} \in \Omega \subset E^3$, then a function $f(\boldsymbol{x}) = y \in \mathcal{T}_0$ is a *scalar field*, a function $f(\boldsymbol{x}) = \boldsymbol{y} \in \mathcal{T}_1$ is a *vector field*, $f(\boldsymbol{x}) = \boldsymbol{Y} \in \mathcal{T}_2$—a *second order tensor field*, etc., all defined on the domain Ω.

3.4.1 Integration and Differentiation of Tensor Fields

Upon fulfillment of appropriate standard conditions, tensor fields may be integrated over selected subdomains of the space E^3. The result of integration of an nth order tensor field is a tensor of the same order. A volume integral over a domain $\Omega \in E^3$ of a tensor field $\boldsymbol{Y}(\boldsymbol{x}) = Y_{ij\dots k}(\boldsymbol{x})\, \boldsymbol{e}_i \otimes \boldsymbol{e}_j \otimes \cdots \otimes \boldsymbol{e}_k$ has the form, for instance,

$$\int_\Omega \boldsymbol{Y}\, \mathrm{d}V = \left(\int_\Omega Y_{ij\dots k}\, \mathrm{d}V \right) \boldsymbol{e}_i \otimes \boldsymbol{e}_j \otimes \cdots \otimes \boldsymbol{e}_k\,, \tag{3.96}$$

while a surface integral over the area $\Gamma \in E^3$ has the form

$$\int_\Gamma \boldsymbol{Y}\, \mathrm{d}A = \left(\int_\Gamma Y_{ij\dots k}\, \mathrm{d}A \right) \boldsymbol{e}_i \otimes \boldsymbol{e}_j \otimes \cdots \otimes \boldsymbol{e}_k\,. \tag{3.97}$$

A sufficiently smooth tensor field may also be differentiated with respect to $\boldsymbol{x}$. Let us introduce the following vector operator,

$$\nabla = \boldsymbol{e}_i \frac{\partial}{\partial x_i}\,, \tag{3.98}$$

and define the gradient of a tensor field $\boldsymbol{Y}(\boldsymbol{x})$ of an arbitrary order as

$$\operatorname{grad} \boldsymbol{Y} \overset{\text{def}}{=} \boldsymbol{Y} \otimes \nabla = Y_{ij\dots k,l} \boldsymbol{e}_i \otimes \boldsymbol{e}_j \otimes \cdots \otimes \boldsymbol{e}_k \otimes \boldsymbol{e}_l \tag{3.99}$$

where the comma preceding an index is an abbreviated notation for the partial derivative with respect to the corresponding coordinate, i.e.

$$(\cdot)_{,i} \overset{\text{def}}{=} \frac{\partial(\cdot)}{\partial x_i}\,, \qquad (\cdot)_{,ij} \overset{\text{def}}{=} \frac{\partial^2(\cdot)}{\partial x_i \partial x_j}\,, \qquad \text{etc.}$$

Exercise 3.21
Show that the differential operator ∇ is a vector.
Solution. Let us examine the transformation conditions of the components $\partial/\partial x_i$ under a change of the coordinate system. Translation of the system does not obviously change the partial derivative of any function of $\boldsymbol{x}$. In the case of an orthogonal transformation we have

$$\frac{\partial}{\partial x_i'} = \frac{\partial}{\partial x_j}\frac{\partial x_j}{\partial x_i'} = \frac{\partial}{\partial x_j} Q_{ij} ,$$

because, according to Eq. (3.4),

$$\frac{\partial x_j}{\partial x_i'} = Q_{ij} .$$

It is thus clear that the components $\nabla_i' = \partial/\partial x_i'$ in the transformed coordinate system are expressed by the components in the old system in the form

$$\nabla_i' = Q_{ij}\nabla_j ,$$

identical to that of the transformation rule (3.13).

The gradient of a tensor field of the order n is thus a tensor field of the order $n+1$, frequently denoted in the literature as $\nabla \boldsymbol{Y}$. However, in view of the assumed notation, this symbol is not compatible with the definition (3.99) and may be a reason of misunderstandings.

Let us additionally define three other gradient-related operations on tensor fields. The divergence of a tensor field $\boldsymbol{Y}(\boldsymbol{x})$ of the order $n > 0$, div $\boldsymbol{Y}$, is the gradient contraction on the last two indices. This is a tensor field of the order $n-1$. The rotation of a tensor field $\boldsymbol{Y}(\boldsymbol{x})$ of the order $n > 0$, rot $\boldsymbol{Y}$, is the field defined as $\nabla \times \boldsymbol{Y}$ of the same order n. The Laplacian of a field $\boldsymbol{Y}(\boldsymbol{x})$ of an arbitrary order is the divergence of its gradient, div (grad $\boldsymbol{Y}$) $\equiv (\boldsymbol{y} \otimes \nabla) \cdot \nabla$. The definition will be explained in detail on the examples below.

- Gradient of a scalar field $f \in \mathcal{T}_0$ is a vector field:

$$\text{grad}\, f \equiv \nabla f = f_{,i}\, \boldsymbol{e}_i .$$

- Laplacian of a scalar field $f \in \mathcal{T}_0$ is a scalar field:

$$\nabla^2 f = f_{,ii} .$$

- Gradient of a vector field $\boldsymbol{y} \in \mathcal{T}_1$ is a second order tensor field:

$$\text{grad}\, \boldsymbol{y} \equiv \nabla \boldsymbol{y} = \boldsymbol{y} \otimes \boldsymbol{\nabla} = y_{i,j}\, \boldsymbol{e}_i \otimes \boldsymbol{e}_j.$$

- Divergence of a vector field $\boldsymbol{y} \in \mathcal{T}_1$ is a scalar field:

$$\text{div}\, \boldsymbol{y} = \boldsymbol{y} \cdot \boldsymbol{\nabla} = \boldsymbol{y}\boldsymbol{\nabla} = y_{i,i}$$

(note that $\text{div}\, \boldsymbol{y} = \text{tr}\,(\text{grad}\, \boldsymbol{y})$).
- Laplacian of a vector field $\boldsymbol{y} \in \mathcal{T}_1$ is a vector field:

$$\nabla^2 \boldsymbol{y} = \text{div}\,(\text{grad}\, \boldsymbol{y}) = y_{i,jj}\, \boldsymbol{e}_i.$$

- Rotation of a vector field $\boldsymbol{y} \in \mathcal{T}_1$ is a vector field:

$$\text{rot}\, \boldsymbol{y} = \boldsymbol{\nabla} \times \boldsymbol{y} = y_{j,i}\epsilon_{ijk}\boldsymbol{e}_k.$$

- Gradient of a tensor field $\boldsymbol{Y} \in \mathcal{T}_2$ is a third order tensor field:

$$\text{grad}\, \boldsymbol{Y} \equiv \nabla \boldsymbol{Y} = \boldsymbol{Y} \otimes \boldsymbol{\nabla} = Y_{ij,k}\, \boldsymbol{e}_i \otimes \boldsymbol{e}_j \otimes \boldsymbol{e}_k.$$

- Divergence of a tensor field $\boldsymbol{Y} \in \mathcal{T}_2$ is a vector field:

$$\text{div}\, \boldsymbol{Y} = \boldsymbol{Y} \cdot \boldsymbol{\nabla} = \boldsymbol{Y}\boldsymbol{\nabla} = Y_{ij,j}\, \boldsymbol{e}_i.$$

- Laplacian of a tensor field $\boldsymbol{Y} \in \mathcal{T}_2$ is a second order tensor field:

$$\nabla^2 \boldsymbol{Y} = \text{div}\,(\text{grad}\, \boldsymbol{Y}) = Y_{ij,kk}.$$

- Rotation of a tensor field $\boldsymbol{Y} \in \mathcal{T}_2$ is a second order tensor field:

$$\text{rot}\, \boldsymbol{Y} = \boldsymbol{\nabla} \times \boldsymbol{Y} = Y_{jk,i}\epsilon_{ijl}\, \boldsymbol{e}_l \otimes \boldsymbol{e}_k.$$

Exercise 3.22
For a scalar field $\varphi(\boldsymbol{x}) \in \mathcal{T}_0$, show that

$$\text{rot}\,(\text{grad}\, \varphi) = \boldsymbol{0}.$$

Solution. Let us compute

$$\text{rot}\,(\text{grad}\, \varphi) = \boldsymbol{\nabla} \times (\boldsymbol{\nabla}\varphi) = (\varphi_{,j})_{,i}\epsilon_{ijk}\, \boldsymbol{e}_k = \varphi_{,ji}\epsilon_{ijk}\, \boldsymbol{e}_k$$

Note that the tensor with components $\varphi_{,ji}$ is symmetric since the definition of partial differentiation implies the equality $\partial^2\varphi/\partial x_i\partial x_j \equiv \partial^2\varphi/\partial x_j\partial x_i$. The full inner product of a symmetric tensor and the antisymmetric symbol ϵ_{ijk} yields zero, see Eq. (3.66):

$$\varphi_{,ij}\epsilon_{ijk}\,\boldsymbol{e}_k = (\varphi_{,23} - \varphi_{,32})\,\boldsymbol{e}_1 + (\varphi_{,31} - \varphi_{,13})\,\boldsymbol{e}_2 + (\varphi_{,12} - \varphi_{,21})\,\boldsymbol{e}_3 = \boldsymbol{0}.$$

Exercise 3.23
For a vector field $\boldsymbol{v}(\boldsymbol{x}) \in \mathcal{T}_1$, show that

$$\operatorname{div}(\operatorname{rot}\boldsymbol{v}) = 0.$$

Solution. Let us rewrite the left-hand side in the index notation as

$$\begin{aligned}\operatorname{div}(\operatorname{rot}\boldsymbol{v}) = (\boldsymbol{\nabla}\times\boldsymbol{v})\boldsymbol{\nabla} &= \frac{\partial}{\partial x_k}\left(\left(\frac{\partial}{\partial x_i}\boldsymbol{e}_i\right)\times(v_j\boldsymbol{e}_j)\right)\boldsymbol{e}_k = \frac{\partial}{\partial x_k}(v_{j,i}\,\epsilon_{ijl}\,\boldsymbol{e}_l)\,\boldsymbol{e}_k\\ &= v_{j,ik}\,\epsilon_{ijl}\,\delta_{lk} = v_{j,ik}\,\epsilon_{ijk}\,.\end{aligned}$$

Since the partial derivative $v_{j,ik}$ is symmetric with respect to the indices ik, its product with the antisymmetric symbol ϵ_{ijk} yields zero, cf. Exercise 3.22.

Exercise 3.24
For vector fields $\boldsymbol{v}(\boldsymbol{x}), \boldsymbol{w}(\boldsymbol{x}) \in \mathcal{T}_1$, show that

$$\operatorname{div}(\boldsymbol{v}\times\boldsymbol{w}) = \boldsymbol{w}\cdot\operatorname{rot}\boldsymbol{v} = -\boldsymbol{v}\cdot\operatorname{rot}\boldsymbol{w}.$$

Solution. Express the divergence as

$$\operatorname{div}(\boldsymbol{v}\times\boldsymbol{w}) = \boldsymbol{\nabla}\cdot(\boldsymbol{v}\times\boldsymbol{w}).$$

Since the operator $\boldsymbol{\nabla}$ is a vector (see Exercise 3.21), let us benefit from the result of Exercise 3.5 in which we substitute $\boldsymbol{u} = \boldsymbol{\nabla}$,

$$\boldsymbol{\nabla}\cdot(\boldsymbol{v}\times\boldsymbol{w}) = \boldsymbol{w}\cdot(\boldsymbol{\nabla}\times\boldsymbol{v}) = \boldsymbol{v}\cdot(\boldsymbol{w}\times\boldsymbol{\nabla}).$$

In view of the equality $\boldsymbol{w}\times\boldsymbol{\nabla} = -\boldsymbol{\nabla}\times\boldsymbol{w}$, this may be rewritten as

$$\boldsymbol{\nabla}\cdot(\boldsymbol{v}\times\boldsymbol{w}) = \boldsymbol{w}\cdot\operatorname{rot}\boldsymbol{v} = -\boldsymbol{v}\cdot\operatorname{rot}\boldsymbol{w}.$$

Exercise 3.25
For a vector field $\boldsymbol{w}(\boldsymbol{x}) \in \mathcal{T}_1$, show that

$$\text{rot}\,(\text{rot}\,\boldsymbol{w}) = \text{grad}\,(\text{div}\,\boldsymbol{w}) - \nabla^2 \boldsymbol{w}.$$

Solution. Express the rotation as

$$\text{rot}\,(\text{rot}\,\boldsymbol{w}) = \boldsymbol{\nabla} \times (\boldsymbol{\nabla} \times \boldsymbol{w})$$

and make use of the result of Exercise 3.7 in which we substitute $\boldsymbol{u} = \boldsymbol{v} = \boldsymbol{\nabla}$, i.e.

$$\boldsymbol{\nabla} \times (\boldsymbol{\nabla} \times \boldsymbol{w}) = \boldsymbol{\nabla}(\boldsymbol{\nabla} \cdot \boldsymbol{w}) - (\boldsymbol{\nabla} \cdot \boldsymbol{\nabla})\boldsymbol{w} = \text{grad}\,(\text{div}\,\boldsymbol{w}) - \nabla^2 \boldsymbol{w}.$$

3.4.2 *Gauss–Ostrogradski Theorem*

The Gauss–Ostrogradski theorem (also known as the divergence theorem or, in the 2D case, as the Green theorem) allows to replace integration over an open domain $\Omega \in E^3$ by integration over a closed, piece-wise smooth surface $\partial\Omega$ constituting the domain's boundary. For a sufficiently smooth vector field $\boldsymbol{y} = \boldsymbol{y}(\boldsymbol{x})$ the following equality holds true:

$$\int_\Omega \text{div}\,\boldsymbol{y}\,\mathrm{d}V = \int_{\partial\Omega} \boldsymbol{y}\boldsymbol{n}\,\mathrm{d}A, \quad \text{i.e.} \quad \int_\Omega y_{i,i}\,\mathrm{d}V = \int_{\partial\Omega} y_i n_i\,\mathrm{d}A, \tag{3.100}$$

where $\boldsymbol{n} = \boldsymbol{n}(\boldsymbol{x})$ is a local external normal versor to the surface $\partial\Omega$. This theorem may be generalized and the following equalities shown to hold true:

$$\int_\Omega \text{grad}\,\boldsymbol{y}\,\mathrm{d}V = \int_{\partial\Omega} \boldsymbol{y} \otimes \boldsymbol{n}\,\mathrm{d}A, \quad \text{i.e.} \quad \int_\Omega y_{i,j}\,\mathrm{d}V = \int_{\partial\Omega} y_i n_j\,\mathrm{d}A,$$
$$\int_\Omega \text{rot}\,\boldsymbol{y}\,\mathrm{d}V = \int_{\partial\Omega} \boldsymbol{n} \times \boldsymbol{y}\,\mathrm{d}A, \quad \text{i.e.} \quad \int_\Omega \epsilon_{ijk} y_{j,i}\,\mathrm{d}V = \int_{\partial\Omega} \epsilon_{ijk} y_j n_i\,\mathrm{d}A,$$

and, for a sufficiently smooth scalar field $f(\boldsymbol{x})$,

$$\int_\Omega \text{grad}\,f\,\mathrm{d}V = \int_{\partial\Omega} f\boldsymbol{n}\,\mathrm{d}A, \quad \text{i.e.} \quad \int_\Omega f_{,i}\,\mathrm{d}V = \int_{\partial\Omega} f n_i\,\mathrm{d}A.$$

Recalling the definitions of the gradient, divergence and rotation operations, one can easily recognize a common rule that governs the above formulae: volume integration over a domain Ω of any kind of product of an arbitrary field and the gradient operator is equivalent to the surface integration over the boundary $\partial\Omega$ of the product in which the gradient operator is replaced by the local normal versor $\boldsymbol{n}$. General relations for an arbitrary, sufficiently smooth tensor field $\boldsymbol{Y} = \boldsymbol{Y}(\boldsymbol{x}) \in \mathcal{T}_N$ can be derived from this rule as

$$\int_\Omega \operatorname{grad} \boldsymbol{Y} \, \mathrm{d}V \equiv \int_\Omega \boldsymbol{Y} \otimes \boldsymbol{\nabla} \, \mathrm{d}V = \int_{\partial\Omega} \boldsymbol{Y} \otimes \boldsymbol{n} \, \mathrm{d}A, \tag{3.101a}$$

$$\int_\Omega Y_{ij\ldots k,l} \, \mathrm{d}V = \int_{\partial\Omega} Y_{ij\ldots k} n_l \, \mathrm{d}A. \tag{3.101b}$$

$$\int_\Omega \operatorname{div} \boldsymbol{Y} \, \mathrm{d}V \equiv \int_\Omega \boldsymbol{Y} \boldsymbol{\nabla} \, \mathrm{d}V = \int_{\partial\Omega} \boldsymbol{Y} \boldsymbol{n} \, \mathrm{d}A, \tag{3.102a}$$

$$\int_\Omega Y_{ij\ldots k,k} \, \mathrm{d}V = \int_{\partial\Omega} Y_{ij\ldots k} n_k \, \mathrm{d}A. \tag{3.102b}$$

$$\int_\Omega \operatorname{rot} \boldsymbol{Y} \, \mathrm{d}V \equiv \int_\Omega \boldsymbol{\nabla} \times \boldsymbol{Y} \, \mathrm{d}V = \int_{\partial\Omega} \boldsymbol{n} \times \boldsymbol{Y} \, \mathrm{d}A, \tag{3.103a}$$

$$\int_\Omega \epsilon_{lim} Y_{ij\ldots k,l} \, \mathrm{d}V = \int_{\partial\Omega} \epsilon_{lim} Y_{ij\ldots k} n_l \, \mathrm{d}A. \tag{3.103b}$$

In the exercises below, V_Ω and A_Ω denote the volume of an arbitrary domain $\Omega \in E^3$ and the area of its boundary surface $\partial\Omega$, respectively.

Exercise 3.26
For the scalar field $f(\boldsymbol{x}) = ax_1 + bx_2 + cx_3$, compute the integral

$$\int_{\partial\Omega} f \boldsymbol{n} \, \mathrm{d}A.$$

Solution. The Gauss–Ostrogradski theorem allows to transform

$$\int_{\partial\Omega} f \boldsymbol{n} \, \mathrm{d}A = \int_\Omega \operatorname{grad} f \, \mathrm{d}V = \int_\Omega (a\boldsymbol{e}_1 + b\boldsymbol{e}_2 + c\boldsymbol{e}_3) \, \mathrm{d}V = (a\boldsymbol{e}_1 + b\boldsymbol{e}_2 + c\boldsymbol{e}_3) V_\Omega .$$

In other words, the solution is the vector with the components $\{aV_\Omega, bV_\Omega, cV_\Omega\}$.

Exercise 3.27
For the vector field $\boldsymbol{v}(\boldsymbol{x}) \in \mathcal{T}_1$, show that

$$\int_{\partial\Omega} (\text{rot}\, \boldsymbol{v}) \cdot \boldsymbol{n}\, \mathrm{d}A = 0.$$

Solution. From the Gauss–Ostrogradski theorem we have

$$\int_{\partial\Omega} (\text{rot}\, \boldsymbol{v}) \cdot \boldsymbol{n}\, \mathrm{d}A = \int_{\Omega} \text{div}\,(\text{rot}\, \boldsymbol{v})\, \mathrm{d}V.$$

The integrand is a special case of a quantity whose vanishing has been shown in Exercise 3.23. Thus, the above integral is zero, too.

Exercise 3.28
Compute the following integrals:

$$\int_{\partial\Omega} \boldsymbol{x}\boldsymbol{n}\, \mathrm{d}A, \qquad \int_{\partial\Omega} \boldsymbol{x}\boldsymbol{B}\boldsymbol{n}\, \mathrm{d}A, \qquad \int_{\partial\Omega} (\boldsymbol{B}\boldsymbol{x} \times \boldsymbol{n})\, \mathrm{d}A,$$

where $\boldsymbol{B}$ is a constant second order tensor. What can be said about the last result if $\boldsymbol{B}$ is symmetric?

Solution. Switching for more convenience to the index notation and taking into account the obvious relation $x_{i,j} \equiv \partial x_i / \partial x_j = \delta_{ij}$, let us make use of the Gauss–Ostrogradski theorem and perform the following transformations:

$$\int_{\partial\Omega} \boldsymbol{x}\boldsymbol{n}\, \mathrm{d}A = \int_{\partial\Omega} x_i n_i\, \mathrm{d}A = \int_{\Omega} x_{i,i}\, \mathrm{d}V = \int_{\Omega} \delta_{ii}\, \mathrm{d}V = 3V_{\Omega},$$

$$\int_{\partial\Omega} \boldsymbol{x}\boldsymbol{B}\boldsymbol{n}\, \mathrm{d}A = \int_{\partial\Omega} x_i B_{ij} n_j\, \mathrm{d}A = \int_{\Omega} (x_i B_{ij})_{,j}\, \mathrm{d}V = \int_{\Omega} x_{i,j} B_{ij}\, \mathrm{d}V$$

$$= \int_{\Omega} \delta_{ij} B_{ij}\, \mathrm{d}V = V_{\Omega} B_{ii} = V_{\Omega} \mathrm{tr}\, \boldsymbol{B},$$

$$\int_{\partial\Omega} (\boldsymbol{B}\boldsymbol{x} \times \boldsymbol{n})\, \mathrm{d}A = \int_{\partial\Omega} B_{il} x_l n_j \epsilon_{ijk} \boldsymbol{e}_k\, \mathrm{d}A = \int_{\Omega} (B_{il} x_l \epsilon_{ijk} \boldsymbol{e}_k)_{,j}\, \mathrm{d}V$$

$$= \int_{\Omega} B_{il} x_{l,j} \epsilon_{ijk} \boldsymbol{e}_k\, \mathrm{d}V = \int_{\Omega} \epsilon_{ijk} B_{ij} \boldsymbol{e}_k\, \mathrm{d}V = V_{\Omega} B_{ij} \epsilon_{ijk} \boldsymbol{e}_k.$$

If the tensor $\boldsymbol{B}$ is symmetric, the last integral vanishes because ϵ_{ijk} is anti-symmetric with respect to ij, cf. Eq. (3.66).

Exercise 3.29
Let $\boldsymbol{v}(\boldsymbol{x})$ denote an arbitrary vector field that—at all points of the closed surface $\partial\Omega$—coincides with the field of its normal versors $\boldsymbol{n}(\boldsymbol{x})$. Compute the integrals

$$\int_{\Omega} \operatorname{div} \boldsymbol{v} \, \mathrm{d}V, \qquad \int_{\Omega} \operatorname{rot} \boldsymbol{v} \, \mathrm{d}V.$$

Solution. Let us apply the Gauss–Ostrogradski theorem and transform the first integral as

$$\int_{\Omega} \operatorname{div} \boldsymbol{v} \, \mathrm{d}V = \int_{\partial\Omega} \boldsymbol{v} \cdot \boldsymbol{n} \, \mathrm{d}A = \int_{\partial\Omega} \boldsymbol{n} \cdot \boldsymbol{n} \, \mathrm{d}A = \int_{\partial\Omega} \mathrm{d}A = A_{\Omega} \, .$$

By transforming the second integral we come to the conclusion that

$$\int_{\Omega} \operatorname{rot} \boldsymbol{v} \, \mathrm{d}V = \int_{\partial\Omega} \boldsymbol{n} \times \boldsymbol{v} \, \mathrm{d}A = \int_{\partial\Omega} \boldsymbol{n} \times \boldsymbol{n} \, \mathrm{d}A = \boldsymbol{0}.$$

3.5 Curvilinear Coordinate Systems

Up to now, in our description of objects in the Euclidean space E^3, we have limited ourselves to Cartesian coordinate systems only. These are generally sufficient as the basis for all theoretical considerations presented in this book—relative simplicity of the formalism associated with the Cartesian description makes it easier to concentrate on physical aspects of the problems discussed. However, one should be aware of the fact that in several practical situations the use of a curvilinear, orthogonal or not, coordinate system may be required as the most natural in formulating the particular equations for the given case.

Fortunately, a formal generalization of the Cartesian formalism to the case of curvilinear systems is not very difficult. Let us briefly point out some basic rules of doing it.

Consider a selected Cartesian coordinate system $\{x_i\}$, $i = 1, 2, 3$, in the Euclidean space E^3. The location of a point $X \in E^3$ with respect to the system's origin is expressed by Eq. (3.2). Let us introduce in this space a curvilinear coordinate system $\{y^i\}$ defined by the mapping

$$y^i = y^i(x_j), \qquad i, j = 1, 2, 3. \tag{3.104}$$

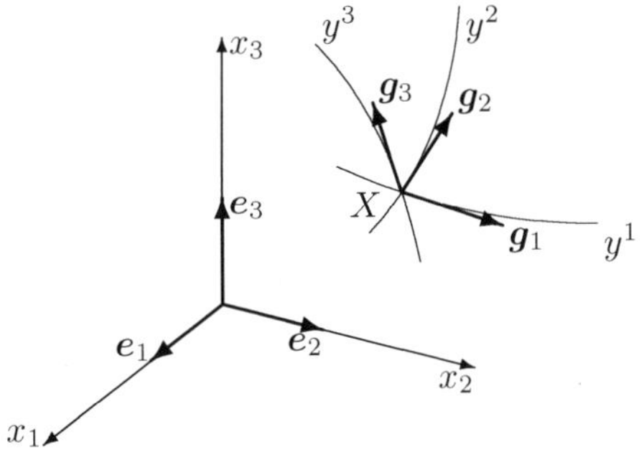

Fig. 3.6 Curvilinear coordinate system

The Jacobian of the transformation (3.104) is assumed non-zero,

$$J = \det\left[\frac{\partial y^i}{\partial x_j}\right] \neq 0. \tag{3.105}$$

Hence, the inverse relation exists and is written here as

$$x_i = x_i(y^j), \tag{3.106}$$

with the unique relationships between triplets of numbers (x_1, x_2, x_3) and (y^1, y^2, y^3).

The vectors

$$\boldsymbol{g}_i = \frac{\partial \boldsymbol{x}}{\partial y^i} = \frac{\partial x_k}{\partial y^i}\,\boldsymbol{e}_k\,. \tag{3.107}$$

are locally tangent at a given point to the lines of constant coordinate y^i of the curvilinear system (Fig. 3.6). The vectors constitute the *local basis* of the system $\{y^i\}$. Contrary to the Cartesian basis vectors $\boldsymbol{e}_k$, the vectors $\boldsymbol{g}_i$ are neither constant in space ($\boldsymbol{g}_i = \boldsymbol{g}_i(y^k)$), nor are they orthonormal. According to Eq. (3.107), the quantity $\frac{\partial x_k}{\partial y^i}$ can be treated as the kth component of the vector $\boldsymbol{g}_i$ in the Cartesian coordinate system (x_1, x_2, x_3) with the basis $\boldsymbol{e}_k$.

Remark. The indices corresponding to the curvilinear coordinates are intentionally upper ones, i.e. y^i. For reasons that will soon become clear, the summation convention has to be modified in the case of such coordinates, so that summation over a pair of repeated indices is only performed if one of them is an upper and the other a lower index, e.g. $a^k b_k \stackrel{\text{def}}{=} \sum_{k=1}^{3} a^k b_k$.

Scalar products of the local basis vectors are components of the metric tensor

$$g_{ij} = \boldsymbol{g}_i \cdot \boldsymbol{g}_j\,. \tag{3.108}$$

For an orthogonal coordinate system, $g_{kl} = 0$ for $k \neq l$. In general, the system $\{y^i\}$ is not orthogonal—for each three local basis vectors $\boldsymbol{g}_i$, $i = 1, 2, 3$ one can thus define three other vectors $\boldsymbol{g}^i$, such that for arbitrary two vectors $\boldsymbol{g}_i, \boldsymbol{g}_j$ $(i \neq j)$ the vector $\boldsymbol{g}^k$ $(k \neq i, k \neq j)$ is orthogonal to each of them. The so constructed vectors $\boldsymbol{g}^i$ constitute the *dual basis* to the basis $\boldsymbol{g}_i$, also called the cobasis. The lengths of the dual basis vectors are selected so that the following equalities hold true

$$\boldsymbol{g}^k \cdot \boldsymbol{g}_l = \delta^k{}_l\,, \tag{3.109}$$

where $\delta^k{}_l$ denotes the Kronecker symbol. Defining the tensor g^{kl} as the inverse to the metric tensor g_{kl},

$$g^{kl} = \frac{\bar{g}_{kl}}{\det(g_{kl})}\,, \qquad g^{km} g_{ml} = \delta^k{}_l\,, \tag{3.110}$$

where $\bar{g}_{kl}$ is the algebraic complement (cofactor) of g_{kl}, one can show that the cobasis vectors fulfilling the above relations are expressed as

$$\boldsymbol{g}^k = g^{kl}\,\boldsymbol{g}_l \tag{3.111}$$

Indeed, $\boldsymbol{g}^k \cdot \boldsymbol{g}_l = (g^{km}\,\boldsymbol{g}_m) \cdot \boldsymbol{g}_l = g^{km}(\boldsymbol{g}_m \cdot \boldsymbol{g}_l) = g^{km} g_{ml} = \delta^k{}_l$. Besides, it is easy to show that

$$g^{kl} = \boldsymbol{g}^k \cdot \boldsymbol{g}^l\,, \tag{3.112}$$

i.e. the quantities g^{kl} are the metric tensor components in the dual basis.

The cobasis $\boldsymbol{g}^k$, similarly as the basis $\boldsymbol{g}_k$, is generally not orthonormal. An arbitrary vector $\boldsymbol{v} \in V^3$ may be expressed in each of the bases as

$$\boldsymbol{v} = v^k \boldsymbol{g}_k = v_k \boldsymbol{g}^k\,, \tag{3.113}$$

where the quantities v^k are called its contravariant components while v_k its covariant components. The components v^k and v_k are generally different (they are only equal in the case of an orthogonal Cartesian system basis $\boldsymbol{g}_k = \boldsymbol{e}_k$ since we have then $\boldsymbol{g}^k = \boldsymbol{e}_k$ and distinguishing between the components v^k and v_k is irrelevant). Taking the scalar product of Eq. (3.113) and $\boldsymbol{g}_l$ and $\boldsymbol{g}^l$, respectively, one obtains

$$v_k = g_{kl} v^l\,, \qquad v^k = g^{kl} v_l\,, \tag{3.114}$$

which means that, given the components v_k and the metric tensor, it is possible to compute the components v^k and vice versa. Such operations are somewhat informally called index raising and lowering, respectively.

Contravariant and covariant vector components are subject to the following transformation at the coordinate system change $\{y^i\} \longrightarrow \{y'^j\}$:

$$v'^i = v^j \frac{\partial y'^i}{\partial y^j}, \qquad v'_i = v_j \frac{\partial y^j}{\partial y'^i}. \tag{3.115}$$

For example, an infinitesimal element $\mathrm{d}y^k$ is a contravariant vector, because

$$\mathrm{d}y'^k = \frac{\partial y'^k}{\partial y^m}\,\mathrm{d}y^m, \tag{3.116}$$

while the partial derivative $\frac{\partial \phi}{\partial y^k}$ of a scalar function $\phi(\boldsymbol{y})$ is a covariant vector, because

$$\frac{\partial \phi}{\partial y'^k} = \frac{\partial \phi}{\partial y^l}\frac{\partial y^l}{\partial y'^k}. \tag{3.117}$$

Remark. It is sometimes convenient to operate on vector components expressed in a basis that coincides with one of the local bases (say, the contravariant basis) but with the normalized basis vectors (versors). We have then

$$\boldsymbol{v} = v^k \boldsymbol{g}_k = v^{(k)} \hat{\boldsymbol{g}}_k, \tag{3.118}$$

where the versors $\hat{\boldsymbol{g}}_k$ are defined as

$$\hat{\boldsymbol{g}}_k = \frac{\boldsymbol{g}_k}{\sqrt{\boldsymbol{g}_k \cdot \boldsymbol{g}_k}} = \frac{\boldsymbol{g}_k}{\sqrt{g_{kk}}} \quad \text{(no summation over } k\text{)}, \tag{3.119}$$

while the components $v^{(k)}$ are called physical components of the vector $\boldsymbol{v}$. The latter are computed as

$$v^{(k)} = v^k \sqrt{g_{kk}}. \tag{3.120}$$

An arbitrary second order tensor $\boldsymbol{A} \in \mathcal{T}_2$ may also be expressed in the curvilinear coordinate system with the help of the two vector bases $\boldsymbol{g}_k$ and $\boldsymbol{g}^k$. Four alternative decompositions are possible,

$$\boldsymbol{A} = A^{ij}\, \boldsymbol{g}_i \otimes \boldsymbol{g}_j = A_{ij}\, \boldsymbol{g}^i \otimes \boldsymbol{g}^j = A^i{}_j\, \boldsymbol{g}_i \otimes \boldsymbol{g}^j = A_i{}^j\, \boldsymbol{g}^i \otimes \boldsymbol{g}_j, \tag{3.121}$$

where A^{ij} and A_{ij} are called the contravariant and covariant components of the tensor $\boldsymbol{A}$, respectively, while $A^i{}_j$ and $A_i{}^j$ its mixed components. The following relations hold true:

$$A_{ij} = A^{pr}\, g_{pi} g_{rj} = A^p{}_j\, g_{pi} = A_i^r\, g_{rj} \tag{3.122}$$

and so do the transformation rules at a change of the coordinate system:

$$
\begin{aligned}
A'^{kl} &= A^{ij}\,\frac{\partial y'^k}{\partial y^i}\,\frac{\partial y'^l}{\partial y^j}\,,\\
A'_{kl} &= A_{ij}\,\frac{\partial y^i}{\partial y'^k}\,\frac{\partial y^j}{\partial y'^l}\,,\\
A'_l{}^k &= A_j{}^i\,\frac{\partial y'^k}{\partial y^i}\,\frac{\partial y^j}{\partial y'^l}\,.
\end{aligned}
\tag{3.123}
$$

Examples of such tensor representations are g_{kl}, g^{kl} and $\delta^k{}_l$.

Similar transformation rules hold true for higher order tensors.

Operations on tensor representations in non-Cartesian coordinate systems are appropriate generalizations of the operations on Cartesian representations defined earlier in this chapter. For example, the outer product of two tensors

$$
\boldsymbol{A} = A^k{}_{lm}\,\boldsymbol{g}_k \otimes \boldsymbol{g}^l \otimes \boldsymbol{g}^m\,, \qquad \boldsymbol{B} = B^k\,\boldsymbol{g}_k
$$

has the form

$$
\boldsymbol{A} \otimes \boldsymbol{B} = A^k{}_{lm}\,B^n\,\boldsymbol{g}_k \otimes \boldsymbol{g}^l \otimes \boldsymbol{g}^m \otimes \boldsymbol{g}_n\,, \tag{3.124}
$$

while the contraction of the tensor

$$
A_{mkl}{}^n\,\boldsymbol{g}^m \otimes \boldsymbol{g}^k \otimes \boldsymbol{g}^l \otimes \boldsymbol{g}_n
$$

on the first and the fourth index has the form

$$
A_{mkl}{}^n\,(\boldsymbol{g}^m \cdot \boldsymbol{g}_n)\,\boldsymbol{g}^k \otimes \boldsymbol{g}^l = A_{mkl}{}^n\,\delta^m{}_n\,\boldsymbol{g}^k \otimes \boldsymbol{g}^l = A_{mkl}{}^m\,\boldsymbol{g}^k \otimes \boldsymbol{g}^l\,. \tag{3.125}
$$

Unfortunately, operations of differentiation of tensor fields with respect to curvilinear coordinates are substantially more complicated than they were in the Cartesian coordinate systems. In the latter case, differentiation concerned the tensor components only (as the basis vectors $\boldsymbol{e}_k$ were constant). Here, the basis $\boldsymbol{g}_k$ is a function of the spatial location and thus is subject to differentiation, too. Consider for instance a vector field $\boldsymbol{v} \in \mathcal{T}_1$. Its gradient in Cartesian basis can be expressed as

$$
\operatorname{grad} \boldsymbol{v} = \boldsymbol{v} \otimes \boldsymbol{\nabla} = \frac{\partial \boldsymbol{v}}{\partial x_k} \otimes \boldsymbol{e}_k = \frac{\partial (v_i \boldsymbol{e}_i)}{\partial x_k} \otimes \boldsymbol{e}_k = v_{i,k}\,\boldsymbol{e}_i \otimes \boldsymbol{e}_k\,. \tag{3.126}
$$

Passing to curvilinear coordinates with the basis $\boldsymbol{g}_i$ and defining, in analogy to Eq. (3.98), the covariant differentiation operator

$$
\boldsymbol{\nabla} = \boldsymbol{g}^i\,\frac{\partial}{\partial y^i} \tag{3.127}
$$

we may write

$$\operatorname{grad} \boldsymbol{v} = \boldsymbol{v} \otimes \nabla = \frac{\partial \boldsymbol{v}}{\partial y_k} \otimes \boldsymbol{g}^k = \frac{\partial (v^i \boldsymbol{g}_i)}{\partial y_k} \otimes \boldsymbol{g}^k = (v^i{}_{,k} \boldsymbol{g}_i + v^i \boldsymbol{g}_{i,k}) \otimes \boldsymbol{g}^k \tag{3.128}$$

where the simplified notation $(\cdot)_{,k}$ refers now to differentiation with respect to y^k. The derivatives of the basis vectors $\boldsymbol{g}_{i,k}$ can be computed with the use of Eq. (3.107) and a similar inverse relation $\boldsymbol{e}_k = \frac{\partial y_l}{\partial x_k} \boldsymbol{g}_l$ as

$$\frac{\partial \boldsymbol{g}_i}{\partial y^k} = \frac{\partial}{\partial y^k}\left(\frac{\partial x_m}{\partial y^i} \boldsymbol{e}_m\right) = \frac{\partial^2 x_m}{\partial y^i \partial y^k} \boldsymbol{e}_m = \frac{\partial^2 x_m}{\partial y^i \partial y^k} \frac{\partial y^l}{\partial x_m} \boldsymbol{g}_l \,. \tag{3.129}$$

Defining the quantity

$$\Gamma^l_{ik} \stackrel{\text{def}}{=} \frac{\partial^2 x_m}{\partial y^i \partial y^k} \frac{\partial y^l}{\partial x_m} \,, \tag{3.130}$$

called the Christoffel symbol of the second kind,[8] one may rewrite the relation (3.129) more concisely as

$$\boldsymbol{g}_{i,k} = \Gamma^l_{ik} \boldsymbol{g}_l \,, \tag{3.131}$$

while Eqs. (3.128) as

$$\operatorname{grad} \boldsymbol{v} = v^i{}_{;k} \boldsymbol{g}_i \otimes \boldsymbol{g}^k \,, \qquad v^i{}_{;k} \stackrel{\text{def}}{=} v^i{}_{,k} + v^j \Gamma^i_{jk} \,. \tag{3.132}$$

The Christoffel symbols are symmetric with respect to the two lower indices. The operation denoted by ";" is called covariant differentiation. In other words, $v^m{}_{;k}$ is the matrix of covariant derivatives of the contravariant components of the vector $\boldsymbol{v}$. Clearly, in the orthogonal Cartesian coordinate systems, all the Christoffel symbol components are zero and the covariant differentiation is equivalent to ordinary partial differentiation.

It can be shown that covariant derivatives of the cobasis vectors $\boldsymbol{g}^i$ may also be expressed with the Christoffel symbols as

$$\boldsymbol{g}^i{}_{,k} = \frac{\partial \boldsymbol{g}^i}{\partial y^k} = -\Gamma^i_{kl} \boldsymbol{g}^l \,. \tag{3.133}$$

Hence, expressing the vector $\boldsymbol{v}$ alternatively in its contravariant components, cf. Eq. (3.114), one can write after series of analogous transformations:

---

[8]Not to be confused with the Christoffel symbol of the first kind, defined as $\Gamma_{n,ik} = g_{nl} \Gamma^l_{ik}$, which is not going to be used in our considerations.

$$\operatorname{grad} \boldsymbol{v} = \frac{\partial(v_i \boldsymbol{g}^i)}{\partial y_k} \otimes \boldsymbol{g}^k = v_{i:k}\, \boldsymbol{g}^i \otimes \boldsymbol{g}^k, \qquad v_{i:k} \overset{\text{def}}{=} v_{i,k} - v_j \Gamma^j_{ki}. \tag{3.134}$$

Note that Eqs. (3.131) and (3.133) imply in particular that the basis vector gradients are expressed as

$$\operatorname{grad} \boldsymbol{g}_i = \Gamma^l_{ik}\, \boldsymbol{g}_l \otimes \boldsymbol{g}_k, \qquad \operatorname{grad} \boldsymbol{g}^i = -\Gamma^i_{kl}\, \boldsymbol{g}^l \otimes \boldsymbol{g}_k. \tag{3.135}$$

The operations of divergence and rotation of a vector field $\boldsymbol{v}$ can be written down in curvilinear coordinates as

$$\operatorname{div} \boldsymbol{v} = v^k_{,k}, \qquad \operatorname{rot} \boldsymbol{v} = v_{m:k}\, \boldsymbol{g}^k \times \boldsymbol{g}^m. \tag{3.136}$$

Analogous transformations can be performed for higher order tensor fields. If $\mathbf{A} \in \mathbf{\mathcal{T}}_2$, for instance, then, recalling the four possible representations of this tensor in the bases $\boldsymbol{g}_k$ and $\boldsymbol{g}^k$, cf. Eq. (3.121), its gradient may also be expressed in four ways:

$$\begin{aligned}
\operatorname{grad} \mathbf{A} = \mathbf{A} \otimes \nabla &= \frac{\partial(A^{ij} \boldsymbol{g}_i \otimes \boldsymbol{g}_j)}{\partial y^k} \otimes \boldsymbol{g}^k = A^{ij}{}_{:k}\, \boldsymbol{g}_i \otimes \boldsymbol{g}_j \otimes \boldsymbol{g}^k && (3.137a)\\
&= \frac{\partial(A_{ij}\, \boldsymbol{g}^i \otimes \boldsymbol{g}^j)}{\partial y^k} \otimes \boldsymbol{g}^k = A_{ij:k}\, \boldsymbol{g}^i \otimes \boldsymbol{g}^j \otimes \boldsymbol{g}^k && (3.137b)\\
&= \frac{\partial(A^i{}_j\, \boldsymbol{g}_i \otimes \boldsymbol{g}^j)}{\partial y^k} \otimes \boldsymbol{g}^k = A^i{}_{j:k}\, \boldsymbol{g}_i \otimes \boldsymbol{g}^j \otimes \boldsymbol{g}^k && (3.137c)\\
&= \frac{\partial(A_i{}^j\, \boldsymbol{g}^i \otimes \boldsymbol{g}_j)}{\partial y^k} \otimes \boldsymbol{g}^k = A_i{}^j{}_{:k}\, \boldsymbol{g}^i \otimes \boldsymbol{g}_j \otimes \boldsymbol{g}^k, && (3.137d)
\end{aligned}$$

where

$$\begin{aligned}
A^{ij}{}_{:k} &\overset{\text{def}}{=} A^{ij}{}_{,k} + A^{lj}\Gamma^i_{lk} + A^{il}\Gamma^j_{lk}, && (3.138a)\\
A_{ij:k} &\overset{\text{def}}{=} A_{ij,k} - A_{lj}\Gamma^l_{ki} - A_{il}\Gamma^l_{kj}, && (3.138b)\\
A^i{}_{j:k} &\overset{\text{def}}{=} A^i{}_{j,k} + A^l{}_j\Gamma^i_{lk} - A^i{}_l\Gamma^l_{kj}, && (3.138c)\\
A_i{}^j{}_{:k} &\overset{\text{def}}{=} A_i{}^j{}_{,k} - A_l{}^j\Gamma^l_{ki} + A_i{}^l\Gamma^j_{lk}. && (3.138d)
\end{aligned}$$

Finally, it is noteworthy that covariant derivatives like $v^i{}_{:k}$, $A^{ij}{}_{:k}$ etc. are representations of tensors of respective orders (i.e. they are subject to the transformation rules (3.50)) while the ordinary derivatives like $v^i{}_{,k}$, $A^{ij}{}_{,k}$ as well as the Christoffel symbols are not. It is interesting to note that

$$g_{ij:k} = 0, \qquad g^{ij}{}_{:k} = 0 \tag{3.139}$$

(the so-called Ricci theorem) which implies the following, very convenient relation:

$$v^i{}_{;k} = (g^{il} v_l)_{;k} = g^{il} v_{l;k} \,. \tag{3.140}$$

Exercise 3.30
Prove Eq. (3.133).

Solution. Derivatives of the cobasis vectors $\boldsymbol{g}^i{}_{,k}$ are vectors which allows to express them as combinations of the cobasis vectors, i.e.

$$\boldsymbol{g}^i{}_{,k} = \Theta^i_{kl} \boldsymbol{g}^l,$$

where Θ^i_{kl} is a set of unknown coefficients. They can be determined by differentiating Eq. (3.109) with respect to the coordinates y^k as

$$(\boldsymbol{g}^i \cdot \boldsymbol{g}_j)_{,k} \equiv \boldsymbol{g}^i{}_{,k} \cdot \boldsymbol{g}_j + \boldsymbol{g}^i \cdot \boldsymbol{g}_{j,k} = 0.$$

In view of this assumption and the relations (3.131) and (3.109), the left-hand side in the above equations can be transformed as follows,

$$\Theta^i_{kl} \boldsymbol{g}^l \cdot \boldsymbol{g}_j + \boldsymbol{g}^i \cdot (\Gamma^l_{jk} \boldsymbol{g}_l) = \Theta^i_{kl} \delta^l{}_j + \Gamma^l_{jk} \delta^i_l = \Theta^i_{kj} + \Gamma^i_{jk} = 0,$$

which due to the symmetry of the Christoffel symbol yields the result $\Theta^i_{kl} = -\Gamma^i_{kl}$.

Exercise 3.31
Consider the following curvilinear coordinate system $\{y^i\}$

$$x_1 = y^1 \cos y^2, \qquad x_2 = y^1 \sin y^2, \qquad x_3 = y^3. \tag{3.141}$$

These are the so-called cylindrical coordinates (y^1—radius, y^2—angle). Determine the basis an cobasis vectors $\boldsymbol{g}^k$, $\boldsymbol{g}_k$, the matrices $[g_{ij}]$ and $[g^{ij}]$, the Christoffel symbols Γ^i_{kl} an the formulae for covariant coordinates of the components $v^i = v^i(y^j)$ of an arbitrary vector field $\boldsymbol{v}$.

Solution. The Jacobian matrix of derivatives of the mapping (3.141) has the form

$$\left[\frac{\partial x_i}{\partial y^j}\right] = \begin{bmatrix} \cos y^2 & -y^1 \sin y^2 & 0 \\ \sin y^2 & y^1 \cos y^2 & 0 \\ 0 & 0 & 1 \end{bmatrix}, \tag{3.142}$$

hence, according to Eq. (3.107), the local basis vectors have the form

$$\boldsymbol{g}_1 = \cos y^2\, \boldsymbol{e}_1 + \sin y^2\, \boldsymbol{e}_2\,, \qquad \boldsymbol{g}_2 = -y^1 \sin y^2\, \boldsymbol{e}_1 + y^1 \cos y^2\, \boldsymbol{e}_2\,, \qquad \boldsymbol{g}_3 = \boldsymbol{e}_3\,.$$

The matrix $[g_{ij}]$ can be determined from Eq. (3.108) as:

$$[g_{ij}] = \begin{bmatrix} 1 & 0 & 0 \\ 0 & (y^1)^2 & 0 \\ 0 & 0 & 1 \end{bmatrix},$$

and, since it is diagonal (the basis $\boldsymbol{g}_k$ is orthogonal here), its inverse can be expressed in quite a trivial form as

$$[g^{ij}] = \begin{bmatrix} 1 & 0 & 0 \\ 0 & 1/(y^1)^2 & 0 \\ 0 & 0 & 1 \end{bmatrix}.$$

The cobasis vectors are determined from Eq. (3.111) as

$$\boldsymbol{g}^1 = \boldsymbol{g}_1\,, \qquad \boldsymbol{g}^2 = -\frac{1}{y^1} \sin y^2\, \boldsymbol{e}_1 + \frac{1}{y^1} \cos y^2\, \boldsymbol{e}_2\,, \qquad \boldsymbol{g}^3 = \boldsymbol{g}_3\,.$$

To determine the Christoffel symbols, we need the inverse of the matrix (3.142), i.e.

$$\left[\frac{\partial y^i}{\partial x_j}\right] = \begin{bmatrix} \cos y^2 & \sin y^2 & 0 \\ -\frac{1}{y^1} \sin y^2 & \frac{1}{y^1} \cos y^2 & 0 \\ 0 & 0 & 1 \end{bmatrix}. \tag{3.143}$$

Its derivatives with respect to y^k become

$$\left[\frac{\partial^2 x_i}{\partial y^j y^1}\right] = \begin{bmatrix} 0 & -\sin y^2 & 0 \\ 0 & \cos y^2 & 0 \\ 0 & 0 & 0 \end{bmatrix}, \qquad \left[\frac{\partial^2 x_i}{\partial y^j y^2}\right] = \begin{bmatrix} -\sin y^2 & -y^1 \cos y^2 & 0 \\ \cos y^2 & -y^1 \sin y^2 & 0 \\ 0 & 0 & 0 \end{bmatrix},$$

$$\frac{\partial^2 x_i}{\partial y^j y^3} \equiv 0. \tag{3.144}$$

Remembering Eq. (3.130) let us multiply rows of the matrix (3.143) by columns of each of the three matrices (3.144) to get the matrices Γ^i_{j1}, Γ^i_{j2} and Γ^i_{j3}, respectively. The only non-zero components in the matrices appear to be the following:

$$\Gamma^2_{21} = \frac{1}{y^1}\,, \qquad \Gamma^2_{12} = \frac{1}{y^1}\,, \qquad \Gamma^1_{22} = -y^1\,.$$

Finally, the covariant coordinates of the vector field components $v^i{}_{;j}$ can be determined from Eq. (3.132) as:

$$\begin{array}{lll} v^1{}_{;1} = v^1{}_{,1}\,, & v^1{}_{;2} = v^1{}_{,2} - y^1 v^2\,, & v^1{}_{;3} = v^1{}_{,3}\,, \\ v^2{}_{;1} = v^2{}_{,1} + \frac{1}{y^1} v^2\,, & v^2{}_{;2} = v^2{}_{,2} + \frac{1}{y^1} v^1\,, & v^2{}_{;3} = v^2{}_{,3}\,, \\ v^3{}_{;1} = v^3{}_{,1}\,, & v^3{}_{;2} = v^3{}_{,2}\,, & v^3{}_{;3} = v^3{}_{,3}\,. \end{array}$$

Note: For $y^1 = 0$, i.e. on the central axis of the cylindrical coordinate system, the Jacobian matrix (3.142) is singular and some of the above quantities are undefined.

3.6 Notations Used in Tensor Description

In the discussion in this chapter, the following two notation systems were alternatively used to write down tensors of an order greater than 1:

- the *absolute notation*, in which tensors are expressed independently of any selected coordinate system so that vectors (first order tensors) are expressed as $\boldsymbol{v}$, $\boldsymbol{w}$, ..., while higher order tensors as $\boldsymbol{A}$, $\boldsymbol{B}$, ...,
- the *index notation*, requiring selection of a coordinate system, in which one refers to the components of vectors (say, v_k, w_l, ...), second order tensors (say, A_{ij}, B_{kl}, ...), etc., in this coordinate system.

Among advantages of the absolute notation one may point out:

- concise form,
- more clear reference to physical aspects of the problem considered.

On the other hand, such features of the other, index notation as

- convenience of complex algebraic transformations,
- the need of use for specific boundary value problems,

are important advantages of this description.

These two formalisms are not the only possible choices, though. Another frequently used notation system is the *matrix notation*. It is particularly convenient in designing computational procedures (i.e. computer implementation of various operations on tensors) and thus it surely deserves our careful attention. Since the readers are probably most familiar with this notation, as it is a subject of most basic courses in mathematics, we limit ourselves to its only short characteristics.

A fundamental object of the matrix notation is a rectangular matrix of the size $M \times N$ being a table of $M \cdot N$ numbers (matrix components) arranged in M rows and N columns:

$$\mathbf{A} = [A_{ij}]_{M\times N} = \begin{bmatrix} A_{11} & \cdots\cdots\cdots & A_{1N} \\ \vdots & \vdots & \vdots \\ \vdots & \cdots\cdots A_{ij} \cdots & \vdots \\ \vdots & \vdots & \vdots \\ A_{M1} & \cdots\cdots\cdots & A_{MN} \end{bmatrix} \begin{matrix} 1 \\ \vdots \\ i \\ \vdots \\ M \end{matrix} \quad . \qquad (3.145)$$

$$1 \quad \cdots\cdots \quad j \quad \cdots \quad N$$

The component A_{ij} is located at the intersection of the ith row and the jth column. If $M = N$ then $\mathbf{A}$ is called a square matrix.

A $M \times 1$ matrix is called a column matrix (or a column vector) and denoted as

$$\mathbf{A} = [A_{ij}]_{M\times 1} = \begin{bmatrix} A_{11} \\ \vdots \\ A_{M1} \end{bmatrix} \stackrel{\text{def}}{=} \{a_i\}_{M\times 1} = \mathbf{a} ,$$

i.e. $a_i = A_{i1}$, $i = 1, 2, \ldots, M$. A $1 \times N$ matrix is called a row matrix (or a row vector) and denoted as

$$\mathbf{B} = [B_{ij}]_{1\times N} = \begin{bmatrix} B_{11} \cdots B_{1N} \end{bmatrix} \stackrel{\text{def}}{=} \lfloor b_j \rfloor_{1\times N} = \mathbf{b}^{\mathrm{T}} ,$$

i.e. $b_j = B_{1j}$, $j = 1, 2, \ldots, N$.

A diagonal matrix is a square matrix $\mathbf{A}$ in which the only non-zero components A_{ij} are located on its main diagonal, $i = j$, i.e. $\forall_{i\neq j}\ A_{ij} = 0$. A special case of a diagonal matrix is the unit matrix $\mathbf{I} = [\delta_{ij}]_{N\times N}$, where δ_{ij} is the Kronecker symbol,

$$\delta_{ij} = \begin{cases} 1 & \text{for } i = j, \\ 0 & \text{for } i \neq j. \end{cases}$$

Among widely used operations on matrices are:

- product of a matrix and a number

$$\lambda\mathbf{A} = \lambda\left[A_{ij}\right]_{M\times N} = \left[\lambda A_{ij}\right]_{M\times N} ,$$

- product of two matrices

$$\mathbf{C} = \mathbf{AB}, \qquad [C_{ij}]_{M\times K} = [A_{ik}]_{M\times K}\,[B_{kj}]_{K\times N} ,$$

  only defined for two compatible matrices, i.e. such that the number of columns of $\mathbf{A}$ equals the number of rows of $\mathbf{B}$, because

$$C_{ij} = A_{ik}B_{kj} = A_{i1}b_{1j} + A_{i2}B_{2j} + \ldots + A_{iK}B_{Kj}$$

(generalization of the summation convention widely used in this book has been employed),

- trace of a square $N \times N$ matrix

$$\mathrm{tr}\,\mathbf{A} = A_{11} + A_{22} + \cdots + A_{NN}\,,$$

- transposition of a matrix

$$\mathbf{A} = \left[A_{ij}\right]_{M\times N} \quad \Longrightarrow \quad \mathbf{A}^{\mathrm{T}} = \left[A_{ji}\right]_{N\times M}\,,$$

for which the following relations also hold true:

$$(\lambda\mathbf{A})^{\mathrm{T}} = \lambda\mathbf{A}^{\mathrm{T}}, \qquad (\mathbf{A}+\mathbf{B})^{\mathrm{T}} = \mathbf{A}^{\mathrm{T}} + \mathbf{B}^{\mathrm{T}}, \qquad (\mathbf{AB})^{\mathrm{T}} = \mathbf{B}^{\mathrm{T}}\mathbf{A}^{\mathrm{T}},$$

- inverse of a square matrix (only defined for a class of so-called non-singular matrices)

$$\mathbf{A}^{-1}\mathbf{A} = \mathbf{A}\mathbf{A}^{-1} = \mathbf{I}\,;$$

for which the following relations also hold true:

$$(\lambda\mathbf{A})^{-1} = \frac{1}{\lambda}\mathbf{A}^{-1}, \qquad (\mathbf{AB})^{-1} = \mathbf{B}^{-1}\mathbf{A}^{-1}, \qquad (\mathbf{A}^{\mathrm{T}})^{-1} = (\mathbf{A}^{-1})^{\mathrm{T}}.$$

In view of the fundamental definitions related to the matrix algebra presented above, several analogies with definitions and operations on first and second order tensors become natural. If the representation of a second order tensor $\boldsymbol{A} \in \mathcal{T}_2$ in a selected Cartesian coordinate system is a matrix of its components $\mathbf{A} = [A_{ij}]_{3\times3}$, then its transpose $\boldsymbol{A}^{\mathrm{T}}$, inverse $\boldsymbol{A}^{-1}$ or trace $\mathrm{tr}\,\boldsymbol{A}$ are in this coordinate system represented by the matrices $\mathbf{A}^{\mathrm{T}}$, $\mathbf{A}^{-1}$ and the number $\mathrm{tr}\,\mathbf{A}$, respectively (the latter value is in fact independent of the coordinate system's choice). The inner product of two tensors $\boldsymbol{AB}$ is represented by a matrix being the result of the matrix product $\mathbf{AB}$ of their corresponding component matrices. Similar analogies may be found between vectors (i.e. the first order tensors) and column or row matrices. The product $\boldsymbol{Av}$ of a tensor $\boldsymbol{A} \in \mathcal{T}_2$ and a vector $\boldsymbol{v} \in \mathcal{T}_1$ is a vector represented by a column matrix being the result of the matrix product of the tensor's component matrix $\mathbf{A} = [A_{ij}]_{3\times3}$ and the column matrix $\mathbf{v} = \{v_i\}_{3\times1}$ containing the vector's components. A similar rule holds true for the product with reversed order of factors, $\boldsymbol{vA}$, however, the vector components have to be now arranged in the form of a row matrix $\lfloor v_i \rfloor_{1\times3}$, i.e., in the matrix notation, the product $\mathbf{v}^{\mathrm{T}}\mathbf{A}$ should be computed.

Table 3.1 illustrates equivalence of the three discussed formalisms in the description of several fundamental operations on tensors. Let us once again underline that vectors and second order tensors cannot be simply identified with their matrix representations ($\boldsymbol{v} \neq \mathbf{v}$, $\boldsymbol{A} \neq \mathbf{A}$)! The latter may only be used to represent tensors

Table 3.1 Equivalence of notations in description of tensor objects and operations

Absolute tensor notation	Index notation	Matrix notation
$\alpha = \boldsymbol{v}\boldsymbol{w}$	$\alpha = v_i w_i$	$\alpha = \mathbf{v}^{\mathrm{T}}\mathbf{w}$
$\boldsymbol{A} = \boldsymbol{v} \otimes \boldsymbol{w}$	$A_{ij} = v_i w_j$	$\mathbf{A} = \mathbf{v}\mathbf{w}^{\mathrm{T}}$
$\boldsymbol{w} = \boldsymbol{A}\boldsymbol{v}$	$w_i = A_{ij} v_j$	$\mathbf{w} = \mathbf{A}\mathbf{v}$
$\boldsymbol{w} = \boldsymbol{v}\boldsymbol{A}$	$w_j = v_i A_{ij}$	$\mathbf{w} = \mathbf{v}^{\mathrm{T}}\mathbf{A}$
$\alpha = \boldsymbol{v}\boldsymbol{A}\boldsymbol{w}$	$\alpha = v_i A_{ij} w_j$	$\alpha = \mathbf{v}^{\mathrm{T}}\mathbf{A}\mathbf{w}$
$\alpha = \operatorname{tr} \boldsymbol{A}$	$\alpha = A_{ii}$	$\alpha = \operatorname{tr} \mathbf{A}$
$\boldsymbol{C} = \boldsymbol{A}\boldsymbol{B}$	$C_{ij} = A_{ik} B_{kj}$	$\mathbf{C} = \mathbf{A}\mathbf{B}$
$\boldsymbol{C} = \boldsymbol{A}\boldsymbol{B}^{\mathrm{T}}$	$C_{ij} = A_{ik} B_{jk}$	$\mathbf{C} = \mathbf{A}\mathbf{B}^{\mathrm{T}}$
$\boldsymbol{C} = (\boldsymbol{A}\boldsymbol{B})^{\mathrm{T}} = \boldsymbol{B}^{\mathrm{T}}\boldsymbol{A}^{\mathrm{T}}$	$C_{ij} = A_{jk} B_{ki}$	$\mathbf{C} = (\mathbf{A}\mathbf{B})^{\mathrm{T}} = \mathbf{B}^{\mathrm{T}}\mathbf{A}^{\mathrm{T}}$
$\boldsymbol{D} = \boldsymbol{A}\boldsymbol{B}\boldsymbol{C}$	$D_{ij} = A_{ik} B_{kl} C_{lj}$	$\mathbf{D} = \mathbf{A}\mathbf{B}\mathbf{C}$
$\alpha = \boldsymbol{A} \cdot \boldsymbol{B}$	$\alpha = A_{ij} B_{ij}$	see text
$\boldsymbol{A} = \boldsymbol{G} \cdot \boldsymbol{B}$	$A_{ij} = G_{ijkl} B_{kl}$	see text
$\alpha \in \mathcal{T}_0,\ \boldsymbol{v}, \boldsymbol{w} \in \mathcal{T}_1,\ \boldsymbol{A}, \boldsymbol{B}, \boldsymbol{C}, \boldsymbol{D} \in \mathcal{T}_2,\ \boldsymbol{G} \in \mathcal{T}_4$		

in a selected coordinate system. However, it should be remembered that the notions of vector $\boldsymbol{v}$ and tensor $\boldsymbol{A}$ are independent of any coordinate system in the space.

As we can see from the table, there is a problem with writing down the full inner product of second order tensors in the matrix notation. For obvious reasons, the notations appears also insufficient for consistent description of higher order tensors which usually appear in problems of mechanics in the context of their full inner products with second order tensors, as in the last row of Table 3.1. To overcome this difficulty, a modification of the matrix notation may be proposed as illustrated below on an example of the operations $\alpha = A_{ij} B_{ij}$ and $A_{ij} = G_{ijkl} B_{kl}$.

Let us arrange the components A_{ij} and B_{ij} in the form of 9×1 column matrices $\tilde{\mathbf{A}}$ and $\tilde{\mathbf{B}}$, respectively. The ordering of components in the matrices is arbitrary provided consistency is preserved when applying it to all tensors. It is usually done in such a way that the first three matrix components are the diagonal tensor components, i.e.

$$\tilde{\mathbf{A}} = \left\{\tilde{A}_i\right\}_{9\times 1} = \{\ A_{11}\ A_{22}\ A_{33}\ A_{12}\ A_{23}\ A_{13}\ A_{21}\ A_{32}\ A_{31}\ \},$$
$$\tilde{\mathbf{B}} = \left\{\tilde{B}_i\right\}_{9\times 1} = \{\ B_{11}\ B_{22}\ B_{33}\ B_{12}\ B_{23}\ B_{13}\ B_{21}\ B_{32}\ B_{31}\ \}. \qquad (3.146)$$

It is obvious that the full inner product operation corresponds in such notation to pair-wise multiplication of subsequent components of the matrices $\tilde{\mathbf{A}}$ and $\tilde{\mathbf{B}}$ and summing up the results, i.e.

$$\alpha = \boldsymbol{A} \cdot \boldsymbol{B} = A_{ij} B_{ij} = \tilde{\mathbf{A}}^{\mathrm{T}}\,\tilde{\mathbf{B}}\,. \qquad (3.147)$$

The operation $\boldsymbol{A} = \boldsymbol{G} \cdot \boldsymbol{B}$, i.e. $A_{ij} = G_{ijkl} B_{kl}$, can be presented in this notation in the form

$$\tilde{\mathbf{A}} = \tilde{\mathbf{G}} \, \tilde{\mathbf{B}}, \qquad \tilde{\mathbf{G}} = \left[\tilde{G}_{ij} \right]_{9\times 9}, \tag{3.148}$$

in which 81 components of the matrix $\tilde{G}_{ij}$ contain 81 components G_{ijkl} representing the tensor $\boldsymbol{G}$, arranged as follows:

$$\left[\tilde{G}_{ij} \right] = \begin{bmatrix} G_{1111} & G_{1122} & G_{1133} & G_{1112} & G_{1123} & G_{1113} & G_{1121} & G_{1132} & G_{1131} \\ G_{2211} & G_{2222} & G_{2233} & G_{2212} & G_{2223} & G_{2213} & G_{2221} & G_{2232} & G_{2231} \\ G_{3311} & G_{3322} & G_{3333} & G_{3312} & G_{3323} & G_{3313} & G_{3321} & G_{3332} & G_{3331} \\ G_{1211} & G_{1222} & G_{1233} & G_{1212} & G_{1223} & G_{1213} & G_{1221} & G_{1232} & G_{1231} \\ G_{2331} & G_{2322} & G_{2333} & G_{2312} & G_{2323} & G_{2313} & G_{2321} & G_{2332} & G_{2331} \\ G_{1311} & G_{1322} & G_{1333} & G_{1312} & G_{1323} & G_{1313} & G_{1321} & G_{1332} & G_{1331} \\ G_{2111} & G_{2122} & G_{2133} & G_{2112} & G_{2123} & G_{2113} & G_{2121} & G_{2132} & G_{2131} \\ G_{3211} & G_{3222} & G_{3233} & G_{3212} & G_{3223} & G_{3213} & G_{3221} & G_{3232} & G_{3231} \\ G_{3111} & G_{3122} & G_{3133} & G_{3112} & G_{3123} & G_{3113} & G_{3121} & G_{3132} & G_{3131} \end{bmatrix}. \tag{3.149}$$

Note that in the presented notation one may also write:

$$A_{ij} G_{ijkl} B_{kl} = \tilde{\mathbf{A}}^{\mathrm{T}} \, \tilde{\mathbf{G}} \, \tilde{\mathbf{B}}. \tag{3.150}$$

In solid mechanics applications, tensors being subjects of the above operations are frequently symmetric, i.e.

$$A_{ij} = A_{ji}, \qquad B_{ij} = B_{ji}, \qquad G_{ijkl} = G_{jikl} = G_{ijlk}. \tag{3.151}$$

In such a case, the matrices $\tilde{\mathbf{A}}$ and $\tilde{\mathbf{B}}$ contain repeated components and it is natural to simplify the notation proposed by removing the latter three of them and reduce the matrices' size to 6×1. In an analogous way, the matrix $\tilde{\mathbf{G}}$ may be reduced to the size 6×6. Let us point out, however, that the result of the full inner product $A_{ij} B_{ij}$ upon such simplification will no longer be equal to $\tilde{\mathbf{A}}^{\mathrm{T}} \, \tilde{\mathbf{B}}$ because the removed terms $A_{21} B_{21} + A_{32} B_{32} + A_{31} B_{31}$ are missing. To preserve the unchanged form of Eq. (3.147), the following modifications of the reduced-dimension column matrices $\tilde{\mathbf{A}}$ and $\tilde{\mathbf{B}}$ may be introduced, for instance:

$$\tilde{\mathbf{A}} = \left\{ \tilde{A}_i \right\}_{6\times 1} = \{ \, A_{11} \;\; A_{22} \;\; A_{33} \;\; A_{12} \;\; A_{23} \;\; A_{13} \, \}, \tag{3.152a}$$

$$\tilde{\mathbf{B}} = \left\{ \tilde{B}_i \right\}_{6\times 1} = \{ \, B_{11} \;\; B_{22} \;\; B_{33} \;\; 2B_{12} \;\; 2B_{23} \;\; 2B_{13} \, \}, \tag{3.152b}$$

or, alternatively:

$$\tilde{\mathbf{A}} = \left\{ \tilde{A}_i \right\}_{6\times 1} = \left\{ A_{11} \; A_{22} \; A_{33} \; \sqrt{2}A_{12} \; \sqrt{2}A_{23} \; \sqrt{2}A_{13} \right\},$$
$$\tilde{\mathbf{B}} = \left\{ \tilde{B}_i \right\}_{6\times 1} = \left\{ B_{11} \; B_{22} \; B_{33} \; \sqrt{2}B_{12} \; \sqrt{2}B_{23} \; \sqrt{2}B_{13} \right\}. \tag{3.153}$$

The first definition, looking more simple at first glance, brings some inconvenience as it requires introduction of an additional convention to determine which tensors are subject to the notation (3.152a) or (3.152b), respectively.

Let us now try to employ the reduced-dimension formulation to write down the full inner product $A_{ij} = G_{ijkl}B_{kl}$ for the symmetries (3.151), in the reduced-dimension matrix notation. It is easy to verify that the matrix product $\tilde{\mathbf{A}} = \tilde{\mathbf{G}}\tilde{\mathbf{B}}$ yields the correct result for the definition (3.152) and for the matrix $\tilde{\mathbf{G}}$ defined as

$$\left[\tilde{G}_{ij} \right]_{6\times 6} = \begin{bmatrix} G_{1111} & G_{1122} & G_{1133} & G_{1112} & G_{1123} & G_{1113} \\ G_{2211} & G_{2222} & G_{2233} & G_{2212} & G_{2223} & G_{2213} \\ G_{3311} & G_{3322} & G_{3333} & G_{3312} & G_{3323} & G_{3313} \\ G_{1211} & G_{1222} & G_{1233} & G_{1212} & G_{1223} & G_{1213} \\ G_{2331} & G_{2322} & G_{2333} & G_{2312} & G_{2323} & G_{2313} \\ G_{1311} & G_{1322} & G_{1333} & G_{1312} & G_{1323} & G_{1313} \end{bmatrix}. \tag{3.154}$$

For the definition (3.153), the latter matrix must be additionally modified by multiplying some of its components by $\sqrt{2}$ or 2:

$$\left[\tilde{G}_{ij} \right]_{6\times 6} = \begin{bmatrix} G_{1111} & G_{1122} & G_{1133} & \sqrt{2}G_{1112} & \sqrt{2}G_{1123} & \sqrt{2}G_{1113} \\ G_{2211} & G_{2222} & G_{2233} & \sqrt{2}G_{2212} & \sqrt{2}G_{2223} & \sqrt{2}G_{2213} \\ G_{3311} & G_{3322} & G_{3333} & \sqrt{2}G_{3312} & \sqrt{2}G_{3323} & \sqrt{2}G_{3313} \\ \sqrt{2}G_{1211} & \sqrt{2}G_{1222} & \sqrt{2}G_{1233} & 2G_{1212} & 2G_{1223} & 2G_{1213} \\ \sqrt{2}G_{2331} & \sqrt{2}G_{2322} & \sqrt{2}G_{2333} & 2G_{2312} & 2G_{2323} & 2G_{2313} \\ \sqrt{2}G_{1311} & \sqrt{2}G_{1322} & \sqrt{2}G_{1333} & 2G_{1312} & 2G_{1323} & 2G_{1313} \end{bmatrix}. \tag{3.155}$$

However, to correctly reproduce the equality (3.150), i.e. to express the quantity $A_{ij}G_{ijkl}B_{kl}$ with symmetries (3.151) in the form of a reduced-dimension matrix operation $\tilde{\mathbf{A}}^{\mathrm{T}}\,\tilde{\mathbf{G}}\,\tilde{\mathbf{B}}$, the correct result is only obtained for the definitions (3.153) and (3.155). Application of the definitions (3.152) and (3.154) would require additional modifications during computations, e.g. doubling the last three components of the column matrix $\mathbf{A}$ before multiplying its transpose by $\tilde{\mathbf{G}}\tilde{\mathbf{B}}$.

Reference

1. Korn G.A., Korn T.M., 1961. *Mathematical Handbook for Scientists and Engineers*. McGraw-Hill.

Chapter 4
Motion, Deformation and Strain in Material Continuum

4.1 Motion of Bodies

A fundamental property of material bodies is their spatial location, i.e. the fact that they occupy certain domains in the Euclidean space E^3. Since each body may occupy different domains at different time instants, none of the domains can be uniquely identified with the body itself. Nevertheless, it is convenient to select one of them as a "reference configuration" of the body; this can be either one of actual geometric configurations assumed by the body during its motion or a hypothetical configuration, potentially possible for this body.

Let us denote the considered material body by the symbol $\mathcal{B}$ and the domain occupied by the body in a selected reference configuration by $\Omega^r \subset E^3$. The domain Ω^r is an open set in E^3 and its boundary $\partial\Omega^r$ is a closed, piece-wise smooth surface. Let $P \in \mathcal{B}$ be a typical material point, i.e. an infinitely small particle of the body. Let us also denote by $\boldsymbol{x}^r \in \Omega^r \subset E^3$ a typical geometrical point in the reference domain. The definition of material body implies the following unique mapping $\boldsymbol{f} : \mathcal{B} \rightarrow \Omega^r \subset E^3$,

$$\boldsymbol{x}^r = \boldsymbol{f}(P)\,, \qquad P \in \mathcal{B}\,. \tag{4.1}$$

Let us consider motion of the body $\mathcal{B}$ within the time interval $[0, \bar{t}]$. Spatial locations of all material points at subsequent time instants $\tau \in [0, \bar{t}]$ will be called instantaneous configurations of $\mathcal{B}$ and denoted by C^τ, Fig. 4.1. In the configuration C^τ material points of the body occupy the domain $\Omega^\tau \subset E^3$. According to the assumed convention, if the distinguished reference configuration C^r coincides with one of the actual configurations of the body $\mathcal{B}$ during its motion, then the upper index r may be referred to as a time instant $\tau = r \in [0, \bar{t}]$ at which the configuration was assumed. In fact, such a choice this is the most natural in practical analyses—the reference configuration in our further discussion will be usually identified with either the initial configuration C^0 or a configuration C^t corresponding to a typical time

M. Kleiber and P. Kowalczyk, *Introduction to Nonlinear Thermomechanics of Solids*, Lecture Notes on Numerical Methods in Engineering and Sciences,
DOI 10.1007/978-3-319-33455-4_4

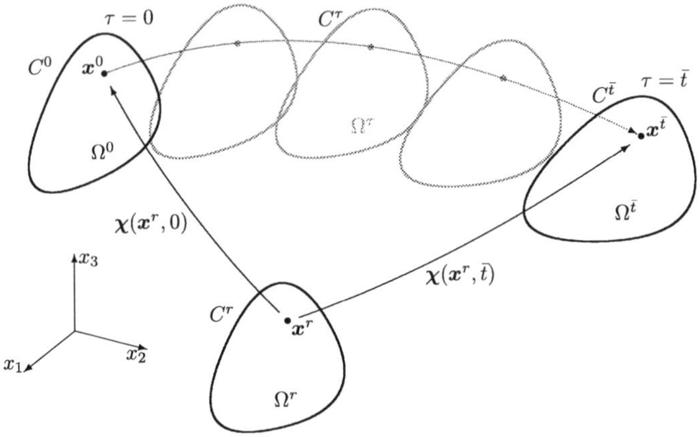

Fig. 4.1 Configurations of a body in motion

instant $\tau = t$. The way the configuration C^r has been depicted in Fig. 4.1 is only intended to underline arbitrary character of its choice.

To make our considerations as clear as possible, only one preselected Cartesian coordinate system $\{x_k\}$ in the Euclidean space will be subsequently employed. All tensor objects in further considerations will be represented by their components in this coordinate system. The summation convention holds true for double indices appearing in monomials.

Denote by $\boldsymbol{x}^\tau = (x_k^\tau)$ the location of the considered particle P at $\tau \in [0, \bar{t}]$. At this time instant, in its geometric configuration C^τ, the body $\mathcal{B}$ occupies the domain Ω^τ, i.e. $\boldsymbol{x}^\tau \in \Omega^\tau \subset E^3$. In particular, at the initial time instant $\tau = 0$, the body's material points coincide with the geometrical points $\boldsymbol{x}^0 \in \Omega^0 \subset E^3$. Let us recall that in the reference configuration $\boldsymbol{x}^r \in \Omega^r \subset E^3$.

Motion of the particle P in the space E^3 is described by the mapping

$$\boldsymbol{x}^\tau = \boldsymbol{\chi}(\boldsymbol{x}^r, \tau) \overset{\text{def}}{=} \boldsymbol{x}^\tau(\boldsymbol{x}^r), \qquad x_k^\tau = x_k(x_l^r, \tau) = x_k^\tau(x_l^r) \tag{4.2}$$

which for each $\tau \in [0, \bar{t}]$ is a unique mapping $E^3 \to E^3$. In other words, the field $\boldsymbol{x}^\tau(\boldsymbol{x}^r)$ parameterized by τ and defined on the set of all material point coordinates of $\mathcal{B}$ in the reference configuration describes the motion of the material points in time. It is assumed that the mapping $\boldsymbol{\chi}(\boldsymbol{x}^r, \tau)$ has continuous partial derivatives with respect to both its arguments up to the order required in the subsequent discussion (i.e. the second order). Apart from the uniqueness and differentiability, no other requirements will be postulated for the mapping, such as limitations on distance change between points (which may vary in time as the body is assumed deformable).

An example of motion is the the *rigid motion* described by the following mapping

$$\boldsymbol{\chi}(\boldsymbol{x}^r, \tau) = \boldsymbol{R}(\tau) \cdot \boldsymbol{x}^r + \hat{\boldsymbol{u}}(\tau), \tag{4.3}$$

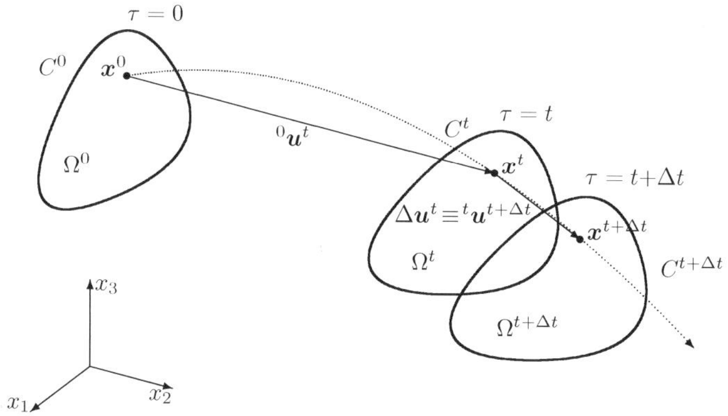

Fig. 4.2 Configurations of a body at a typical time instant and in a typical time increment

where $\boldsymbol{R}(\tau)$ denotes an arbitrary time-dependent rotation tensor while $\hat{\boldsymbol{u}}(\tau)$ a time-dependent vector of the rotation centre displacement. It will be shown that the so-defined mapping preserves distances between material points at all time instances τ, i.e. the body does not change its shape. Another example of motion is *inflation*, i.e. uniform volume deformation defined as

$$\chi(\boldsymbol{x}^r, \tau) = k(\tau)\boldsymbol{x}^r + \hat{\boldsymbol{u}}(\tau), \tag{4.4}$$

where $k(\tau)$ denotes an arbitrary scalar function with positive values denoting time-dependent scaling factor of the body's linear dimensions while $\hat{\boldsymbol{u}}(\tau)$ describes displacement of the body's centre.

Our subsequent discussion will concentrate on motion from an initial configuration C^0 (assumed at $\tau = 0$) to another configuration, further called current, corresponding to a selected, typical time instant $\tau = t$, Fig. 4.2. Our objective is to find the solution to the problem of mechanics (i.e. to determine, among other unknowns, locations of the body's points) at each of the time instants $\tau \in [0, \bar{t}]$. Looking for it at an arbitrary instant $\tau = t$ seems obviously a step in the good direction—having a method of the problem solution at $\tau = t$ on hand, one can in analogy apply it to any other time instant and in this way solve the problem in the entire time interval $[0, \bar{t}]$.

In practice, the motion of the body will be analysed at subsequent discrete time instants with interpolation of solution within short time intervals between them. Since many problems in mechanics have an evolutionary character and thus the solution has to be searched step-by-step in subsequent time instants, let us define the notion of *time increment* that separates the time instant $\tau = t$ from the next "close" time instant $\tau = t + \Delta t$, Fig. 4.2.

The vector

$$ {}^{t_1}\boldsymbol{u}^{t_2} \overset{\text{def}}{=} \boldsymbol{x}^{t_2} - \boldsymbol{x}^{t_1}, \qquad {}^{t_1}u_i^{t_2} = x_i^{t_2} - x_i^{t_1}. \tag{4.5}$$

is called the *displacement* of a particle between the two configurations C^{t_1} and C^{t_2}. In particular, total displacement of the particle from the beginning of its motion at $\tau = 0$ to its current location at $\tau = t$ is denoted as ${}^{0}\boldsymbol{u}^{t} = \boldsymbol{x}^{t} - \boldsymbol{x}^{0}$, Fig. 4.2, while for its increment in the interval $[t, t + \Delta t]$ the following notation is introduced:

$$\Delta \boldsymbol{u}^{t} \stackrel{\text{def}}{=} {}^{t}\boldsymbol{u}^{t+\Delta t} = \boldsymbol{x}^{t+\Delta t} - \boldsymbol{x}^{t} . \tag{4.6}$$

Displacements between various configurations are additive according to the rule ${}^{t_1}\boldsymbol{u}^{t_2} + {}^{t_2}\boldsymbol{u}^{t_3} = {}^{t_1}\boldsymbol{u}^{t_3}$. Obviously, the relation ${}^{t_1}\boldsymbol{u}^{t_2} = -{}^{t_2}\boldsymbol{u}^{t_1}$ holds true while the displacement between two identical locations is zero, e.g. ${}^{t}\boldsymbol{u}^{t} \equiv \boldsymbol{0}$.

Since the particle locations at various time instants $\boldsymbol{x}^{\tau}$ are transformations of their locations in the reference configuration $\boldsymbol{x}^{r}$, cf. Eq. (4.2), all the above defined displacements and their increments are vector fields defined on $\boldsymbol{x}^{r} \in \Omega^{r}$. In particular, displacement with respect to the reference configuration, e.g. ${}^{r}\boldsymbol{u}^{t} = \boldsymbol{x}^{t} - \boldsymbol{x}^{r}$ is such a field, which is a generalization of the definition (4.5).

In cases there is no doubt about the choice of the limit time instants the corresponding upper indices at the displacement symbol will be skipped.

If the body does not move and deform, than the displacement field of its points vanishes, $\boldsymbol{u} \equiv \boldsymbol{0}$.

The *deformation gradient* at the time instant $\tau = t$ is the second order tensor,

$${}^{r}\boldsymbol{F}^{t} \stackrel{\text{def}}{=} \nabla \boldsymbol{x}^{t}(\boldsymbol{x}^{r}) = \nabla \boldsymbol{\chi}(\boldsymbol{x}^{r}, \tau), \qquad {}^{r}F^{t}_{ij} = \frac{\partial x^{t}_{i}}{\partial x^{r}_{j}} = x^{t}_{i,j} , \tag{4.7}$$

where the operator $\nabla(\cdot)$ is the gradient operator grad $(\cdot)$, cf. Eq. (3.99), and the notation $(\cdot)_{,i}$ will be consistently used for simplified description of partial differentiation with respect to x^{r}_{i}. Since the deformation gradient may be defined at each material point of the body $\mathcal{B}$, the deformation gradient is a second order tensor field on $\boldsymbol{x}^{r} \in \Omega^{r}$.

The notion of deformation gradient may be generalized to various other reference configurations,

$${}^{t_1}F^{t_2}_{ij} = \frac{\partial x^{t_2}_{i}}{\partial x^{t_1}_{j}} ,$$

since, if the mapping $\boldsymbol{x}^{\tau}(\boldsymbol{x}^{r})$ is unique for each τ, there are no formal limitations in expressing coordinates at an arbitrary time instant as functions of coordinates at another arbitrary time instant. The following multiplicative rule of combining gradients holds true:

$${}^{r}\boldsymbol{F}^{t_2} = {}^{t_1}\boldsymbol{F}^{t_2}\, {}^{r}\boldsymbol{F}^{t_1} , \qquad {}^{r}F^{t_2}_{ij} = {}^{t_1}F^{t_2}_{ik}\, {}^{r}F^{t_1}_{kj} . \tag{4.8}$$

In particular, for two identical configurations we have ${}^{t}\boldsymbol{F}^{t} \equiv \boldsymbol{I}$. Hence, in view of Eq. (4.8) we can conclude that ${}^{t_1}\boldsymbol{F}^{t_2}\, {}^{t_2}\boldsymbol{F}^{t_1} = {}^{t_2}\boldsymbol{F}^{t_1}\, {}^{t_1}\boldsymbol{F}^{t_2} = \boldsymbol{I}$, i.e.

$$\left({}^{t_1}\boldsymbol{F}^{t_2}\right)^{-1} = {}^{t_2}\boldsymbol{F}^{t_1} . \tag{4.9}$$

In cases when there is no doubt about the choice of the limit time instants, the corresponding upper indices at the deformation gradient symbol $\boldsymbol{F}$ will be skipped.

If the body does not move, then the deformation gradient field equals ${}^r\boldsymbol{F}^\tau \equiv \boldsymbol{I}$. If the motion is described by Eq. (4.3) or (4.4), then the corresponding deformation gradient is expressed as ${}^r\boldsymbol{F}^\tau = \boldsymbol{R}(\tau)$ or ${}^r\boldsymbol{F}^\tau = k(\tau)\,\boldsymbol{I}$, respectively. It is noteworthy that in both the cases the translation $\hat{u}(\tau)$ does not affect the deformation gradient—at pure translation the latter is the same as at the no motion case.

Since, in view of the definition of displacement, $\boldsymbol{x}^t = \boldsymbol{x}^r + {}^r\boldsymbol{u}^t$, the deformation gradient defined by Eq. (4.7) may be expressed by the displacement gradient:

$$ {}^r\boldsymbol{F}^t = \boldsymbol{I} + \nabla\,{}^r\boldsymbol{u}^t, \qquad {}^rF^t_{ij} = \delta_{ij} + \frac{\partial\,{}^ru^t_i}{\partial x^r_j} = \delta_{ij} + {}^ru^t_{i,j}\,, \tag{4.10} $$

which can be generalized to any other two configurations of the body in motion as

$$ {}^{t_1}F^{t_2}_{ij} = \delta_{ij} + \frac{\partial\,{}^{t_1}u^{t_2}_i}{\partial x^{t_1}_j}\,. \tag{4.11} $$

Remark. While introducing the inverse deformation gradient in Eq. (4.9) we have not mentioned the important question of its existence, i.e. the nonsingularity of the tensor $\boldsymbol{F}$ (whichever two configurations might be involved here), $\det \boldsymbol{F} \neq 0$. In fact, this holds true which is implied by the assumption of reversibility of the differentiated transformation $\boldsymbol{x}^\tau(\boldsymbol{x}^r)$. Moreover, an even stronger assumption will be made in further considerations, i.e. $\det \boldsymbol{F} > 0$. This means that our discussion will not cover mappings $\boldsymbol{x}^\tau(\boldsymbol{x}^r)$ that include local symmetric flips of material particles. This may be justified as a consequence of the assumed continuity of the mapping in time τ—nonsingularity of $\boldsymbol{F}$ implies that its determinant $\det \boldsymbol{F}$ cannot change its sign during the motion and since it is positive for one of the configurations C^r, it must remain positive for all other configurations.

4.2 Strain

4.2.1 Definitions

In Sect. 2.1, an example of 1D bar extended along its axis was discussed. Two notions were defined there: deformation λ and strain ε. The first of them, the ratio of the current (deformed) to the initial bar length, is—as it is easy to observe—a one-dimensional counterpart of the deformation gradient tensor (more specifi-

Fig. 4.3 Deformation of an infinitesimal linear element

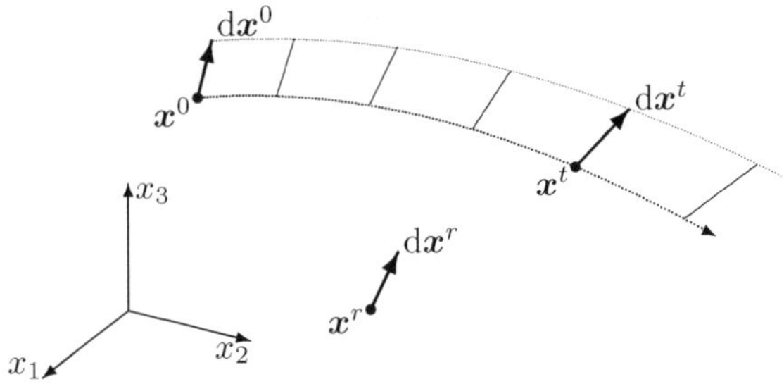

cally, its ${}^{r}F^{t}_{11}$ component), under the assumption that the initial bar's configuration is assumed as the reference configuration. We may also introduce a tensor quantity as the 3D counterpart of 1D strain defined as the ratio of the extension length Δl to the bar's initial length l. As we are going to see soon, even in the simple 1D case the so-defined strain is only an approximate of the rigorously defined strain measure, valid only for $\Delta l \ll l$.

Consider an infinitesimal vector $\mathrm{d}\boldsymbol{x}^r$ at a given point $\boldsymbol{x}^r \in \Omega^r$ that can be imagined as a connector between the material particle located at $\boldsymbol{x}^r$ and a "neighbouring" particle in the configuration C^r (Fig. 4.3). Such a vector will be called a linear material element. A square of its length is $(\mathrm{d}l^r)^2 = \mathrm{d}\boldsymbol{x}^r \cdot \mathrm{d}\boldsymbol{x}^r$. In an arbitrary current configuration C^τ this vector takes the form

$$\mathrm{d}\boldsymbol{x}^\tau = {}^{r}\boldsymbol{F}^\tau \, \mathrm{d}\boldsymbol{x}^r , \qquad \mathrm{d}x^\tau_i = \frac{\partial x^\tau_i}{\partial x^r_j} \, \mathrm{d}x^r_j ,$$

so that

$$(\mathrm{d}l^\tau)^2 = \left({}^{r}\boldsymbol{F}^\tau \, \mathrm{d}\boldsymbol{x}^r\right)\left({}^{r}\boldsymbol{F}^\tau \, \mathrm{d}\boldsymbol{x}^r\right) = \mathrm{d}\boldsymbol{x}^r \, {}^{r}\boldsymbol{C}^\tau \, \mathrm{d}\boldsymbol{x}^r , \tag{4.12}$$

where

$${}^{r}\boldsymbol{C}^\tau \overset{\text{def}}{=} ({}^{r}\boldsymbol{F}^\tau)^{\mathrm{T}} \, {}^{r}\boldsymbol{F}^\tau , \qquad {}^{r}C^\tau_{ij} = {}^{r}F^\tau_{ki} \, {}^{r}F^\tau_{kj} , \tag{4.13}$$

is a symmetric tensor called the *right Cauchy–Green deformation tensor*.

Let us now consider the motion of the body between the initial configuration C^0 and the current configuration C^t. At a given material point let us define the *strain* ${}^{0}_{r}\varepsilon^t$ from the configuration C^0 to C^t as a second order tensor that, for each $\mathrm{d}\boldsymbol{x}^r$, fulfills the following condition:

$$\mathrm{d}\boldsymbol{x}^r \, {}^{0}_{r}\varepsilon^t \, \mathrm{d}\boldsymbol{x}^r = \frac{1}{2}\left[(\mathrm{d}l^t)^2 - (\mathrm{d}l^0)^2\right] . \tag{4.14}$$

Substituting Eq. (4.12) for $\tau = 0$ and $\tau = t$ we can derive the following form of this tensor:

$$ {}^{0}_{r}\varepsilon^{t} = \frac{1}{2}\left({}^{r}\boldsymbol{C}^{t} - {}^{r}\boldsymbol{C}^{0}\right) = \frac{1}{2}\left[({}^{r}\boldsymbol{F}^{t})^{\mathrm{T}}\,{}^{r}\boldsymbol{F}^{t} - ({}^{r}\boldsymbol{F}^{0})^{\mathrm{T}}\,{}^{r}\boldsymbol{F}^{0}\right], \tag{4.15a} $$

$$ {}^{0}_{r}\varepsilon^{t}_{ij} = \frac{1}{2}\left({}^{r}F^{t}_{ki}\,{}^{r}F^{t}_{kj} - {}^{r}F^{0}_{ki}\,{}^{r}F^{0}_{kj}\right). \tag{4.15b} $$

Recalling now the relation (4.10), one may also write after transformation:

$$ \begin{aligned} {}^{0}_{r}\varepsilon^{t}_{ij} &= \frac{1}{2}\left(\frac{\partial\,{}^{0}u^{t}_{i}}{\partial x^{r}_{j}} + \frac{\partial\,{}^{0}u^{t}_{j}}{\partial x^{r}_{i}} + \frac{\partial\,{}^{r}u^{t}_{k}}{\partial x^{r}_{i}}\frac{\partial\,{}^{r}u^{t}_{k}}{\partial x^{r}_{j}} - \frac{\partial\,{}^{r}u^{0}_{k}}{\partial x^{r}_{i}}\frac{\partial\,{}^{r}u^{0}_{k}}{\partial x^{r}_{j}}\right). \\ &= \frac{1}{2}\left({}^{0}u^{t}_{i,j} + {}^{0}u^{t}_{j,i} + {}^{r}u^{t}_{k,i}\,{}^{r}u^{t}_{k,j} - {}^{r}u^{0}_{k,i}\,{}^{r}u^{0}_{k,j}\right). \end{aligned} \tag{4.16} $$

Let us note that the strain tensor is defined with respect to the predefined reference configuration which means that it describes relations between geometric dimensions in three configurations: initial C^0, current C^t and reference C^r (this explains the presence of three indices at the tensor symbol). The tensor is symmetric which can be easily verified by mutual replacement of the indices i, j in the above formulae. Finally, let us note that in the case of no motion ($\boldsymbol{u} \equiv \boldsymbol{0}$) the strain tensor vanishes as well.

Consider now two special cases of the reference configuration's choice, i.e. $C^r = C^0$ and $C^r = C^t$.

The description of motion in which the reference configuration is identified with the initial configuration of the body is called the *Lagrangian description*. In this case we have ${}^{r}\boldsymbol{u}^{0} = \boldsymbol{0}$ and ${}^{r}\boldsymbol{F}^{0} = \boldsymbol{I}$. Skipping upper indices at the symbols of displacement and deformation gradient, i.e. assuming $\boldsymbol{u} \stackrel{\text{def}}{=} {}^{0}\boldsymbol{u}^{t}$ and $\boldsymbol{F} \stackrel{\text{def}}{=} {}^{0}\boldsymbol{F}^{t}$, we can express the strain tensor as

$$ {}^{0}_{0}\varepsilon^{t} = \frac{1}{2}\left(\boldsymbol{F}^{\mathrm{T}}\boldsymbol{F} - \boldsymbol{I}\right) = \frac{1}{2}(\boldsymbol{C} - \boldsymbol{I}), \tag{4.17a} $$

$$ {}^{0}_{0}\varepsilon^{t}_{ij} = \frac{1}{2}\left(\frac{\partial u_i}{\partial x^0_j} + \frac{\partial u_j}{\partial x^0_i} + \frac{\partial u_k}{\partial x^0_i}\frac{\partial u_k}{\partial x^0_j}\right). \tag{4.17b} $$

This tensor, describing the total (initial-to-current) strain with respect to the initial configuration is called the *Green strain tensor* and denoted as ε^G.

The description of motion in which the reference configuration is identified with the current configuration of the body, i.e. $C^r = C^t$, is called the *Eulerian description*. In this case we have ${}^{r}\boldsymbol{u}^{t} = \boldsymbol{0}$ and ${}^{r}\boldsymbol{F}^{t} = \boldsymbol{I}$. Recalling additionally that ${}^{t}\boldsymbol{F}^{0} = ({}^{0}\boldsymbol{F}^{t})^{-1}$ we can again simplify the form of the strain tensor as

$$ {}^{0}_{t}\varepsilon^{t} = \frac{1}{2}\left[\boldsymbol{I} - \left(\boldsymbol{F}\boldsymbol{F}^{\mathrm{T}}\right)^{-1}\right] = \frac{1}{2}\left(\boldsymbol{I} - \boldsymbol{B}^{-1}\right), \tag{4.18a} $$

$$ {}^{0}_{t}\varepsilon^{t}_{ij} = \frac{1}{2}\left(\frac{\partial u_i}{\partial x^t_j} + \frac{\partial u_j}{\partial x^t_i} - \frac{\partial u_k}{\partial x^t_i}\frac{\partial u_k}{\partial x^t_j}\right), \tag{4.18b} $$

where $\boldsymbol{B} = \boldsymbol{F}\boldsymbol{F}^{\mathrm{T}}$ denotes the *left Cauchy–Green deformation tensor* which, as it is easy to verify, is a symmetric tensor again. It describes the total strain with respect to the current configuration and it is called the *Almansi strain tensor* denoted as $\boldsymbol{\varepsilon}^{A}$.

In many problems of mechanics, displacements values of material points are much (by, say, 3 orders of magnitude) smaller than the body's geometric dimensions. In such a case, changes of the overall geometric configuration of the body are insignificant and the displacement gradient components $|\partial u_i/\partial x_j| \ll 1$. Looking at Eqs. (4.16), (4.17b) and (4.18b) we can propose the following simplifications in their form that, under the above assumptions, may be assumed not to result in significant numerical errors:

- skipping all the products of the form $\frac{\partial u_k}{\partial x_i}\frac{\partial u_k}{\partial x_j}$ as negligible in comparison to the gradients $\frac{\partial u_i}{\partial x_j}$ and
- neglecting differences between differentiation with respect to the initial or current coordinates of the particle (as the choice of the reference configuration becomes of little importance in this kind of deformation).

Under such assumptions, all the strain measures defined above reduce to the following linearized approximate measure called the *Cauchy strain tensor*:

$$ \boldsymbol{\varepsilon} = \mathrm{sym}\,(\nabla\boldsymbol{u}) \equiv \frac{1}{2}\left[\nabla\boldsymbol{u} + (\nabla\boldsymbol{u})^{\mathrm{T}}\right], \qquad \varepsilon_{ij} = \frac{1}{2}\left(\frac{\partial u_i}{\partial x_j} + \frac{\partial u_j}{\partial x_i}\right), \tag{4.19} $$

which may also be denoted as $\boldsymbol{\varepsilon}^{C}$ since the difference between the Lagrangian and Eulerian description disappears.

A strain tensor defined with respect to a given reference configuration can be redefined with respect to another reference configuration by simple algebraic operations. Assume the strain state is described with respect to a configuration C^{t_1} by the tensor

$$ {}^{0}_{t_1}\varepsilon^{t} = \frac{1}{2}\left[({}^{t_1}\boldsymbol{F}^{t})^{\mathrm{T}}\,{}^{t_1}\boldsymbol{F}^{t} - ({}^{t_1}\boldsymbol{F}^{0})^{\mathrm{T}}\,{}^{t_1}\boldsymbol{F}^{0}\right]. \tag{4.20} $$

The same state in reference to another configuration C^{t_2} is expressed by the tensor

$$ {}^{0}_{t_2}\varepsilon^{t} = \frac{1}{2}\left[({}^{t_2}\boldsymbol{F}^{t})^{\mathrm{T}}\,{}^{t_2}\boldsymbol{F}^{t} - ({}^{t_2}\boldsymbol{F}^{0})^{\mathrm{T}}\,{}^{t_2}\boldsymbol{F}^{0}\right]. \tag{4.21} $$

Making use of the multiplicative gradient combination rule (4.8), the above formula may be transformed as

$$
\begin{aligned}
{}^{0}_{t_2}\varepsilon^{t} &= \frac{1}{2}\left[({}^{t_1}F^{t}\,{}^{t_2}F^{t_1})^{\mathrm{T}}\,{}^{t_1}F^{t}\,{}^{t_2}F^{t_1} - ({}^{t_1}F^{0}\,{}^{t_2}F^{t_1})^{\mathrm{T}}\,{}^{t_1}F^{0}\,{}^{t_2}F^{t_1}\right] \\
&= ({}^{t_2}F^{t_1})^{\mathrm{T}}\,{}^{0}_{t_1}\varepsilon^{t}\,{}^{t_2}F^{t_1}\,.
\end{aligned}
\tag{4.22}
$$

In particular, the following relationships hold true for the Green and Almansi strain measures ($\boldsymbol{F}$ stands here for the gradient ${}^{0}\boldsymbol{F}^{t}$):

$$
\varepsilon^{G} = \boldsymbol{F}^{\mathrm{T}}\,\varepsilon^{A}\,\boldsymbol{F}\,, \qquad \varepsilon^{A} = \boldsymbol{F}^{-\mathrm{T}}\,\varepsilon^{G}\,\boldsymbol{F}^{-1}\,, \tag{4.23}
$$

where the notation $\boldsymbol{F}^{-\mathrm{T}}$ has been used to denote the inverse of the transpose gradient $(\boldsymbol{F}^{-1})^{\mathrm{T}}$.

4.2.2 *Physical Meaning of Strain in One Dimension*

Let us now go back to the 1D example of an extended bar. Assume that the bar length in the reference configuration is l^{r} while the process of its deformation runs from the initial length l^{0} to current l^{t} (Fig. 4.4). If the deformation is uniform along the bar's length, then the gradients ${}^{r}\boldsymbol{F}^{0}$ and ${}^{r}\boldsymbol{F}^{t}$ correspond to scalar constant quantities l^{0}/l^{r} and l^{t}/l^{r}. It is easy to verify that the one-dimensional strain in the bar, cf. Eq. (4.15), can be expressed as

$$
{}^{0}_{r}\varepsilon^{t} = \frac{1}{2}\,\frac{(l^{t})^2-(l^{0})^2}{(l^{r})^2}\,. \tag{4.24}
$$

Identifying the reference configuration with the initial or the current one, one can write

$$
{}^{0}_{0}\varepsilon^{t} = \varepsilon^{G} = \frac{1}{2}\,\frac{(l^{t})^2-(l^{0})^2}{(l^{0})^2} = \frac{1}{2}\left[\left(\frac{l^{t}}{l^{0}}\right)^2 - 1\right], \tag{4.25}
$$

$$
{}^{0}_{t}\varepsilon^{t} = \varepsilon^{A} = \frac{1}{2}\,\frac{(l^{t})^2-(l^{0})^2}{(l^{t})^2} = \frac{1}{2}\left[1 - \left(\frac{l^{0}}{l^{t}}\right)^2\right], \tag{4.26}
$$

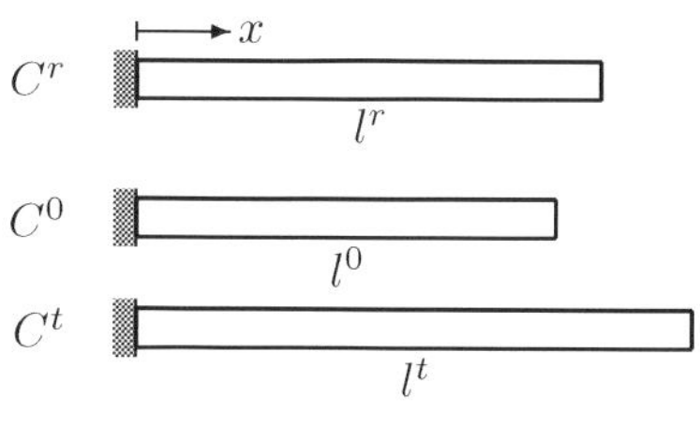

Fig. 4.4 Extended bar—geometric configurations

respectively. Assuming additionally $|l^t - l^0| \ll l^0$ and denoting $l = l^0 \approx l^t$, $\Delta l = l^t - l^0$, one may transform each of the above equations, e.g. Eq. (4.25), as

$$\varepsilon^G = \frac{1}{2}\frac{(l^t)^2 - (l^0)^2}{(l^0)^2} = \frac{1}{2}\frac{(l^t+l^0)(l^t-l^0)}{(l^0)^2} = \frac{1}{2}\frac{(2l^0+\Delta l)\Delta l}{(l^0)^2} \approx \frac{1}{2}\frac{2l\Delta l}{l^2} = \frac{\Delta l}{l},$$

which leads to a conclusion that the linearized Cauchy strain is

$$\varepsilon^C = \frac{\Delta l}{l} \approx \varepsilon^G \approx \varepsilon^A . \qquad (4.27)$$

This is the same quantity that was assigned the name of strain in the initial discussion of Sect. 2.1.

4.2.3 Physical Meaning of Strain Components

Consider two-dimensional motion in which points of a unit square domain $\Omega^0 = [-1, 1] \times [-1, 1]$ move at $\tau = t$ to the following locations:

$$x_1^t = (1 + k_1)x_1^0 + k_3 x_2^0 , \qquad x_2^t = k_2 x_1^0 + (1 + k_4)x_2^0 , \qquad (4.28)$$

whose locus (domain Ω^t in the deformed configuration) is the parallelogram shown in Fig. 4.5.

Displacement field components ${}^0u_i^t = x_i^t - x_i^0$ can be expressed as

$${}^0u_1^t = k_1 x_1^0 + k_3 x_2^0 , \qquad {}^0u_2^t = k_2 x_1^0 + k_4 x_2^0 .$$

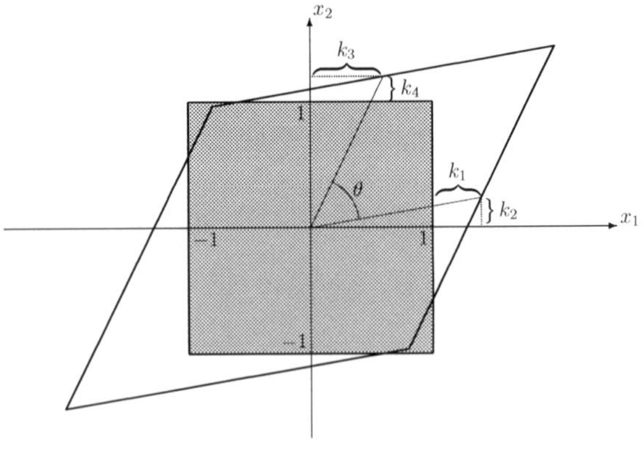

Fig. 4.5 Example of deformation in the 2D space

Assuming the initial configuration as the reference configuration, one can then compute components of displacement gradient and the Green strain tensors as

$$\begin{aligned}\left[\frac{\partial\,{}^0u_i^t}{\partial x_j^0}\right] &= \begin{bmatrix} k_1 & k_3 \\ k_2 & k_4 \end{bmatrix},\\ \left[{}^0_0\varepsilon^t_{ij}\right] = \left[\varepsilon^G_{ij}\right] &= \begin{bmatrix} k_1 + \frac{1}{2}(k_1^2 + k_3^2) & \frac{1}{2}(k_2 + k_3 + k_1k_3 + k_2k_4) \\ \frac{1}{2}(k_2 + k_3 + k_1k_3 + k_2k_4) & k_4 + \frac{1}{2}(k_2^2 + k_4^2) \end{bmatrix}.\end{aligned}$$

They turn out to be uniform, i.e. independent of x_i^0 in this case. If $k_i \ll 1$, strain may be approximated by the linearized Cauchy tensor,

$$\left[\varepsilon^C_{ij}\right] = \begin{bmatrix} k_1 & \frac{1}{2}(k_2 + k_3) \\ \frac{1}{2}(k_2 + k_3) & k_4 \end{bmatrix}.$$

The geometrical meaning of the latter tensor components can easily be read from Fig. 4.5. The diagonal components ε_{11} and ε_{22} denote the relative increase of dimensions along the corresponding two coordinate system axes (negative values would correspond to shortening). The remaining two (equal) components $\varepsilon_{12} = \varepsilon_{21}$ describe the shear strain—they are an indirect measure of the angle change between straight lines initially parallel to the coordinate axes. For sufficiently small k_i we have $\varepsilon_{12} \approx \frac{1}{2}\cos\theta$.

The above conclusions remain valid for nonuniform deformation (such that the displacement gradient is a function of location $\boldsymbol{x}^0$). The Cauchy strain components, different at each material point, describe then relative changes in dimensions and shape of an infinitesimal rectangular element $\mathrm{d}x_1^0\mathrm{d}x_2^0$ at each material point of the body. In various configurations during its motion, such a rectangle gets deformed to a parallelogram $\mathrm{d}x_1^\tau\mathrm{d}x_2^\tau$ (Fig. 4.6). The above interpretation of strain components may obviously be generalized to the 3D case—the diagonal components describe

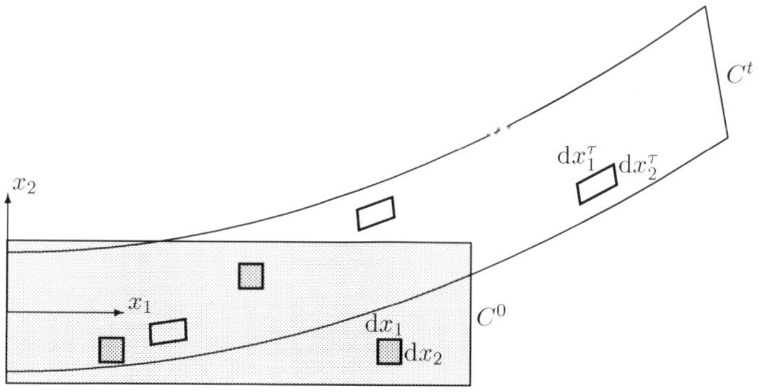

Fig. 4.6 Example of nonuniform deformation in the 2D space. Local values of strain components describe relative changes of dimensions of an infinitesimal rectangle $\mathrm{d}x_1^0\mathrm{d}x_2^0$

then relative extensions/contractions of an elementary cuboid $\mathrm{d}x_1^0\mathrm{d}x_2^0\mathrm{d}x_3^0$ while the remaining ones changes of angles between its walls while it gets deformed to a parallelepiped $\mathrm{d}x_1^T\mathrm{d}x_2^T\mathrm{d}x_3^T$.

In cases of finite deformation, i.e. when the Cauchy tensor cannot be considered a good approximate of the strain tensor, interpretation of the strain components is no more that intuitive. Let us anyway try to look into the precise meaning of the Green strain tensor components. For this reason, let us consider two infinitesimal vectors $\mathrm{d}\boldsymbol{x}^0$ and $\mathrm{d}\boldsymbol{y}^0$ in the initial configuration, parallel to the coordinate system axes x_1 and x_2, respectively. We can express them as $\mathrm{d}\boldsymbol{x}^0 = \boldsymbol{e}_1\mathrm{d}x^0$ and $\mathrm{d}\boldsymbol{y}^0 = \boldsymbol{e}_2\mathrm{d}y^0$ where $\mathrm{d}x^0$ and $\mathrm{d}y^0$ are the vectors' lengths, respectively. In the current, deformed configuration C^t, the vectors assume the form $\mathrm{d}\boldsymbol{x}^t = {}^0\boldsymbol{F}^t\,\mathrm{d}\boldsymbol{x}^0$ and $\mathrm{d}\boldsymbol{y}^t = {}^0\boldsymbol{F}^t\,\mathrm{d}\boldsymbol{y}^0$, respectively, and their lengths $\mathrm{d}x^t$, $\mathrm{d}y^t$, as well as directions are obviously different from what they were in the initial configuration.

Since the versors $\boldsymbol{e}_1$ and $\boldsymbol{e}_2$ have the simple zero/one representation in the considered coordinate system, one may write, for instance

$$\mathrm{d}\boldsymbol{x}^0\,{}_0^0\boldsymbol{\varepsilon}^t\,\mathrm{d}\boldsymbol{x}^0 = \boldsymbol{e}_1\,{}_0^0\boldsymbol{\varepsilon}^t\,\boldsymbol{e}_1\,(\mathrm{d}x^0)^2 = {}_0^0\varepsilon_{11}^t\,(\mathrm{d}x^0)^2 = \varepsilon_{11}^G\,(\mathrm{d}x^0)^2.$$

In view of the definition (4.14) it can be concluded that the Green tensor component ε_{11}^G describes the relative change of the square length in the direction x_1,

$$\varepsilon_{11}^G = \frac{1}{2}\,\frac{(\mathrm{d}x^t)^2 - (\mathrm{d}x^0)^2}{(\mathrm{d}x^0)^2} = \frac{1}{2}\left[\left(\frac{\mathrm{d}x^t}{\mathrm{d}x^0}\right)^2 - 1\right]. \tag{4.29}$$

An analogous geometrical interpretation can be assigned to the remaining diagonal components ε_{22}^G and ε_{33}^G.

Consider now the following expression,

$$\mathrm{d}\boldsymbol{x}^0\,{}_0^0\boldsymbol{\varepsilon}^t\,\mathrm{d}\boldsymbol{y}^0 = \boldsymbol{e}_1\,{}_0^0\boldsymbol{\varepsilon}^t\,\boldsymbol{e}_2\,\mathrm{d}x^0\mathrm{d}y^0 = \varepsilon_{12}^G\,\mathrm{d}x^0\mathrm{d}y^0.$$

In view of the definition (4.17) this expression may also be written as

$$\mathrm{d}\boldsymbol{x}^0\,{}_0^0\boldsymbol{\varepsilon}^t\,\mathrm{d}\boldsymbol{y}^0 = \frac{1}{2}\left[\left({}^0\boldsymbol{F}^t\mathrm{d}\boldsymbol{x}^0\right)\left({}^0\boldsymbol{F}^t\mathrm{d}\boldsymbol{y}^0\right) - \mathrm{d}\boldsymbol{x}^0\mathrm{d}\boldsymbol{y}^0\right] = \frac{1}{2}\,\mathrm{d}\boldsymbol{x}^t\mathrm{d}\boldsymbol{y}^t,$$

where the orthogonality of the vectors $\mathrm{d}\boldsymbol{x}^0$ and $\mathrm{d}\boldsymbol{y}^0$ has been used. Denoting by θ_{12}^t the angle between the vectors $\mathrm{d}\boldsymbol{x}^t$ and $\mathrm{d}\boldsymbol{y}^t$ in the current configuration and recalling Eq. (3.18), one can also write

$$\mathrm{d}\boldsymbol{x}^t\mathrm{d}\boldsymbol{y}^t = \cos\theta_{12}^t\,\mathrm{d}x^t\mathrm{d}y^t.$$

Collecting the above equalities, we finally arrive at the result

$$\varepsilon_{12}^{G} = \frac{1}{2}\cos\theta_{12}^{t}\,\frac{\mathrm{d}x^{t}}{\mathrm{d}x^{0}}\,\frac{\mathrm{d}y^{t}}{\mathrm{d}y^{0}}\,. \tag{4.30}$$

This allows to interpret the component ε_{12}^{G} as a measure of change of the initially right angle between infinitesimal segments parallel to the axes x_1 and x_2, additionally scaled by the length change factors of these segments. An analogous geometrical interpretation can be assigned to the remaining off-diagonal components ε_{23}^{G} and ε_{13}^{G}.

A similar discussion can be carried out for the Almansi tensor components, $\varepsilon_{ij}^{A} = {}_{t}^{0}\varepsilon_{ij}^{t}$. This time the infinitesimal orthogonal vectors have to be selected in the current, deformed configuration, i.e. $\mathrm{d}\boldsymbol{x}^{t} = \boldsymbol{e}_1\mathrm{d}x^{t}$ and $\mathrm{d}\boldsymbol{y}^{t} = \boldsymbol{e}_2\mathrm{d}y^{t}$. In the initial configuration their counterparts have the form $\mathrm{d}\boldsymbol{x}^{0} = ({}^{0}\boldsymbol{F}^{t})^{-1}\mathrm{d}\boldsymbol{x}^{t}$ and $\mathrm{d}\boldsymbol{y}^{0} = ({}^{0}\boldsymbol{F}^{t})^{-1}\mathrm{d}\boldsymbol{y}^{t}$ while their lengths equal $\mathrm{d}x^{0}$ and $\mathrm{d}y^{0}$, respectively, After analogous transformations one arrives at the following interpretation of the Almansi strain components. For the diagonal components we have, for instance

$$\varepsilon_{11}^{A} = \frac{1}{2}\,\frac{(\mathrm{d}x^{t})^2 - (\mathrm{d}x^{0})^2}{(\mathrm{d}x^{t})^2} = \frac{1}{2}\left[1 - \left(\frac{\mathrm{d}x^{0}}{\mathrm{d}x^{t}}\right)^2\right]. \tag{4.31}$$

For the remaining, off-diagonal components we have, for instance

$$\varepsilon_{12}^{A} = -\frac{1}{2}\cos\theta_{12}^{0}\,\frac{\mathrm{d}x^{0}}{\mathrm{d}x^{t}}\,\frac{\mathrm{d}y^{0}}{\mathrm{d}y^{t}}\,, \tag{4.32}$$

where θ_{12}^{0} denotes the angle between $\mathrm{d}\boldsymbol{x}^{0}$ and $\mathrm{d}\boldsymbol{y}^{0}$ in the initial configuration.

4.2.4 Some Other Strain Tensor Properties

As we already know, the strain tensor is a measure of relative change in shape and dimension of an infinitesimal element of a deformable body. The tensor is symmetric, assumes the zero value while the body is at rest, and its components' values depend on the choice of the reference configuration.

Consider a rigid motion of a body defined by Eq. (4.3). The deformation gradient at an arbitrary time instant τ equals the instantaneous value of the rotation tensor $\boldsymbol{R}^{\tau}$. In particular, we can write

$$^{r}\boldsymbol{F}^{0} = \boldsymbol{R}^{0}\,, \qquad ^{r}\boldsymbol{F}^{t} = \boldsymbol{R}^{t}\,.$$

The strain tensor from the configuration C^0 to C^t is thus given as

$${}_{r}^{0}\boldsymbol{\varepsilon}^{t} = \frac{1}{2}\left[(\boldsymbol{R}^{t})^{\mathrm{T}}\boldsymbol{R}^{t} - (\boldsymbol{R}^{0})^{\mathrm{T}}\boldsymbol{R}^{0}\right] = \boldsymbol{0},$$

and its vanishing is implied by orthogonality of the rotation tensor. Hence, strain is zero not only if there is no motion at all but also in the case of any rigid motion.

Let us now analyse what features of the body's motion affect the strain components and which do not? The polar decomposition theorem (cf. Sect. 3.3.4) applied to the deformation gradient tensor $\boldsymbol{F}$, whose determinant is assumed positive (upper indices are skipped as the following discussion and its conclusions are valid for deformation between any two configurations of the body), allows to present $\boldsymbol{F}$ as a product of a rotation tensor $\boldsymbol{R}$ and a symmetric tensor $\boldsymbol{U} = \sqrt{\boldsymbol{F}^{\mathrm{T}}\boldsymbol{F}}$, i.e.

$$\boldsymbol{F} = \boldsymbol{R}\boldsymbol{U}. \tag{4.33}$$

In other words, considering an infinitesimal spherical surrounding of a selected material point, described by vectors $\mathrm{d}\boldsymbol{x}^r$ of unit lengths and all possible directions, and interpreting the tensor $\boldsymbol{F}$ as a transformation operator in the mapping $\mathrm{d}\boldsymbol{x} = \boldsymbol{F}\,\mathrm{d}\boldsymbol{x}^r$, we can conclude (see Fig. 3.4 and the discussion which followed) that the deformed surrounding of the point is an ellipsoid obtained by (i) 3D stretching the sphere by the transformation described by the symmetric tensor $\boldsymbol{U}$ and (ii) rotating the result with the use of the rotation operator $\boldsymbol{R}$. Recalling the definitions of the right Cauchy–Green deformation tensor $\boldsymbol{C}$ (4.13) and the strain tensor $\boldsymbol{\varepsilon}$ (4.15) or (4.17), we can also see that $\boldsymbol{U} = \sqrt{\boldsymbol{C}}$, i.e. it is the very tensor $\boldsymbol{U}$ (more precisely — its square $\boldsymbol{C}$) that actually affects the strain state $\boldsymbol{\varepsilon}$. What can be concluded from the above discussion is that the strain tensor describes solely the "symmetric" part of local deformation, i.e. the shape change of the material particle's surrounding, while it remains insensitive to any rigid motions of the surrounding.

Remark. The above issue may also be formulated in another way. If the given deformation state, described by the gradient $\boldsymbol{F}$, is modified by imposing additional rigid rotation $\boldsymbol{R}'$, i.e. $\boldsymbol{F}' = \boldsymbol{R}'\boldsymbol{F}$, then the modified deformation tensor $\boldsymbol{C}' = \boldsymbol{F}'^{\mathrm{T}}\boldsymbol{F}' = \boldsymbol{F}^{\mathrm{T}}\boldsymbol{R}'^{\mathrm{T}}\boldsymbol{R}'\boldsymbol{F} = \boldsymbol{F}^{\mathrm{T}}\boldsymbol{F} = \boldsymbol{C}$, i.e. it does not change, which implies that no change affects the strain tensor $\boldsymbol{\varepsilon}$, too. This feature of the tensors $\boldsymbol{C}$ and $\boldsymbol{\varepsilon}$ is called their *rotational invariance*.

Figure 4.7 presents graphical illustration of the deformation gradient's polar decomposition in which the numerical data from the example discussed in Sect. 4.2.3 have been used. Apart from the initial and the current configurations, an "intermediate" configuration has been additionally depicted in which the considered domain is only "reshaped" by the tensor $\boldsymbol{U}$ but not yet rotated by the tensor $\boldsymbol{R}$. Although this configuration is different from C^t, the strain tensors computed for the two configurations are identical.

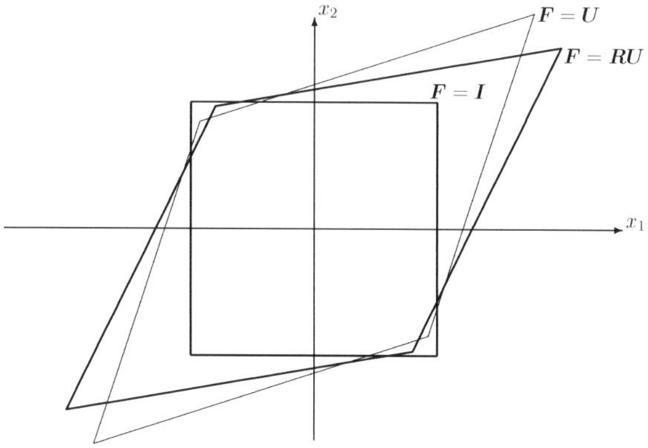

Fig. 4.7 Polar decomposition of deformation in the 2D space

Remark. The fact that the tensor defined by Eq. (4.15) vanishes under rigid motion does not mean that its linearized approximate, i.e. the Cauchy tensor (4.19) is zero. Let us verify this by taking as an example a 2D rigid motion in which, at a selected time instant $\tau = t$,

$$\boldsymbol{x}^t = \boldsymbol{R}\boldsymbol{x}^0, \qquad [R_{ij}] = \begin{bmatrix} \cos\varphi & -\sin\varphi \\ \sin\varphi & \cos\varphi \end{bmatrix}.$$

The displacement gradient with respect to the initial configuration is expressed as

$$[u_{i,j}] = \left[\frac{\partial u_i}{\partial x_j^0}\right] = [R_{ij} - \delta_{ij}] = \begin{bmatrix} \cos\varphi - 1 & -\sin\varphi \\ \sin\varphi & \cos\varphi - 1 \end{bmatrix},$$

and thus the Cauchy strain

$$[\varepsilon_{ij}^C] = \frac{1}{2}[u_{i,j} + u_{j,i}] = \begin{bmatrix} \cos\varphi - 1 & 0 \\ 0 & \cos\varphi - 1 \end{bmatrix}$$

is non-zero, although for small φ it assumes very small values. On the other hand, it is easy to show by adding the higher order terms that the Green tensor is exactly zero in this case.

This example provides some clues about applicability limits of strain linearization. Let us estimate how large the rotation angle φ might be so as to preserve sufficient accuracy of the linear strain approximation. Expanding the function $\cos\varphi$ as a Taylor series we obtain

$$\varepsilon_{11} = \cos\varphi - 1 = 1 - \frac{\varphi^2}{2} + \cdots - 1 \approx 0\left(\frac{\varphi^2}{2}\right),$$

i.e. we can conclude that the error of the linear approximation is of the order of the square rotation angle. Therefore, if expected strains in an analysed problem are of the order of 10^{-2} and the accepted error in their evaluation is 1 % (in many practical applications these assumptions are realistic), then the rotation angles should not exceed the order of 10^{-2} rad ($\varphi^2 \approx 10^{-4}$). If the expected strains were lower then—at the same acceptable relative error—the admissible rotation limits should be lower, too.

Another property of the strain tensor is its symmetry. Recalling considerations in Sect. 3.3.2, each symmetric tensor has three orthogonal principal directions and transformation of its components to the coordinate system based on these orthogonal directions results in a diagonal matrix representation in which the subsequent diagonal elements are the tensor's eigenvalues (principal strains),

$$[\varepsilon'_{ij}] = \begin{bmatrix} \varepsilon_{\rm I} & 0 & 0 \\ 0 & \varepsilon_{\rm II} & 0 \\ 0 & 0 & \varepsilon_{\rm III} \end{bmatrix} .$$

The physical interpretation of the components is that they describe longitudinal strains (relative extensions) at a material particle in the principal directions. We may thus conclude that, for each strain state at a point $\boldsymbol{x}$, there exists an orthogonal triplet of directions $\{x'_1, x'_2, x'_3\}$ in which an infinitesimal cuboid $\mathrm{d}x'_1\mathrm{d}x'_2\mathrm{d}x'_3$ is extended/contracted along its axes and subject to no shear deformation (i.e. it remains a cuboid). This is illustrated in Fig. 4.8 where, for the considered example of 2D motion, such a square with sides parallel to the principal Green strain directions has been depicted in the initial configuration. Its deformed shape in the current configuration is a rectangle, too. It is noteworthy that rotated sides of the deformed rectangle are parallel to the principal directions of the Almansi strain tensor.

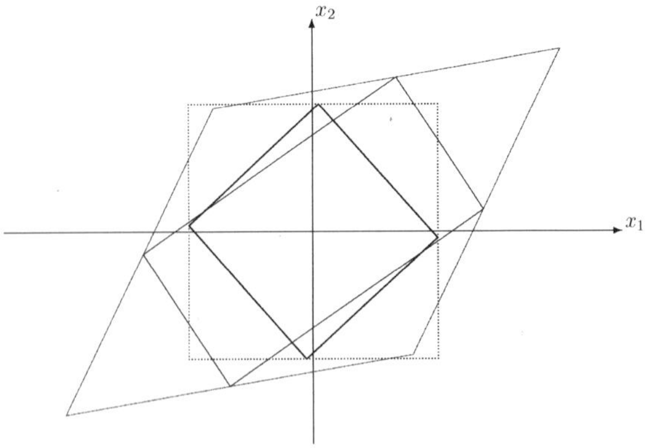

Fig. 4.8 Principal directions of strain in the 2D space

Exercise 4.1
The motion of a deformable body is given by the equations

$$
\begin{aligned}
x_1^\tau &= x_1^0 + 0.02\tau\left[(x_1^0)^2 - (x_2^0)^2\right],\\
x_2^\tau &= x_2^0 + 0.01\tau\left[(x_1^0)^2 - (x_2^0)^2\right],\\
x_3^\tau &= x_3^0 .
\end{aligned}
\tag{4.34}
$$

Express components of the deformation gradient ${}^0\boldsymbol{F}^\tau$ as functions of $\boldsymbol{x}^0$ and τ. Besides, at $\tau = t = 1$ and at the material point with initial coordinates $x_1^0 = x_2^0 = 1$, compute components of the Green and the Almansi strain tensors.

Solution. Consider the initial configuration C^0 as the reference configuration. The deformation gradient can be derived by differentiation of Eq. (4.34) with respect to x_j^0 as

$$
\left[{}^0F_{ij}^\tau\right] = \left[\frac{\partial x_i^\tau}{\partial x_j^0}\right] = \begin{bmatrix} 1 + 0.04\tau x_1^0 & -0.04\tau x_2^0 & 0 \\ 0.02\tau x_1^0 & 1 - 0.02\tau x_2^0 & 0 \\ 0 & 0 & 1 \end{bmatrix}.
$$

In particular, at $\tau = t = 1$ and at $x_1^0 = x_2^0 = 1$ the gradient components are

$$
\left[{}^0F_{ij}^t\right] = \begin{bmatrix} 1.04 & -0.04 & 0 \\ 0.02 & 0.98 & 0 \\ 0 & 0 & 1 \end{bmatrix},
$$

while the displacement gradient components are

$$
\left[{}^0u_{i,j}^t\right] = \left[\frac{\partial\, {}^0u_i^t}{\partial x_j^0}\right] = \left[{}^0F_{ij}^t - \delta_{ij}\right] = \begin{bmatrix} 0.04 & -0.04 & 0 \\ 0.02 & -0.02 & 0 \\ 0 & 0 & 0 \end{bmatrix}.
$$

The Green strain components can be computed from Eq. (4.17b) as

$$
\begin{aligned}
\varepsilon_{11}^G &= 0.04 + \tfrac{1}{2}\left[0.04^2 + 0.02^2\right] = 0.041,\\
\varepsilon_{22}^G &= -0.02 + \tfrac{1}{2}\left[(-0.04)^2 + (-0.02)^2\right] = -0.019,\\
\varepsilon_{12}^G &= \tfrac{1}{2}\left[-0.04 + 0.02 + 0.04(-0.04) + 0.02(-0.02)\right] = -0.011,\\
\varepsilon_{21}^G &= \varepsilon_{12}^G
\end{aligned}
$$

(all remaining components are zero).

Consider now the current configuration C^t as the reference configuration. To determine the Almansi strain components, the displacement gradients with respect to x_i^t are needed. Instead of inverting the relations (4.34) let us benefit from the relation ${}^t\boldsymbol{F}^0 = ({}^0\boldsymbol{F}^t)^{-1}$ and simply invert the deformation gradient's component matrix computed above. Its determinant is $1.04 \cdot 0.98 + 0.02 \cdot 0.04 = 1.02$, thus the inverse gradient components are expressed as

$$[{}^tF_{ij}^0] = \frac{1}{102}\begin{bmatrix} 98 & 4 & 0 \\ -2 & 104 & 0 \\ 0 & 0 & 102 \end{bmatrix},$$

while the displacement gradient components in the current configuration becomes

$$\left[\frac{\partial\, {}^0u_i^t}{\partial x_j^t}\right] = [\delta_{ij} - {}^tF_{ij}^0] = \frac{1}{102}\begin{bmatrix} 4 & -4 & 0 \\ 2 & -2 & 0 \\ 0 & 0 & 0 \end{bmatrix}.$$

The Almansi strain components can be computed from Eq. (4.18b) as

$$\varepsilon_{11}^A = \tfrac{4}{102} - \tfrac{1}{2}\left[(\tfrac{4}{102})^2 + (\tfrac{2}{102})^2\right] = \tfrac{398}{10404} \approx 0.038255,$$
$$\varepsilon_{22}^A = -\tfrac{2}{102} - \tfrac{1}{2}\left[(-\tfrac{4}{102})^2 + (-\tfrac{2}{102})^2\right] = -\tfrac{214}{10404} \approx -0.020569,$$
$$\varepsilon_{12}^A = \tfrac{1}{2}\left[-\tfrac{4}{102} + \tfrac{2}{102} - \tfrac{4}{102}(-\tfrac{4}{102}) - \tfrac{2}{102}(-\tfrac{2}{102})\right] = -\tfrac{92}{10404} \approx -0.008843,$$
$$\varepsilon_{21}^A = \varepsilon_{12}^A$$

(all remaining components are zero).

4.3 Area and Volumetric Deformation

Consider a material point P of a deformable body and an infinitesimal surface element at this point with the area $\mathrm{d}A^r$ in the reference configuration. Shape of the element does not really matter, but we assume it for simplicity to be a parallelogram spanned on two non-colinear material elements $\mathrm{d}\boldsymbol{x}^r$ and $\mathrm{d}\boldsymbol{y}^r$, Fig. 4.9. Let $\boldsymbol{n}^r$ be a unit vector normal to this surface element (with the right-hand orientation with respect to $\mathrm{d}\boldsymbol{x}^r$ and $\mathrm{d}\boldsymbol{y}^r$). Making use of the vector product property (3.24) we can write

$$\mathrm{d}\boldsymbol{x}^r \times \mathrm{d}\boldsymbol{y}^r = \boldsymbol{n}^r\,\mathrm{d}A^r. \tag{4.35}$$

The right-hand side of the above equation may be interpreted as the so-called vector surface element $\boldsymbol{n}\mathrm{d}A$ in the configuration C^r.

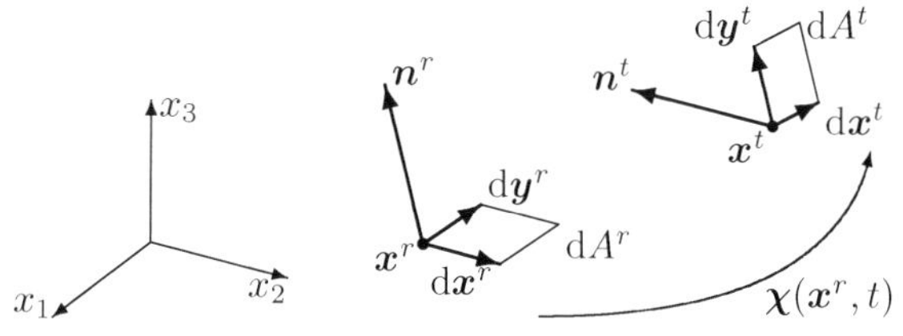

Fig. 4.9 Surface element deformation

At an arbitrary time instant $\tau = t$, the vectors $d\boldsymbol{x}^r$ and $d\boldsymbol{y}^r$ get deformed according to

$$d\boldsymbol{x}^t = {}^r\boldsymbol{F}^t \, d\boldsymbol{x}^r, \qquad d\boldsymbol{y}^t = {}^r\boldsymbol{F}^t \, d\boldsymbol{y}^r, \tag{4.36}$$

while the relation between them and the deformed vector surface element in the current configuration C^t assumes the form

$$d\boldsymbol{x}^t \times d\boldsymbol{y}^t = \boldsymbol{n}^t \, dA^t. \tag{4.37}$$

Substituting Eq. (4.36) into Eq. (4.37) and skipping for clarity the upper indices r and t at the symbol $\boldsymbol{F}$, one can write down in the index notation:

$$\begin{aligned} \boldsymbol{n}^t \, dA^t &= (\boldsymbol{F} d\boldsymbol{x}^r) \times (\boldsymbol{F} d\boldsymbol{y}^r) = (F_{ik} dx_k^r \, \boldsymbol{e}_i) \times (F_{jl} dy_l^r \, \boldsymbol{e}_j) \\ &= F_{ik} F_{jl} \, dx_k^r dy_l^r \, \boldsymbol{e}_i \times \boldsymbol{e}_j = F_{ik} F_{jl} \, dx_k^r dy_l^r \, \epsilon_{ijm} \boldsymbol{e}_m. \end{aligned} \tag{4.38}$$

This expression will now be transformed by using subsequently: the property of the Kronecker symbol (3.20), the inverse tensor definition (3.73), the property (3.28) of the tensor determinant and the orthogonality of the coordinate system versors (3.33) to yield

$$\begin{aligned} F_{ik} F_{jl} \, dx_k^r dy_l^r \, \epsilon_{ijm} \boldsymbol{e}_m &= F_{ik} F_{jl} \, dx_k^r dy_l^r \, \epsilon_{ijm} \underbrace{\delta_{mn} \boldsymbol{e}_n}_{\boldsymbol{e}_m} \\ &= F_{ik} F_{jl} \underbrace{F_{mp} F_{pn}^{-1}}_{\delta_{mn}} dx_k^r dy_l^r \, \epsilon_{ijm} \boldsymbol{e}_n \\ &= \underbrace{\epsilon_{klp} \, \det[F_{rs}]}_{F_{ik} F_{jl} F_{mp} \epsilon_{ijm}} F_{pn}^{-1} \, dx_k^r dy_l^r \, \boldsymbol{e}_n \\ &= \det[F_{rs}] \, F_{pn}^{-1} \, dx_k^r dy_l^r \, \boldsymbol{e}_n \underbrace{\epsilon_{klq} (\boldsymbol{e}_p \boldsymbol{e}_q)}_{\epsilon_{klp}} \\ &= \det[F_{rs}] \, F_{pn}^{-1} \, dx_k^r dy_l^r \epsilon_{klq} \, (\boldsymbol{e}_n \otimes \boldsymbol{e}_p) \boldsymbol{e}_q \\ &= \det[F_{rs}] \, (F_{pn}^{-1} \, \boldsymbol{e}_n \otimes \boldsymbol{e}_p) \, (dx_k^r dy_l^r \epsilon_{klq} \boldsymbol{e}_q) \\ &= (\det \boldsymbol{F}) \, \boldsymbol{F}^{-\mathrm{T}} (d\boldsymbol{x}^r \times d\boldsymbol{y}^r), \end{aligned} \tag{4.39}$$

where again the notation $\boldsymbol{F}^{-\mathrm{T}} \equiv (\boldsymbol{F}^{-1})^{\mathrm{T}}$ has been used for convenience. Introducing a widely adopted in the literature notation for the deformation gradient determinant

$$J \stackrel{\text{def}}{=} \det \boldsymbol{F} \tag{4.40}$$

and making use of Eq. (4.35), one can finally write down the formula for the infinitesimal vector surface element deformation between the configurations C^r and C^t in the following form:

$$\boldsymbol{n}^t\,\mathrm{d}A^t = {}^rJ^t\,({}^r\boldsymbol{F}^t)^{-\mathrm{T}}\,\boldsymbol{n}^r\,\mathrm{d}A^r, \qquad n_i^t\,\mathrm{d}A^t = {}^rJ^t\,{}^tF_{ji}^r\,n_j^r\,\mathrm{d}A^r. \tag{4.41}$$

The relation derived above is known in the literature as the *Nanson formula* which is valid for any other two configurations of the body, e.g. $\boldsymbol{n}^{t_2}\,\mathrm{d}A^{t_2} = {}^{t_1}J^{t_2}\,({}^{t_1}\boldsymbol{F}^{t_2})^{-\mathrm{T}}\,\boldsymbol{n}^{t_1}\,\mathrm{d}A^{t_1}$. It is clear in view of this formula that the vector surface element is not a material element—it does not obey the transformation rule (4.36). In other words, the material particle "pointed" by the normal versor $\boldsymbol{n}^r$ in the surrounding of the point P in the configuration C^r is generally not the same particle as that pointed by the versor $\boldsymbol{n}^t$ in the configuration C^t.

Let us further note that, since $\boldsymbol{n}^t$ is a unit vector,

$$\begin{aligned}(\mathrm{d}A^t)^2 &= (\boldsymbol{n}^t\mathrm{d}A^t)\,(\boldsymbol{n}^t\mathrm{d}A^t)\\ &= ({}^rJ^t)^2\,(\boldsymbol{n}^r\mathrm{d}A^r)\,({}^r\boldsymbol{F}^t)^{-1}({}^r\boldsymbol{F}^t)^{-\mathrm{T}}\,(\boldsymbol{n}^r\mathrm{d}A^r)\\ &= (\boldsymbol{n}^r\mathrm{d}A^r)\left[({}^rJ^t)^2({}^r\boldsymbol{C}^t)^{-1}\right](\boldsymbol{n}^r\mathrm{d}A^r)\\ &= \left[({}^rJ^t)^2\,\boldsymbol{n}^r({}^r\boldsymbol{C}^t)^{-1}\boldsymbol{n}^r\right](\mathrm{d}A^r)^2.\end{aligned} \tag{4.42}$$

Comparing the above formula with Eq. (4.12) one can see that—in determining vector surface element deformation—the tensor $J^2\boldsymbol{C}^{-1}$ plays an analogous role as the tensor $\boldsymbol{C}$ in determining linear material element deformation.

Let us investigate in an analogous way the relative volume change (volumetric deformation) of an infinitesimal volume element $\mathrm{d}V^r$ at a material point P in the reference configuration C^r. Let us define this element as a parallelepiped spanned on three non-colinear material elements $\mathrm{d}\boldsymbol{x}^r$, $\mathrm{d}\boldsymbol{y}^r$ and $\mathrm{d}\boldsymbol{z}^r$ with the right-hand orientation (Fig. 4.10). As has been shown in Exercise 3.5, its volume equals

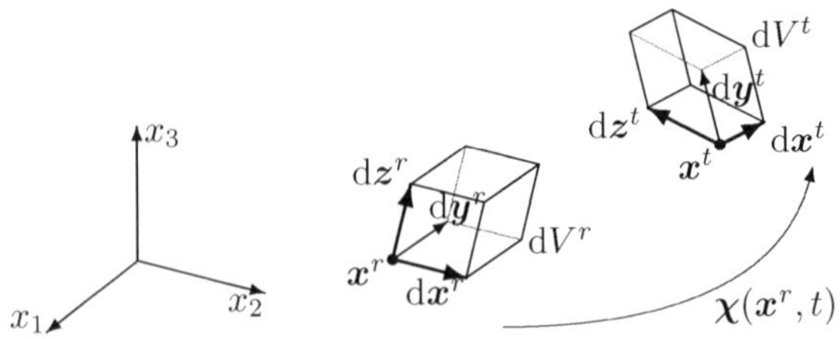

Fig. 4.10 Volume element deformation

$$\mathrm{d}V^r = (\mathrm{d}\boldsymbol{x}^r \times \mathrm{d}\boldsymbol{y}^r) \cdot \mathrm{d}\boldsymbol{z}^r = \epsilon_{ijk}\, \mathrm{d}x_i^r \mathrm{d}y_j^r \mathrm{d}x_k^r \,. \tag{4.43}$$

At a selected time instant $\tau = t$, in the deformed configuration C^t of the body, the three vectors assume the form

$$\mathrm{d}\boldsymbol{x}^t = {}^r\boldsymbol{F}^t\, \mathrm{d}\boldsymbol{x}^r, \qquad \mathrm{d}\boldsymbol{y}^t = {}^r\boldsymbol{F}^t\, \mathrm{d}\boldsymbol{y}^r, \qquad \mathrm{d}\boldsymbol{z}^t = {}^r\boldsymbol{F}^t\, \mathrm{d}\boldsymbol{z}^r, \tag{4.44}$$

while the deformed volume of the considered element

$$\mathrm{d}V^t = (\mathrm{d}\boldsymbol{x}^t \times \mathrm{d}\boldsymbol{y}^t) \cdot \mathrm{d}\boldsymbol{z}^t$$

can be expressed after transformations as (the indices at $\boldsymbol{F}$ are left out again)

$$\begin{aligned} \mathrm{d}V^t &= (\boldsymbol{F}\mathrm{d}\boldsymbol{x}^r \times \boldsymbol{F}\mathrm{d}\boldsymbol{y}^r) \cdot \boldsymbol{F}\mathrm{d}\boldsymbol{z}^r = F_{il}\mathrm{d}x_l^r\, F_{jm}\mathrm{d}y_m^r\, \epsilon_{ijk}\, F_{kn}\mathrm{d}z_n^r \\ &= \underbrace{F_{il}F_{jm}F_{kn}\epsilon_{ijk}}_{\det \boldsymbol{F}\,\epsilon_{lmn}}\, \mathrm{d}x_l^r \mathrm{d}y_m^r \mathrm{d}z_n^r = {}^rJ^t\, \mathrm{d}V^r. \end{aligned} \tag{4.45}$$

Thus, the deformation gradient determinant ${}^rJ^t = \det {}^r\boldsymbol{F}^t$ defines the ratio of the deformed to reference volume of an infinitesimal material volume element. This relation is obviously valid for any other two configurations of the body, i.e. $\mathrm{d}V^{t_2} = {}^{t_1}J^{t_2}\, \mathrm{d}V^{t_1}$.

If the displacement gradient is small enough in the considered motion, one may approximate the result by a linear formula in which terms containing products of displacement gradient components are skipped. In such a case we have, for instance

$${}^rJ^t = \det\left[\delta_{ij} + \frac{\partial\, {}^ru_i^t}{\partial x_j^r}\right] \approx 1 + \frac{\partial\, {}^ru_i^t}{\partial x_i^r}\,, \tag{4.46}$$

i.e.

$$\frac{\mathrm{d}V^t - \mathrm{d}V^r}{\mathrm{d}V^r} = {}^rJ^t - 1 \approx {}^ru_{i,i}^t = \varepsilon_{ii}^C\,. \tag{4.47}$$

Exercise 4.2 For the motion defined in Exercise 4.1, compute at $\tau = t = 1$ and at the material point $x_1^0 = x_2^0 = 1$ the change of area of the element $\mathrm{d}A^0$ whose normal versor is $\{n_i^0\} = \frac{1}{\sqrt{3}}\{1,\ 1,\ 1\}$. Determine the change of the versor's orientation, i.e. the components $\{n_i^t\}$.

Solution. Let us make use of the Nanson formula, taking the initial configuration as the reference configuration,

$$n_i^t\, \mathrm{d}A^t = {}^0J^t\, {}^tF_{ji}^0\, n_j^0\, \mathrm{d}A^0,$$

and use the values of the determinant ${}^0J^t = 1.02$ as well as the inverse tensor components ${}^tF^0_{ij}$ computed before in Exercise 4.1. The transposed tensor components $\left[{}^0J^t\,{}^tF^0_{ji}\right]$ appearing in above formula become

$$[{}^0J^t\,{}^tF^0_{ji}] = \begin{bmatrix} 0.98 & -0.02 & 0 \\ 0.04 & 1.04 & 0 \\ 0 & 0 & 1 \end{bmatrix}.$$

Multiplying them by the appropriate components $n^0_j\,\mathrm{d}A^0$ we obtain

$$\{n^t_i\}\,\mathrm{d}A^t = \frac{1}{\sqrt{3}}\{\,0.96,\ 1.08,\ 1\,\}\,\mathrm{d}A^0.$$

The right-hand side vector length equals approximately $1.014561\,\mathrm{d}A^0$. Since the vector $\{n^t_i\}$ has by definition the unit length, normalization of the right-hand side leads to the result

$$\mathrm{d}A^t = 1.014561\,\mathrm{d}A^0, \qquad \{n^t_i\} = \{\,0.546302,\ 0.614589,\ 0.569064\,\}.$$

4.4 Strain Rate and Strain Increments

4.4.1 Time Derivative of a Tensor Field. Lagrangian and Eulerian Description of Motion

Let $\boldsymbol{a}(\boldsymbol{x}^r, \tau)$ denote an arbitrary tensor field defined on material points of the body $\mathcal{B}$, assumed a smooth function of time. At each material point $\boldsymbol{x}^r$ and each time instant $\tau = t$, the field's time derivative (i.e. its rate) can be defined as

$$\dot{\boldsymbol{a}}^t = \left.\frac{\partial \boldsymbol{a}(\boldsymbol{x}^r, \tau)}{\partial \tau}\right|_{\tau=t}. \tag{4.48}$$

This quantity is a limit value of the quotient

$$\dot{\boldsymbol{a}}^t = \lim_{\Delta t \to 0} \frac{\Delta \boldsymbol{a}(\boldsymbol{x}^r, t)}{\Delta t}, \tag{4.49}$$

where $\Delta \boldsymbol{a}(\boldsymbol{x}^r, t)$ denotes an increment of the field's value at a given point $\boldsymbol{x}^r$ during the finite time interval $[t, t + \Delta t]$, i.e.

$$\Delta \boldsymbol{a}(\boldsymbol{x}^r, t) \stackrel{\text{def}}{=} \boldsymbol{a}(\boldsymbol{x}^r, t + \Delta t) - \boldsymbol{a}(\boldsymbol{x}^r, t) = \boldsymbol{a}^{t+\Delta t}(\boldsymbol{x}^r) - \boldsymbol{a}^t(\boldsymbol{x}^r). \tag{4.50}$$

In particular, for the initial and current configurations taken as reference configurations of the body, we have

$$C^r = C^0: \qquad \Delta \boldsymbol{a}(\boldsymbol{x}^0, t) = \boldsymbol{a}(\boldsymbol{x}^0, t + \Delta t) - \boldsymbol{a}(\boldsymbol{x}^0, t), \tag{4.51}$$

$$C^r = C^t: \qquad \Delta \boldsymbol{a}(\boldsymbol{x}^t, t) = \boldsymbol{a}(\boldsymbol{x}^t, t + \Delta t) - \boldsymbol{a}(\boldsymbol{x}^t, t). \tag{4.52}$$

Let us stress that the argument in the last equation is always $\boldsymbol{x}^t$ and not $\boldsymbol{x}^{t+\Delta t}$—the once selected reference configuration remains constant during the motion, i.e. at $\tau = t + \Delta t$ as well.

Time differentiation of a tensor field defined on the current (transient) coordinates of material points $\boldsymbol{c}(\boldsymbol{x}^\tau, \tau)$ appears to be a somewhat more complex task. The partial derivative of such a function with respect to time $\frac{\partial \boldsymbol{c}}{\partial \tau}$ (computed at $\boldsymbol{x}^\tau$ kept constant) does not describe the function's rate at a selected material point because this point is in motion. Its physical meaning is rather the function's rate at a selected point in the space through which various material points flow during their motion. To compute the so-called *material time derivative* of the field $\boldsymbol{c}$, i.e. the rate of its change at a given material point of the body, one has to express $\boldsymbol{x}^\tau$ as a field defined on material points, i.e. on their coordinates in the reference configuration, cf. Eq. (4.2):

$$\boldsymbol{x}^\tau = \chi(\boldsymbol{x}^r, \tau) = \boldsymbol{x}^\tau(\boldsymbol{x}^r). \tag{4.53}$$

One can then define

$$\dot{\boldsymbol{c}}^t \stackrel{\text{def}}{=} \left.\frac{\mathrm{d}\boldsymbol{c}(\boldsymbol{x}^\tau, \tau)}{\mathrm{d}\tau}\right|_{\tau=t} = \left.\frac{\mathrm{d}\boldsymbol{c}(\boldsymbol{x}^\tau(\boldsymbol{x}^r, \tau), \tau)}{\mathrm{d}\tau}\right|_{\tau=t} = \left.\frac{\partial \boldsymbol{c}^\tau}{\partial \tau} + \frac{\partial \boldsymbol{c}^\tau}{\partial x_k^\tau}\frac{\partial x_k^\tau}{\partial \tau}\right|_{\tau=t}$$
$$= \frac{\partial \boldsymbol{c}^t}{\partial \tau} + \operatorname{grad} \boldsymbol{c}^t \cdot \boldsymbol{v}^t, \tag{4.54}$$

where $\operatorname{grad} \boldsymbol{c}^t$ denotes the spatial gradient operation with respect to coordinates in the current configuration C^t while $\boldsymbol{v}^t$ is the velocity field of material points,

$$\boldsymbol{v}^t = \left.\frac{\partial \chi(\boldsymbol{x}^r, \tau)}{\partial \tau}\right|_{\tau=t} = \dot{\chi}(\boldsymbol{x}^r, t) = \dot{\boldsymbol{u}}^t(\boldsymbol{x}^r), \tag{4.55}$$

which by using the inverse transformation χ^{-1} may obviously be expressed in terms of the coordinates $\boldsymbol{x}^t$, too. The material derivative can be determined as a limit value of the following quotient:

$$\dot{\boldsymbol{c}}^t = \lim_{\Delta t \to 0} \frac{\dot{\Delta} \boldsymbol{c}(\boldsymbol{x}^t, t)}{\Delta t}, \tag{4.56}$$

where $\dot{\Delta}\boldsymbol{c}(\boldsymbol{x}^t, t)$ denotes the increment of $\boldsymbol{c}$ at a material point that occupies the position $\boldsymbol{x}^t$ at the time instant $\tau = t$ in the finite time interval $[t, t + \Delta t]$,

$$\dot{\Delta}\boldsymbol{c}(\boldsymbol{x}^t, t) \stackrel{\text{def}}{=} \frac{\boldsymbol{c}(\boldsymbol{x}^{t+\Delta t}, t + \Delta t) - \boldsymbol{c}(\boldsymbol{x}^t, t)}{\Delta t} . \tag{4.57}$$

The two ways of defining the time derivative (rate) of a tensor field highlight the difference between two alternative ways of description of phenomena related to motion in solid mechanics. The latter, in which physical quantities are attributes of points in space, is called the *Eulerian description*. Apart from rare exceptions, this description will not be a subject of our interest in this book; the reason why the definitions (4.54)–(4.57) have been presented above is just to make our discussion of continuum mechanics equations complete.

Most of the issues discussed in this book will be presented in the *Lagrangian description*, i.e. the description in which physical quantities are attributes of material points subject to motion. In the traditional Lagrangian approach, material points have uniquely assigned coordinates in the initial geometric configuration of the body C^0. In our subsequent discussion, a generalized version of this approach will be employed, in which the fixed reference configuration C^r with spatial coordinates assigned to material points may be selected in an arbitrary way and does not need to be identified with the initial configuration. The essence of the description remains the same, though—material points of the body $\mathcal{B}$ are identified with their locations in a reference configuration which is fixed (does not move with the body). To avoid misunderstandings, let us further refer to the case with initial configuration taken as the reference configuration ($C^r = C^0$) as the *stationary Lagrangian description*.

A specific version of the Lagrangian description called the *updated Lagrangian description* is frequently used in computational pracice. This notion refers to a case when motion of the body is analysed in subsequent time instances upon the recursive rule of searching the solution at the time instance $\tau = t + \Delta t$ once the solution at the instance $\tau = t$ is known. In such a "step-by-step" analysis concept, it is frequently very convenient to consider the "last known" configuration at the beginning of the new time increment as the reference configuration, i.e. $C^r = C^t$. According to the Lagrangian approach, this configuration remains a fixed reference configuration during the body motion within the time interval $[t, t + \Delta t]$. However, when passing to the next time increment, it is necessary to assume the new reference configuration, as the "new" time instant t becomes now the "old" time instant $t + \Delta t$. This requires re-computation of several quantities that were formerly defined in the old reference configuration C^t in order to consequently express them in the new reference configuration. Although the updated Lagrangian description somewhat remains the Eulerian description, it should not be confused with the latter: here we have to do with a stationary (fixed) reference configuration during the motion and its updates are only performed step-by-step at certain analysis stages (and not continuously, as it is in the Eulerian description). For example, Eq. (4.52) defines a tensor field increment according to the updated Lagrangian description while Eq. (4.57) according to the Eulerian description.

4.4.2 Increments and Rates of Strain Tensor Measures

Let us now go back to analysis of strain. Let us recall Eq. (4.16) representing components of strain between the configurations at $\tau = 0$ and at $\tau = t$ with respect to the configuration C^r, expressed as functions of displacement gradient components, i.e.

$$ {}^{0}_{r}\varepsilon^{t}_{ij} = \frac{1}{2}\left(\frac{\partial\, {}^{0}u^{t}_{i}}{\partial x^{r}_{j}} + \frac{\partial\, {}^{0}u^{t}_{j}}{\partial x^{r}_{i}} + \frac{\partial\, {}^{r}u^{t}_{k}}{\partial x^{r}_{i}} \frac{\partial\, {}^{r}u^{t}_{k}}{\partial x^{r}_{j}} - \frac{\partial\, {}^{r}u^{0}_{k}}{\partial x^{r}_{i}} \frac{\partial\, {}^{r}u^{0}_{k}}{\partial x^{r}_{j}} \right). \tag{4.58} $$

Expressing in analogy the tensor components for $\tau = t + \Delta t$ and subtracting the two equations side-wise, we arrive at the following definition of the strain tensor increment:

$$ \begin{aligned} \Delta_r\varepsilon^{t}_{ij} &= {}^{0}_{r}\varepsilon^{t+\Delta t}_{ij} - {}^{0}_{r}\varepsilon^{t}_{ij} \\ &= \frac{1}{2}\left(\frac{\partial \Delta u^{t}_{i}}{\partial x^{r}_{j}} + \frac{\partial \Delta u^{t}_{j}}{\partial x^{r}_{i}} + \frac{\partial \Delta u^{t}_{k}}{\partial x^{r}_{i}} \frac{\partial\, {}^{r}u^{t}_{k}}{\partial x^{r}_{j}} + \frac{\partial\, {}^{r}u^{t}_{k}}{\partial x^{r}_{i}} \frac{\partial \Delta u^{t}_{k}}{\partial x^{r}_{j}} + \frac{\partial \Delta u^{t}_{k}}{\partial x^{r}_{i}} \frac{\partial \Delta u^{t}_{k}}{\partial x^{r}_{j}} \right), \end{aligned} \tag{4.59} $$

where the value of displacement increment $\Delta u^{t}_{k} = {}^{r}u^{t+\Delta t}_{k} - {}^{r}u^{t}_{k} = {}^{0}u^{t+\Delta t}_{k} - {}^{0}u^{t}_{k}$ does not depend on the initial configuration of the displacement itself (thus, the left upper index may be skipped). Let us also note that the right-hand side of Eq. (4.59) is independent of the initial configuration C^0 which allows to skip also the left upper index at the strain increment symbol, i.e. $\Delta_r\varepsilon^{t}_{ij} \equiv \Delta^{0}_{r}\varepsilon^{t}_{ij}$.

It can be seen from the above considerations that the strain increment depends, among other factors, on the displacement increment gradient and that it is the sum of linear and quadratic terms with respect to the gradient components. Let us separate the terms by introducing the notation

$$ \Delta_r\varepsilon^{t}_{ij} = \overline{\Delta_r\varepsilon^{t}_{ij}} + \overline{\overline{\Delta_r\varepsilon^{t}_{ij}}}\,, \tag{4.60a} $$

$$ \overline{\Delta_r\varepsilon^{t}_{ij}} = \frac{1}{2}\left(\frac{\partial \Delta u^{t}_{i}}{\partial x^{r}_{j}} + \frac{\partial \Delta u^{t}_{j}}{\partial x^{r}_{i}} + \frac{\partial \Delta u^{t}_{k}}{\partial x^{r}_{i}} \frac{\partial\, {}^{r}u^{t}_{k}}{\partial x^{r}_{j}} + \frac{\partial\, {}^{r}u^{t}_{k}}{\partial x^{r}_{i}} \frac{\partial \Delta u^{t}_{k}}{\partial x^{r}_{j}} \right), \tag{4.60b} $$

$$ \overline{\overline{\Delta_r\varepsilon^{t}_{ij}}} = \frac{1}{2} \frac{\partial \Delta u^{t}_{k}}{\partial x^{r}_{i}} \frac{\partial \Delta u^{t}_{k}}{\partial x^{r}_{j}}\,. \tag{4.60c} $$

If displacement increments Δu^{t}_{k} are small enough, the quantity $\overline{\overline{\Delta_r\varepsilon^{t}_{ij}}}$ appears to be incomparably smaller than $\overline{\Delta_r\varepsilon^{t}_{ij}}$ and may be neglected.

Passing to the limit as $\Delta t \to 0$, one arrives at the formula for the strain tensor's time derivative, i.e. its rate, in the form

$$ {}_{r}\dot{\varepsilon}^{t}_{ij} = \frac{1}{2}\left(\frac{\partial \dot{u}^{t}_{i}}{\partial x^{r}_{j}} + \frac{\partial \dot{u}^{t}_{j}}{\partial x^{r}_{i}} + \frac{\partial \dot{u}^{t}_{k}}{\partial x^{r}_{i}} \frac{\partial\, {}^{r}u^{t}_{k}}{\partial x^{r}_{j}} + \frac{\partial\, {}^{r}u^{t}_{k}}{\partial x^{r}_{i}} \frac{\partial \dot{u}^{t}_{k}}{\partial x^{r}_{j}} \right). \tag{4.61} $$

Left upper indices at the displacement and strain rate symbols have been removed. They are not necessary since the rate values are independent of the initial (fixed) configuration of motion.

Comparison of Eqs. (4.60) and (4.61) allows to make a general remark on the concept of finite time increments in derivation of equations of mechanics (the so-called incremental approach). Performing numerical computations with the step-by-step method, it is necessary to express the problem equations in increments of certain physical quantities defined on a given time step. The simplest way to do it may appear to be the simple replacement of all time derivatives by their finite time increments. However, the above equations clearly indicate that such a replacement would result in negligence of nonlinear incremental terms which might lead to a significant inaccuracy of the final solution.

Consider two special cases of reference configurations: $C^r = C^0$ and $C^r = C^t$. In the first case, corresponding to the stationary Lagrangian description, the strain tensor is called the Green tensor. Its increment and rate can be derived by replacing in Eqs. (4.59) and (4.61) the index r by 0, i.e.

$$\Delta_0\varepsilon_{ij}^t = \frac{1}{2}\left(\frac{\partial \Delta u_i^t}{\partial x_j^0} + \frac{\partial \Delta u_j^t}{\partial x_i^0} + \frac{\partial \Delta u_k^t}{\partial x_i^0}\frac{\partial\, {}^0u_k^t}{\partial x_j^0} + \frac{\partial\, {}^0u_k^t}{\partial x_i^0}\frac{\partial \Delta u_k^t}{\partial x_j^0} + \frac{\partial \Delta u_k^t}{\partial x_i^0}\frac{\partial \Delta u_k^t}{\partial x_j^0}\right), \tag{4.62}$$

$${}_0\dot{\varepsilon}_{ij}^t = \frac{1}{2}\left(\frac{\partial \dot{u}_i^t}{\partial x_j^0} + \frac{\partial \dot{u}_j^t}{\partial x_i^0} + \frac{\partial \dot{u}_k^t}{\partial x_i^0}\frac{\partial\, {}^0u_k^t}{\partial x_j^0} + \frac{\partial\, {}^0u_k^t}{\partial x_i^0}\frac{\partial \dot{u}_k^t}{\partial x_j^0}\right). \tag{4.63}$$

In the other case, corresponding to the updated Lagrangian description, to derive the formulae for the strain increment and rate one needs to replace the index r by t in Eqs. (4.59) and (4.61). This makes some of the terms vanish (as ${}^tu_i^t \equiv 0$) and leads to the following formulae:

$$\Delta_t\varepsilon_{ij}^t = \frac{1}{2}\left(\frac{\partial \Delta u_i^t}{\partial x_j^t} + \frac{\partial \Delta u_j^t}{\partial x_i^t} + \frac{\partial \Delta u_k^t}{\partial x_i^t}\frac{\partial \Delta u_k^t}{\partial x_j^t}\right), \tag{4.64}$$

$${}_t\dot{\varepsilon}_{ij}^t = \frac{1}{2}\left(\frac{\partial \dot{u}_i^t}{\partial x_j^t} + \frac{\partial \dot{u}_j^t}{\partial x_i^t}\right). \tag{4.65}$$

Let us underline that increments and rates of the displacement gradients at $\tau = t$ are understood in the Lagrangian sense (cf. Eq. (4.52)), i.e. they are taken at the fixed reference configuration, e.g.

$$\frac{\partial \Delta u_i^t}{\partial x_j^t} = \frac{\partial\, {}^0u_i^{t+\Delta t}}{\partial x_j^t} - \frac{\partial\, {}^0u_i^t}{\partial x_j^t}, \qquad \Delta_t\varepsilon_{ij}^t = {}_t^0\varepsilon_{ij}^{t+\Delta t} - {}_t^0\varepsilon_{ij}^t\,.$$

The above formulae cannot thus be identified with the increment and rate of the Almansi strain tensor (4.18) which is always defined with respect to the current (moving) configuration and which at $\tau = t + \Delta t$ depends on displacement gradients with respect to the coordinates $\boldsymbol{x}^{t+\Delta t}$. This is because the Almansi strain tensor definition is compatible with the Eulerian description of the deformable body so that $\Delta \varepsilon^A_{ij} = {}_{t+\Delta t}{}^{0}\varepsilon^{t+\Delta t}_{ij} - {}^{0}_{t}\varepsilon^{t}_{ij}$, while here the Lagrangian concept is consistently observed. We note, however, that for sufficiently small displacement increments the configuration change during the time step, $C^t \mapsto C^{t+\Delta t}$, may be so small that Eqs. (4.64) and (4.65) may turn out good enough approximates of the Almansi strain increment and rate, respectively.

Let us stop for a while at the updated Lagrangian description of motion. Note that in this case the strain rate (4.65) is exactly equal to the linearized Cauchy strain rate (4.19). Denoting by $v_i = \dot{u}_i$ the material point velocity one can write:

$$ {}_t\dot{\varepsilon}^t_{ij} = \frac{1}{2}\,(v^t_{i,j} + v^t_{j,i}), \qquad {}_t\dot{\boldsymbol{\varepsilon}}^t = \frac{1}{2}\left[\nabla \boldsymbol{v}^t + (\nabla \boldsymbol{v}^t)^{\mathrm{T}}\right] = \mathrm{sym}\,(\nabla \boldsymbol{v}^t) \tag{4.66} $$

(since $\boldsymbol{x}^r = \boldsymbol{x}^t$, the simplified notation for partial differentiation in space should be here interpreted as $(\cdot)_{,i} = \partial(\cdot)/\partial x^t_i$). Let us introduce the notation

$$ \boldsymbol{L}^t \stackrel{\mathrm{def}}{=} \nabla \boldsymbol{v}^t, \qquad L^t_{ij} = v^t_{i,j} = \frac{\partial v^t_i}{\partial x^t_j} \tag{4.67} $$

for the velocity gradient tensor with respect to the current configuration. This tensor, as each second order tensor, can be presented as a sum of its symmetric and antisymmetric part, cf. Eqs. (3.63) and (3.64):

$$ \boldsymbol{L} = \boldsymbol{d} + \boldsymbol{\omega}, \qquad \boldsymbol{d} = \frac{1}{2}(\boldsymbol{L} + \boldsymbol{L}^{\mathrm{T}}), \qquad \boldsymbol{\omega} = \frac{1}{2}(\boldsymbol{L} - \boldsymbol{L}^{\mathrm{T}}). \tag{4.68} $$

The first symmetric part equals at $\tau = t$ the strain rate $\boldsymbol{d}^t \equiv {}_t\dot{\boldsymbol{\varepsilon}}^t$ and describes the rate of relative shape change of a material with respect to the current configuration. The antisymmetric tensor $\boldsymbol{\omega}$, called the spin or rotation tensor, describes the velocity of rigid motion of the material particle. This velocity does not affect the strain components values, as was shown in Sect. 4.2.4.

Strain rates defined with respect to two different reference configurations are related to each other by the same formulae as the strains themselves, cf. Eq. (4.22). It is easy to check that for two arbitrary reference configurations C^{t_1} and C^{t_2} we have

$$ {}_{t_2}\dot{\boldsymbol{\varepsilon}}^t = ({}^{t_2}\boldsymbol{F}^{t_1})^{\mathrm{T}}\ {}_{t_1}\dot{\boldsymbol{\varepsilon}}^t\ {}^{t_2}\boldsymbol{F}^{t_1}, \tag{4.69} $$

and, in particular, for C^0 and C^t,

$$ {}_0\dot{\boldsymbol{\varepsilon}}^t = \boldsymbol{F}^{\mathrm{T}}\ {}_0\dot{\boldsymbol{\varepsilon}}^t\ \boldsymbol{F}, \qquad {}_0\dot{\boldsymbol{\varepsilon}}^t = \boldsymbol{F}^{-\mathrm{T}}\ {}_0\dot{\boldsymbol{\varepsilon}}^t\ \boldsymbol{F}^{-1}, \tag{4.70} $$

where $\boldsymbol{F}$ stands for the gradient ${}^0\boldsymbol{F}^t$.

Let us complete our discussion on strain rates by a remark on the volumetric deformation rate. Time differentiation of the scalar equation (4.45) at the current time instant $\tau = t$, i.e. $\mathrm{d}V^t = {}^rJ^t\,\mathrm{d}V^r$, leads to the following relation:

$$\mathrm{d}\dot{V}^t = {}^r\dot{J}^t\,\mathrm{d}V^r. \tag{4.71}$$

Let us transform it using the known formula for the tensor determinant derivative

$$\dot{J} = J\,\mathrm{tr}(\dot{\boldsymbol{F}}\boldsymbol{F}^{-1}), \tag{4.72}$$

which in the particular case of the deformation gradient ${}^rF^t_{ij}$ can be expressed as follows,

$${}^r\dot{J}^t = {}^rJ^t\,{}^r\dot{F}^t_{ij}\,{}^tF^r_{ji} = {}^rJ^t\,\frac{\partial \dot{x}^t_i}{\partial x^r_j}\,\frac{\partial x^r_j}{\partial x^t_i} = {}^rJ^t\,\frac{\partial \dot{x}^t_i}{\partial x^t_i} = {}^rJ^t\,\frac{\partial v^t_i}{\partial x^t_i}\,. \tag{4.73}$$

Taking into account Eqs. (4.45) and (4.67) one can finally express the result as

$$\mathrm{d}\dot{V}^t = \frac{\partial v^t_i}{\partial x^t_i}\,\mathrm{d}V^t = L^t_{ii}\,\mathrm{d}V^t\,, \tag{4.74}$$

where, which is noteworthy, the reference configuration C^r is no longer involved. In the case of small deformations, the volumetric deformation rate is simply expressed as the trace of the Cauchy strain tensor with respect to the current configuration C^t,

$$\mathrm{d}\dot{V}^t \approx \dot{\varepsilon}^C_{ii}\,\mathrm{d}V^t\,, \tag{4.75}$$

cf. Eq. (4.47).

Exercise 4.3
Prove the formula (4.72).

Solution. The definition of tensor determinant (3.69) implies that

$$\dot{J} = \frac{1}{6}\,\epsilon_{ijk}\epsilon_{lmn}\,(\dot{F}_{il}F_{jm}F_{kn} + F_{il}\dot{F}_{jm}F_{kn} + F_{il}F_{jm}\dot{F}_{kn}).$$

Let us transform the first of the three terms, making subsequently use of the Kronecker symbol property (3.20), the inverse tensor definition (3.73), the property of the tensor determinant (3.28) and the property of the permutation symbol (3.26):

$$\frac{1}{6}\epsilon_{ijk}\epsilon_{lmn}\,\dot{F}_{il}F_{jm}F_{kn} = \frac{1}{6}\epsilon_{ijk}\epsilon_{lmn}\,\underbrace{\dot{F}_{ip}\delta_{pl}}_{\dot{F}_{il}}\,F_{jm}F_{kn}$$

$$= \frac{1}{6}\epsilon_{ijk}\epsilon_{lmn}\,\dot{F}_{ip}\,\underbrace{F^{-1}_{pq}F_{ql}}_{\delta_{pl}}\,F_{jm}F_{kn}$$

$$= \frac{1}{6}\epsilon_{ijk}\,\underbrace{\epsilon_{qjk}\det(\boldsymbol{F}^{\mathrm{T}})}_{\epsilon_{lmn}F_{ql}F_{jm}F_{kn}}\,\dot{F}_{ip}F^{-1}_{pq} = \frac{1}{6}\underbrace{2\delta_{iq}}_{\epsilon_{ijk}\epsilon_{qjk}}J\,\dot{F}_{ip}F^{-1}_{pq} = \frac{J}{3}\,\dot{F}_{ip}F^{-1}_{pi}.$$

It is easy to show that transformation of the two remaining terms yields the same results which allows to finally write

$$\dot{J} = J\,\dot{F}_{ip}F^{-1}_{pi} = J\,\mathrm{tr}(\dot{\boldsymbol{F}}\boldsymbol{F}^{-1}).$$

4.4.3 Strain Increments and Rates in One Dimension

Consider once again the problem of uniaxially stretched homogeneous bar shown in Fig. 4.4. Its strain with respect to an arbitrary configuration C^r is given by Eq. (4.24) while for the special cases of $C^r = C^0$ and $C^r = C^t$ by Eqs. (4.25) and (4.26), respectively.

Consider a finite time increment $[t, t + \Delta t]$ and denote by $l^{t+\Delta t}$ the bar length at $\tau = t + \Delta t$. The length change during this increment is thus $\Delta l^t = l^{t+\Delta t} - l^t$ (note that the definition of Δl differs here from that of Sect. 4.2.2). The strain increment with respect to the reference configuration C^r can be computed as

$$\Delta_r\varepsilon^t = {}^0_r\varepsilon^{t+\Delta t} - {}^0_r\varepsilon^t = \frac{1}{2}\,\frac{(l^t + \Delta l^t)^2 - (l^t)^2}{(l^r)^2} = \frac{l^t\Delta l^t}{(l^r)^2} + \frac{1}{2}\,\frac{(\Delta l^t)^2}{(l^r)^2}, \tag{4.76}$$

which can also be written in a form in which linear and quadratic terms with respect to Δl^t are separated, cf. Eqs. (4.60):

$$\Delta_r\varepsilon^t = \overline{\Delta_r\varepsilon}^t + \overline{\overline{\Delta_r\varepsilon}}^t, \qquad \overline{\Delta_r\varepsilon}^t = \frac{l^t\Delta l^t}{(l^r)^2}, \qquad \overline{\overline{\Delta_r\varepsilon}}^t = \frac{1}{2}\,\frac{(\Delta l^t)^2}{(l^r)^2}. \tag{4.77}$$

Considering the two special cases of stationary and updated Lagrangian description, one can write, respectively,

$$\Delta_0\varepsilon^t = \frac{l^t\Delta l^t}{(l^0)^2} + \frac{1}{2}\,\frac{(\Delta l^t)^2}{(l^0)^2}, \tag{4.78}$$

$$\Delta_t \varepsilon^t = \frac{\Delta l^t}{l^t} + \frac{1}{2} \frac{(\Delta l^t)^2}{(l^t)^2} . \tag{4.79}$$

Passing to the limit as $\Delta l^t \to 0$, one arrives at the differential formulae for strain rates expressed, depending on the reference configuration assumed, as

$$_r\dot{\varepsilon}^t = \frac{l^t \, \dot{l}^t}{(l^r)^2} , \qquad _0\dot{\varepsilon}^t = \frac{l^t \, \dot{l}^t}{(l^0)^2} , \qquad _t\dot{\varepsilon}^t = \frac{\dot{l}^t}{l^t} . \tag{4.80}$$

Upon an additional assumption of small deformation, i.e. $|l^t - l^0| \ll l^0$ and $l = l^0 \approx l^t \approx l^t + \Delta t \approx l^r$, one may reduce all the above strain measures to the following linearized Cauchy strain rate tensor:

$$\dot{\varepsilon}^C = \frac{\dot{l}}{l} \tag{4.81}$$

or, written in the differential form,

$$\mathrm{d}\varepsilon^C = \frac{\mathrm{d}l}{l} . \tag{4.82}$$

Remark. Equation (4.82) may suggest careful readers to associate the one-dimensional Cauchy strain ε^C with the logarithmic function $\ln l - a$ (where a stands for an integration constant) which is the solution of the differential equation (4.82). We note that the strain measures discussed in this chapter are not the only ones used in formulating and solving problems of deformable body kinematics. Among other measures we may mention the so-called *natural* or *logarithmic strain*, defined as

$$\varepsilon^L \stackrel{\text{def}}{=} \ln \frac{l^t}{l^r} ,$$

whose linearized approximate is also the Cauchy strain ε^C. The logarithmic strain differential is expressed as

$$\mathrm{d}\varepsilon^L = \frac{\mathrm{d}l^t}{l^t} ,$$

i.e. the length rate is referred to the current length at each time instant. Unfortunately, consistent generalization of such a 1D definition of strain to the 3D case appears a difficult task. In our further discussion, the logarithmic strain and its properties will not be used but it is useful to remember of their existence.

4.5 Strain Compatibility Equations

In the previous sections of this chapter we have focused on equations that allow to uniquely derive the strain tensor field with respect to a selected reference configuration, given a sufficiently smooth displacement vector field at points of a deformable body. Let us now formulate an inverse problem: is it possible to find the displacement field $\boldsymbol{u}(\boldsymbol{x})$ given the strain field $\boldsymbol{\varepsilon}(\boldsymbol{x})$ defined in the space of material points?

Since the symmetric strain tensor $\boldsymbol{\varepsilon}$ has generally six independent components ε_{ij} that depend on three components u_i of the displacement field $\boldsymbol{u}$, it seems intuitively clear that certain additional conditions must be fulfilled by ε_{ij} to guarantee consistency of any of the equation systems (4.16), (4.17b) or (4.18b) (depending on our choice of the reference configuration). In other words, we have to make sure that components u_i can be determined given the components of ε_{ij}. The conditions are called the *strain compatibility equations*.

In the case of infinitesimal (linearized) strain, to which we are going to limit ourselves here, the compatibility equations can be derived in a quite straightforward way. To this aim, one has to twice differentiate the expression defining the Cauchy strain tensor components,

$$\varepsilon_{ij} = \frac{1}{2}(u_{i,j} + u_{j,i}), \tag{4.83}$$

to obtain

$$\varepsilon_{ij,kl} = \frac{1}{2}(u_{i,jkl} + u_{j,ikl}). \tag{4.84}$$

Changing the indices, one may also write:

$$\begin{aligned} \varepsilon_{kl,ij} &= \frac{1}{2}(u_{k,lij} + u_{l,kij}), \\ \varepsilon_{jl,ik} &= \frac{1}{2}(u_{j,lik} + u_{l,jik}), \\ \varepsilon_{ik,jl} &= \frac{1}{2}(u_{i,kjl} + u_{k,ijl}). \end{aligned} \tag{4.85}$$

Since the operation of multiple partial differentiation of a function is insensitive to the ordering of subsequent differentiations, i.e. $u_{i,jkl} = u_{i,kjl} = u_{i,ljk}$ etc., Eqs. (4.84) and (4.85) imply the following relations,

$$\varepsilon_{ij,kl} + \varepsilon_{kl,ij} - \varepsilon_{ik,jl} - \varepsilon_{jl,ik} = 0, \tag{4.86}$$

being in fact a system of $3^4 = 81$ strain compatibility equations. In view of several symmetries with respect to various index pairs, most of the equations are identities and only nine of them turn out non-trivial and can be written as

$$\epsilon_{mik}\,\epsilon_{njl}\,\varepsilon_{ij,kl} = 0 \tag{4.87}$$

(for all $m, n = 1, 2, 3$). It can be further shown that only 6 of them are essentially different, i.e.

$$\begin{aligned}
&\varepsilon_{11,22} + \varepsilon_{22,11} - 2\varepsilon_{12,12} = 0,\\
&\varepsilon_{11,33} + \varepsilon_{33,11} - 2\varepsilon_{13,13} = 0,\\
&\varepsilon_{22,33} + \varepsilon_{33,22} - 2\varepsilon_{23,23} = 0,\\
&\varepsilon_{13,12} + \varepsilon_{12,13} - \varepsilon_{23,11} - \varepsilon_{11,23} = 0,\\
&\varepsilon_{12,23} + \varepsilon_{23,12} - \varepsilon_{13,22} - \varepsilon_{22,13} = 0,\\
&\varepsilon_{23,13} + \varepsilon_{13,23} - \varepsilon_{12,33} - \varepsilon_{33,12} = 0.
\end{aligned} \tag{4.88}$$

Equation (4.88) constitute the system of conditions sufficient for existence of a unique solution u_i to the system of Eq. (4.83) (excluding displacement fields u_i corresponding to rigid motion, for which $\varepsilon_{ij} = 0$).

Exercise 4.4
The linearized Cauchy strain field for a deformable body in the two-dimensional space is given by the equations

$$\varepsilon_{11} = ax_2^2\,, \qquad \varepsilon_{22} = bx_1^2\,, \qquad \varepsilon_{12} = cx_1x_2\,,$$

where x_1, x_2 denote initial coordinates of the body's material points. For which a, b, c the compatibility equations are fulfilled? Compute the displacement field corresponding to the above strains.

Solution. In the 2D case, only the first of Eq. (4.88) is of our interest. Differentiating the given strain components twice with respect to the coordinates and substituting the results into the compatibility equation, we arrive at

$$2a + 2b - 2c = 0,$$

i.e.

$$c = a + b.$$

The Cauchy strain components are given as

$$\varepsilon_{11} = u_{1,1}\,, \qquad \varepsilon_{22} = u_{2,2}\,, \qquad \varepsilon_{12} = \tfrac{1}{2}\left(u_{1,2} + u_{2,1}\right).$$

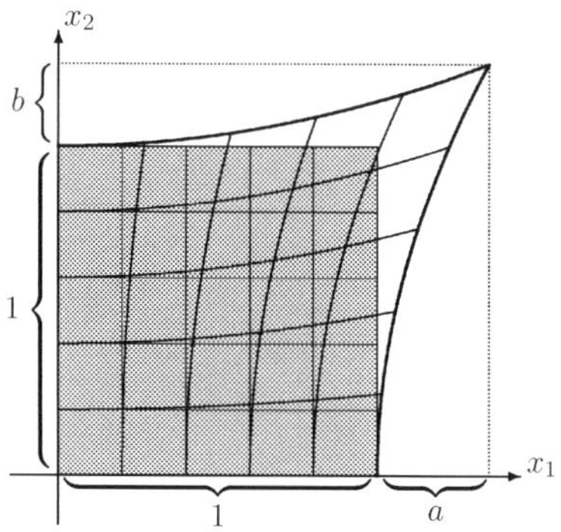

Fig. 4.11 The deformation of a 1 × 1 *square* for the data from Exercise 4.4 at an assumption of zero rigid motion

Integration of the first two formulae leads to the relations

$$u_1 = u_{10}(x_2) + a x_1 x_2^2 \,, \qquad u_2 = u_{20}(x_1) + b x_1^2 x_2 \,,$$

where u_{10} and u_{20} denote arbitrary continuous functions of one variable. Substituting the above results into the third formula, we obtain

$$\begin{aligned} \varepsilon_{12} &= \tfrac{1}{2}\left[2a x_1 x_2 + u'_{10}(x_2) + 2b x_1 x_2 + u'_{20}(x_1)\right] \\ &= (a+b) x_1 x_2 + \tfrac{1}{2}\left[u'_{10}(x_2) + u'_{20}(x_1)\right]. \end{aligned}$$

Since this quantity should equal $c x_1 x_2$ and the derivatives u'_{10} and u'_{20} cannot contain terms with $x_1 x_2$ (because these are one-variable functions), it can be concluded that (i) $c = a + b$ (i.e. the compatibility equation must be fulfilled for existence of a solution) and (ii) the functions u_{10} and u_{20} are linear with opposite slope coefficients, e.g. $u_{10}(x_2) = p x_2 + q$, $u_{20}(x_1) = -p x_1 + r$ (only then their derivatives are constant and add up to zero). Such a choice of the functions u_{10} and u_{20} corresponds, for small deformations, to a rigid motion (rotation combined with translation) that has no influence on strain value.

Figure 4.11 shows the deformation corresponding to zero rigid motion ($u_{10} = u_{20} \equiv 0$) for the example discussed.

Chapter 5
Description of Stress State

5.1 Introduction

5.1.1 Forces, Stress Vectors and Stress Tensor in Continuum

Force is one of the key notions in mechanics. Forces are interactions between bodies. According to laws of mechanics, presented in the subsequent chapters, action of forces is directly related to motion and deformation of bodies.

Forces are vector quantities. They may act on a body from outside but they may also describe mutual interaction between parts (subdomains) of a body. Some forces are concentrated, i.e. acting on particular, separate points of a body while other are continuously distributed on the body's surface or within its volume. Forces are subject to certain rules. The most fundamental of them are (cf. Fig. 5.1) the principle of equilibrium (all forces and their moments, acting on the body or its part, sum up to zero) and the principle of action and reaction (forces describing mutual interactions between two bodies have equal values and opposite signs).

In this chapter, we will focus on forces acting on and within a deformable solid and continuously distributed on its external or internal surfaces. This way of their description is tightly related to our postulate of treating deformable bodies as material continuum.

Let us consider a body $\mathcal{B}$ in the three dimensional space. At the moment, let us neglect the fact that the body is movable and deformable and assume it to have a fixed spatial configuration. Let us define the notion of *stress vector* $\boldsymbol{t}$ at a surface point $P \in \mathcal{B}$ as a density of the resultant force $\boldsymbol{f}$ acting on a surface element $\mathrm{d}A$ containing P, i.e.

$$\boldsymbol{t} = \frac{\mathrm{d}\boldsymbol{f}}{\mathrm{d}A} \,. \tag{5.1}$$

M. Kleiber and P. Kowalczyk, *Introduction to Nonlinear Thermomechanics of Solids*, Lecture Notes on Numerical Methods in Engineering and Sciences,
DOI 10.1007/978-3-319-33455-4_5

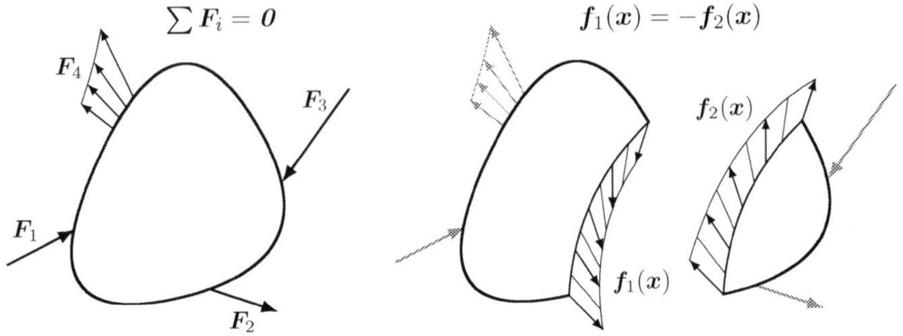

Fig. 5.1 Forces acting on a body; principles of equilibrium and action/reaction

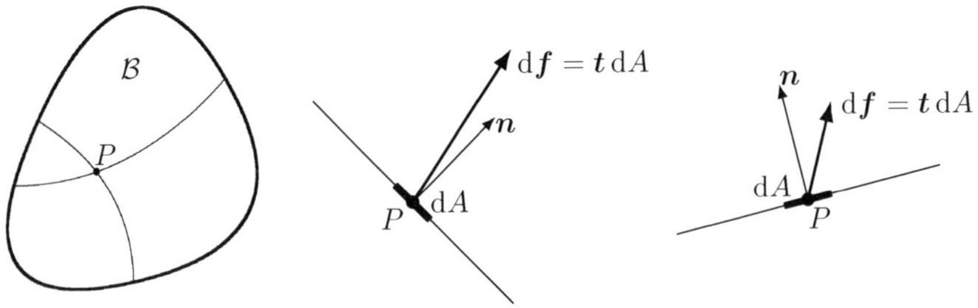

Fig. 5.2 Internal stress vectors revealed by different sectioning surfaces

If A is an external surface of $\mathcal{B}$, we are talking about external stresses acting on the body, otherwise $\boldsymbol{t}$ describes internal forces of interaction between two subdomains of $\mathcal{B}$ divided by the surface. In the latter case, the stress value $\boldsymbol{t}$ at the point P will generally depend on the orientation of the cross-sectional surface: if $\boldsymbol{n}$ denotes the normal unit vector (versor) to the surface element $\mathrm{d}A$ at the point P, then (cf. Fig. 5.2)

$$\boldsymbol{t} = \boldsymbol{t}(\boldsymbol{n}). \tag{5.2}$$

The form of the relationship (5.2) cannot be arbitrary. The principle of action and reaction requires that $\boldsymbol{t}(-\boldsymbol{n}) = -\boldsymbol{t}(\boldsymbol{n})$; each sectioning surface is an external surface for two interacting subdomains of the body $\mathcal{B}$ and the normal versors (by the definition oriented outwards) have opposite signs for each subdomain. On the other hand, the principle of equilibrium requires that the internal forces (as well as their moments with respect to any point) which appear on the external surface of an infinitesimal element $\mathrm{d}V$ of the body's domain must remain in equilibrium, i.e. they must sum up to zero over the entire surface. For instance, if $\mathrm{d}V$ is an infinitesimal polyhedron with N walls, whose areas and normal versors are $\mathrm{d}A_1, \ldots, \mathrm{d}A_N$ and $\boldsymbol{n}_1, \ldots, \boldsymbol{n}_N$, respectively, then the equality $\sum_{s=1}^{N} \boldsymbol{t}(\boldsymbol{n}_s)\,\mathrm{d}A_s = \boldsymbol{0}$ must hold true.

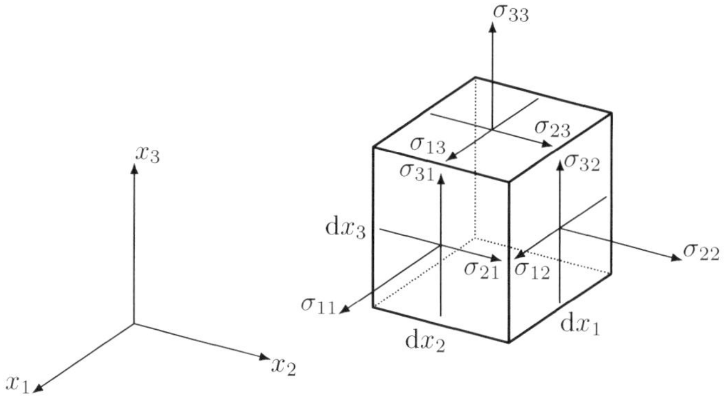

Fig. 5.3 Stress tensor components—physical interpretation

These conditions appear to be fulfilled for the following form of the function (5.2),

$$\boldsymbol{t} = \boldsymbol{\sigma n}, \tag{5.3}$$

where $\boldsymbol{\sigma}$ denotes a symmetric tensor of the second rank, called *stress tensor*. Let us make sure that the so-defined internal forces remain in equilibrium for an infinitesimal cuboid with dimensions dx_1, dx_2, dx_3, and with edges aligned with the axes of a given system of coordinates. Since normal versors to the cuboid walls have in this system simple zero/one representations $\{\pm 1, 0, 0\}$, $\{0, \pm 1, 0\}$, $\{0, 0, \pm 1\}$, the stress vector components $t_i = \sigma_{ij} n_j$ on particular walls are simply equal to appropriate components of the stress tensor σ_{ij}. This is shown in Fig. 5.3, where—for clarity—stress components on the invisible back walls are not displayed (they equal the corresponding displayed front walls components but are directed oppositely due to negative sign of the components $\{n_i\}$).

Resultant forces acting on particular walls of the cuboid are products of stresses and the wall areas. Although stresses are distributed loads, the resultant forces can be treated as concentrated loads acting in the geometrical centres of corresponding walls. They have to obviously be in equilibrium—summing up e.g. components parallel to the x_1 axis one comes at

$$\begin{aligned} \bar{f}_1 = {} & \sigma_{11} dx_2 dx_3 + \sigma_{12} dx_3 dx_1 + \sigma_{13} dx_1 dx_2 \\ & - \sigma_{11} dx_2 dx_3 - \sigma_{12} dx_3 dx_1 - \sigma_{13} dx_1 dx_2 = 0. \end{aligned}$$

Similar results can be obtained for the remaining directions, too.

Internal force moments can be determined from Eqs. (3.29)–(3.30); let us do it with respect to the point $\boldsymbol{y}$ being the geometrical center of the cuboid. Particular components, i.e. moments with respect to the axes containing the cuboid center and parallel to corresponding axes of the coordinate system, can be determined from Eq. (3.30). Summing up moments of particular internal force components with respect to the axis parallel to x_1 (the m_1 component of the moment vector $\boldsymbol{m}$) and noting that the only stress components generating non-zero terms are σ_{23} and σ_{32} (on both front and back cuboid walls), we come at the following moment equilibrium equation:

$$\begin{aligned} m_1 &= \sigma_{32}\mathrm{d}x_3\mathrm{d}x_1 \cdot \frac{\mathrm{d}x_2}{2} - \sigma_{23}\mathrm{d}x_1\mathrm{d}x_2 \cdot \frac{\mathrm{d}x_3}{2} \\ &\quad + \sigma_{32}\mathrm{d}x_3\mathrm{d}x_1 \cdot \frac{\mathrm{d}x_2}{2} - \sigma_{23}\mathrm{d}x_1\mathrm{d}x_2 \cdot \frac{\mathrm{d}x_3}{2} \\ &= (\sigma_{32} - \sigma_{23})\,\mathrm{d}x_1\mathrm{d}x_2\mathrm{d}x_3 = 0. \end{aligned}$$

Analogous equations can be derived for the moment components with respect to the remaining axes. It is thus clear that the necessary condition for the equilibrium is

$$\sigma_{32} - \sigma_{23} = \sigma_{13} - \sigma_{31} = \sigma_{21} - \sigma_{12} = 0,$$

i.e. symmetry of the stress tensor $\boldsymbol{\sigma}$.

Let us note that Fig. 5.3 presents a graphical interpretation of the physical sense of particular stress tensor components in a given coordinate system. Staying still for a while at the example discussed here, diagonal components $\sigma_{11}, \sigma_{22}, \sigma_{33}$ correspond to forces that stretch an elementary "brick" of continuum in the directions of the Cartesian coordinate axes. Negative values of the components mean that the material is compressed, rather than stretched, in this direction. The remaining stress components correspond to tangential forces shearing the cuboid walls.

Remark. Deriving the above formulae for stress equilibrium, we assumed that the stress state in the material is uniform, i.e. gradients $\sigma_{ij,k} = 0$. Otherwise, values of stresses on the reciprocal cuboid walls would not be the same, which would obviously affect the form of the equilibrium equations. The case with non-zero gradients will be discussed in Chap. 6, while introducing the principles of momentum and angular momentum conservation.

Let us recall now—in the context of the above definitions and interpretations—the example of uniaxially stretched bar from Sect. 2.1. Let us compare the stress σ defined there by Eq. (2.3) with the quantities defined above. It is easy to conclude that $\sigma = F/A$ is the one-dimensional counterpart of the vector $\boldsymbol{t}$ which—on the bar cross-section surface—is directed parallelly to the bar axis and whose length is just σ. It is also the value of the axial normal stress component, $\sigma_{11} = \sigma$, which in this case is the only non-zero stress tensor component. We can thus conclude that the notions introduced above are generalization of the simple stress description in a bar for the case of three-dimensional material continuum.

The stress unit in the international system of units SI is pascal, [Pa] = [N/m^2]. In real materials, e.g. metals, the order of stress values usually reaches millions of

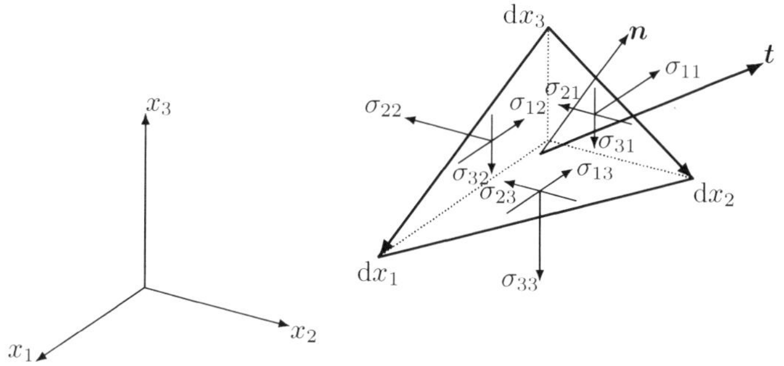

Fig. 5.4 Internal forces acting on an infinitesimal tetrahedron

pascals and thus the derived unit, megapascal (1 MPa $= 10^6$ Pa), is more frequently used in technical applications.

Exercises 5.1
Stress at a continuum point is described by a symmetric tensor $\boldsymbol{\sigma}$. Verify the equilibrium conditions of internal forces and their moments for an infinitesimal tetrahedron spanned on three edges $\mathrm{d}x_1, \mathrm{d}x_2, \mathrm{d}x_3$, parallel to coordinate axes, respectively, Fig. 5.4.

Solution. Let us number the walls S_k, $k = 1, 2, 3, 4$, so that the first three of them are perpendicular to the three axes x_k, $k = 1, 2, 3$, respectively, and S_4 is the largest oblique wall visible in the figure. Internal force vectors acting on the first three walls can be defined in a similar way as in the previously considered cuboid by multiplying the corresponding stress components by the wall areas:

$$\begin{aligned} S_1 &: \quad \{\mathrm{d}f_i^{(1)}\} = -\tfrac{1}{2}\mathrm{d}x_2\mathrm{d}x_3\,\{\sigma_{11}, \sigma_{21}, \sigma_{31}\}, \\ S_2 &: \quad \{\mathrm{d}f_i^{(2)}\} = -\tfrac{1}{2}\mathrm{d}x_3\mathrm{d}x_1\,\{\sigma_{12}, \sigma_{22}, \sigma_{32}\}, \\ S_3 &: \quad \{\mathrm{d}f_i^{(3)}\} = -\tfrac{1}{2}\mathrm{d}x_1\mathrm{d}x_2\,\{\sigma_{13}, \sigma_{23}, \sigma_{33}\}, \end{aligned}$$

(the minus sign means that the forces are directed against the directions of coordinate axes). On the fourth wall we have

$$S_4 : \quad \mathrm{d}f_i = t_i\,\mathrm{d}A = \sigma_{ij}n_j\,\mathrm{d}A\,,$$

where $\mathrm{d}A$ denotes the wall area and n_j are components of its normal versor.

Let us define the auxiliary vectors $\mathrm{d}\boldsymbol{u}$ and $\mathrm{d}\boldsymbol{v}$ whose components are $\{\mathrm{d}u_i\} = \{\mathrm{d}x_1, 0, -\mathrm{d}x_3\}$ and $\{\mathrm{d}v_i\} = \{0, \mathrm{d}x_2, -\mathrm{d}x_3\}$; they are displayed in the figure as arrows starting from the upper vertex of the tetrahedron. S_4 is thus a half of the parallelogram spanned on the two vectors. It comes from the properties of the vector product that $\boldsymbol{n}\,\mathrm{d}A = \frac{1}{2}(\mathrm{d}\boldsymbol{u} \times \mathrm{d}\boldsymbol{v})$ and the components of this vector, see Eq. (3.23), can be expressed as

$$\{n_j\,\mathrm{d}A\} = \left\{\tfrac{1}{2}\mathrm{d}x_2\mathrm{d}x_3,\ \tfrac{1}{2}\mathrm{d}x_3\mathrm{d}x_1,\ \tfrac{1}{2}\mathrm{d}x_1\mathrm{d}x_2\right\}.$$

Hence, on the oblique wall S_4,

$$\mathrm{d}f_i^{(4)} = \tfrac{1}{2}\sigma_{i1}\mathrm{d}x_2\mathrm{d}x_3 + \tfrac{1}{2}\sigma_{i2}\mathrm{d}x_3\mathrm{d}x_1 + \tfrac{1}{2}\sigma_{i3}\mathrm{d}x_1\mathrm{d}x_2\,.$$

Let us now sum up the first direction ($i = 1$) components of forces acting on all the 4 walls:

$$\begin{aligned} \mathrm{d}f_1 &= \mathrm{d}f_1^{(1)} + \mathrm{d}f_1^{(2)} + \mathrm{d}f_1^{(3)} + \mathrm{d}f_1^{(4)} \\ &= -\tfrac{1}{2}\sigma_{11}\mathrm{d}x_2\mathrm{d}x_3 - \tfrac{1}{2}\sigma_{12}\mathrm{d}x_3\mathrm{d}x_1 - \tfrac{1}{2}\sigma_{13}\mathrm{d}x_1\mathrm{d}x_2 \\ &\quad + \tfrac{1}{2}\sigma_{11}\mathrm{d}x_2\mathrm{d}x_3 + \tfrac{1}{2}\sigma_{12}\mathrm{d}x_3\mathrm{d}x_1 + \tfrac{1}{2}\sigma_{13}\mathrm{d}x_1\mathrm{d}x_2 = 0 \end{aligned}$$

The equilibrium condition is thus fulfilled. It is easy to arrive at the same result for the other two directions.

The equilibrium condition for internal force moments can be very easily formulated with respect to the three orthogonal axes parallel to x_1, x_2, x_3, containing the geometrical centres of the walls S_1, S_2, S_3, respectively. Geometrical considerations indicate that all the three axes intersect at the central point of S_4 which will be considered as $\boldsymbol{y}$ in Eq. (3.30). Upon such a choice of axes, majority of resultant stress components (acting at the wall central points) produce zero moments and the equilibrium equation with respect to the first axis can be written in a simple form

$$m_1 = \tfrac{1}{2}\sigma_{32}\mathrm{d}x_3\mathrm{d}x_1 \cdot \tfrac{1}{3}\mathrm{d}x_2 - \tfrac{1}{2}\sigma_{23}\mathrm{d}x_1\mathrm{d}x_2 \cdot \tfrac{1}{3}\mathrm{d}x_3 = 0.$$

This equation holds true for $\sigma_{32} = \sigma_{23}$. Similarly, from the moment equilibrium equations with respect to the other axes one obtains the conditions $\sigma_{13} = \sigma_{31}$ i $\sigma_{21} = \sigma_{12}$.

Exercises 5.2 A stress state in which the component matrix of the tensor $\boldsymbol{\sigma}$ has the form

$$[\sigma_{ij}] = \begin{bmatrix} \sigma_{11} & \sigma_{12} & 0 \\ \sigma_{21} & \sigma_{22} & 0 \\ 0 & 0 & 0 \end{bmatrix}, \tag{5.4}$$

is called *plane stress*. In this state, the stress vector on each cross section plane orthogonal to x_3 ($\{n_i\} = \{0, 0, 1\}$) is zero while stress vectors on cross-section surfaces orthogonal to x_1 and x_2 have zero component t_3. Show that the component $t_3 = 0$ on all cross-section surfaces parallel to x_3 and the zero components of the matrix $[\sigma_{ij}]$ in Eq. (5.4) remain zero in all coordinate systems transformed by rotation around the x_3 axis.

Solution. The normal versor to a surface parallel to x_3 is $\{n_i\} = \{n_1, n_2, 0\}$. The inner product $\boldsymbol{\sigma n}$ results thus in the vector

$$\{t_i\} = \{\sigma_{ij} n_j\} = \{\sigma_{11} n_1 + \sigma_{12} n_2,\ \sigma_{21} n_1 + \sigma_{22} n_2,\ 0\},$$

i.e. the component t_3 remains zero.

Rotation around the x_3 axis by the angle α is described by the following orthogonal matrix:

$$[R_{ij}] = \begin{bmatrix} \cos\alpha & \sin\alpha & 0 \\ -\sin\alpha & \cos\alpha & 0 \\ 0 & 0 & 1 \end{bmatrix}.$$

The components σ'_{ij} in the rotated system can be computed from Eq. (3.44), which, after taking into account symmetry $\sigma_{21} = \sigma_{12}$, yields

$$\begin{aligned}
\sigma'_{11} &= \sigma_{11} \cos^2\alpha + \sigma_{22} \sin^2\alpha + 2\sigma_{12} \sin\alpha \cos\alpha, \\
\sigma'_{22} &= \sigma_{11} \sin^2\alpha + \sigma_{22} \cos^2\alpha - 2\sigma_{12} \sin\alpha \cos\alpha, \\
\sigma'_{12} &= (\sigma_{22} - \sigma_{11}) \sin\alpha \cos\alpha + \sigma_{12} (\cos^2\alpha - \sin^2\alpha), \\
\sigma'_{21} &= \sigma'_{12}, \qquad \sigma'_{i3} = \sigma'_{3i} = 0.
\end{aligned}$$

When collected in a 3×3 matrix, they assume the plane stress form (5.4).

Remark. The components σ'_{ij} determined in Exercise 5.2 may also be expressed as

$$\begin{aligned}
\sigma'_{11} &= \tfrac{1}{2}(\sigma_{11} + \sigma_{22}) + \tfrac{1}{2}(\sigma_{11} - \sigma_{22}) \cos 2\alpha + \sigma_{12} \sin 2\alpha\,, \\
\sigma'_{22} &= \tfrac{1}{2}(\sigma_{11} + \sigma_{22}) - \tfrac{1}{2}(\sigma_{11} - \sigma_{22}) \cos 2\alpha - \sigma_{12} \sin 2\alpha\,, \\
\sigma'_{12} &= \sigma'_{21} = \sigma_{12} \cos 2\alpha - \tfrac{1}{2}(\sigma_{11} - \sigma_{22}) \sin 2\alpha\,.
\end{aligned} \tag{5.5}$$

There is a simple geometric interpretation of the above transformation equations called *Mohr's circle*, Fig. 5.5, which used to help generations of engineers

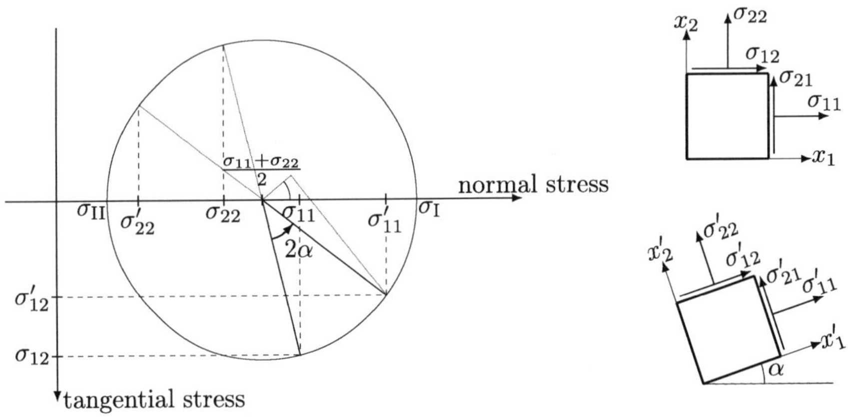

Fig. 5.5 Mohr's circle as graphical method of determination of plane stress components

of the pre-computer era to graphically determine plane stress components σ'_{ij}. The circle's center corresponds to the stress value $\frac{1}{2}(\sigma_{11} + \sigma_{22})$ and its diameter equals $\sqrt{(\sigma_{22} - \sigma_{11})^2 + (2\sigma_{12})^2}$. Stress components revealed on the cross sections orthogonal to x_1 and x_2 axes correspond to two opposite points on the circle. Each other pair of opposite points on the circle determines stress components in a coordinate system rotated by angle α. The horizontal diameter, for which $\sigma_{12} = 0$, corresponds to the coordinate system related to principal direction of the stress tensor, which lie in the x_1, x_2 plane.

5.1.2 *Principal Stress Directions. Extreme Stress Values*

As we have learned from Sect. 3.3.2, each symmetric tensor, in particular the stress tensor $\boldsymbol{\sigma}$, has three orthogonal principal directions. Transformation of the tensor components to the coordinate system defined by these directions leads to a diagonal matrix representation in which the numbers on the main diagonal of the component matrix are the three eigenvalues of the tensor:

$$[\sigma'_{ij}] = \begin{bmatrix} \sigma_I & 0 & 0 \\ 0 & \sigma_{II} & 0 \\ 0 & 0 & \sigma_{III} \end{bmatrix}.$$

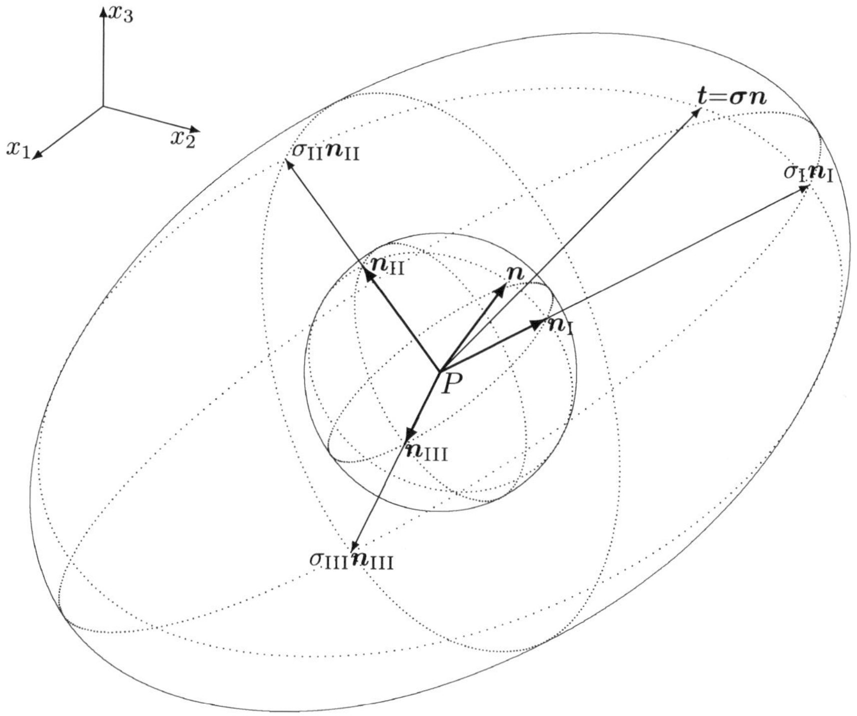

Fig. 5.6 Stress vectors at the point P as functions of cross-section's orientation. Physical interpretation of the principal stresses and their directions

These values are called *principal stresses*. Following the physical interpretation of stress tensor components we can conclude that—for each stress state $\boldsymbol{\sigma}$ at a given point $\boldsymbol{x}$—a set of three orthogonal directions can be found in which an elementary cuboid $\mathrm{d}x_1'\mathrm{d}x_2'\mathrm{d}x_3'$ is only subject to stretching or compressing forces in these directions, with no tangential (shear) forces present.

Recalling Fig. 3.4 and the related discussion we can also conclude that the locus of vector end points of all stress vectors $\boldsymbol{t}$ revealed by different cross-sections at a given material point is an ellipsoid whose major axes are parallel to the principal stress directions and whose semi-axes' lengths (stress values on the cross-sections normal to the directions) equal the principal stresses. The highest and the lowest of the three principal stresses are the extreme values of stress appearing at this point in all possible cross-sections. This is graphically presented in Fig. 5.6, which illustrates the physical interpretation of the principal stresses and their directions.

If all the three principal stress values are equal ($\sigma_{\mathrm{I}} = \sigma_{\mathrm{II}} = \sigma_{\mathrm{III}} = \sigma$), i.e. if $\boldsymbol{\sigma}$ is a spherical tensor, $\boldsymbol{\sigma} = \sigma \boldsymbol{I}$, then the stress vector acting on each cross-section is orthogonal to the section plane and equals $\boldsymbol{t} = \sigma \boldsymbol{n}$. Such a stress state is called hydrostatic—it corresponds to stress that appears in a material particle subject to hydrostatic pressure ($\sigma < 0$) or uniform tension ($\sigma > 0$) acting in all directions, e.g.

submerged deep in water. This is the only stress state in which the material is free of tangential stresses in all of its cross-sections.

The level of stress values in material particles plays an important role in examining the material's strength. Excessive internal forces may result in breaking inter-atomic bounds and either destroy the material or at least deprive it of certain important functional properties. This issue will be discussed in more detail in Chap. 7. Here we only observe that from the strength's point of view the tangential stress components are potentially much more dangerous than the normal ones. Let us thus focus our attention on the former ones to answer the question about their maximum value at a given material point.

Let the stress state be expressed by the tensor $\boldsymbol{\sigma}$ defined by its eigenvalues $\sigma_{\mathrm{I}}, \sigma_{\mathrm{II}}, \sigma_{\mathrm{III}}$ and principal directions $\boldsymbol{n}_{\mathrm{I}}, \boldsymbol{n}_{\mathrm{II}}, \boldsymbol{n}_{\mathrm{III}}$. Tangential stresses on cross-sections orthogonal to the principal directions are zero by the definition. The stress vector $\boldsymbol{t} = \boldsymbol{\sigma}\boldsymbol{n}$ revealed on an arbitrary cross-section defined by the normal versor $\boldsymbol{n}$ is the sum of two orthogonal component vectors: the normal stress $\sigma_n \boldsymbol{n}$ and tangential stress $\boldsymbol{\tau}$. The normal component's length is expressed as

$$\sigma_n = \boldsymbol{t} \cdot \boldsymbol{n}. \tag{5.6}$$

Denoting by τ the value of tangential stress of our interest, i.e. the length of the vector $\boldsymbol{\tau}$, one can find out from the Pythagoras theorem

$$\sigma_n^2 + \tau^2 = |\boldsymbol{t}|^2 = \boldsymbol{t} \cdot \boldsymbol{t}, \tag{5.7}$$

which yields

$$\tau^2 = \boldsymbol{t} \cdot \boldsymbol{t} - \sigma_n^2 = \boldsymbol{t} \cdot \boldsymbol{t} - (\boldsymbol{t} \cdot \boldsymbol{n})^2. \tag{5.8}$$

Consider the Cartesian coordinate system based on the principal stress directions $\boldsymbol{n}_{\mathrm{I}}, \boldsymbol{n}_{\mathrm{II}}, \boldsymbol{n}_{\mathrm{III}}$. Let components of $\boldsymbol{n}$ in this system be $\{n_1, n_2, n_3\}$. The vector $\boldsymbol{t} = \boldsymbol{\sigma}\boldsymbol{n}$ has thus components $\{\sigma_{\mathrm{I}} n_1, \sigma_{\mathrm{II}} n_2, \sigma_{\mathrm{III}} n_3\}$. Substituting this into Eq. (5.8) leads to

$$\begin{aligned} \tau^2 &= t_1^2 + t_2^2 + t_3^2 - (t_1 n_1 + t_2 n_2 + t_3 n_3)^2 \\ &= (\sigma_{\mathrm{I}} n_1)^2 + (\sigma_{\mathrm{II}} n_2)^2 + (\sigma_{\mathrm{III}} n_3)^2 - (\sigma_{\mathrm{I}} n_1^2 + \sigma_{\mathrm{II}} n_2^2 + \sigma_{\mathrm{III}} n_3^2)^2 . \end{aligned} \tag{5.9}$$

Employing the normalizing condition

$$n_1^2 + n_2^2 + n_3^2 = 1, \tag{5.10}$$

one obtains after transformations

$$\tau^2 = (\sigma_{\mathrm{I}} - \sigma_{\mathrm{II}})^2 n_1^2 n_2^2 + (\sigma_{\mathrm{II}} - \sigma_{\mathrm{III}})^2 n_2^2 n_3^2 + (\sigma_{\mathrm{III}} - \sigma_{\mathrm{I}})^2 n_1^2 n_3^2 . \tag{5.11}$$

To find extreme values of this quantity as functions of $\boldsymbol{n}$ at the condition (5.10), let us employ the Lagrange multipliers method. It consists in looking for the values of n_1, n_2, n_3 and the coefficient λ that ensure stationarity of the function

$$\begin{aligned} f(n_1, n_2, n_3, \lambda) &= \tau^2 - \lambda(n_1^2 + n_2^2 + n_3^2 - 1) \\ &= (\sigma_{\mathrm{I}} - \sigma_{\mathrm{II}})^2 n_1^2 n_2^2 + (\sigma_{\mathrm{II}} - \sigma_{\mathrm{III}})^2 n_2^2 n_3^2 + (\sigma_{\mathrm{III}} - \sigma_{\mathrm{I}})^2 n_1^2 n_3^2 \\ &\quad - \lambda\left(n_1^2 + n_2^2 + n_3^2 - 1\right). \end{aligned} \tag{5.12}$$

Differentiating f subsequently with respect to its arguments, one gets

$$\begin{aligned} &(\sigma_{\mathrm{I}} - \sigma_{\mathrm{II}})^2 n_1 n_2^2 + (\sigma_{\mathrm{I}} - \sigma_{\mathrm{III}})^2 n_1 n_3^2 - \lambda n_1 = 0, \\ &(\sigma_{\mathrm{II}} - \sigma_{\mathrm{III}})^2 n_2 n_3^2 + (\sigma_{\mathrm{II}} - \sigma_{\mathrm{I}})^2 n_2 n_1^2 - \lambda n_2 = 0, \\ &(\sigma_{\mathrm{III}} - \sigma_{\mathrm{I}})^2 n_3 n_1^2 + (\sigma_{\mathrm{III}} - \sigma_{\mathrm{II}})^2 n_3 n_2^2 - \lambda n_3 = 0, \\ &n_1^2 + n_2^2 + n_3^2 - 1 = 0, \end{aligned}$$

which is a nonlinear system of four algebraic equations with respect to n_1, n_2, n_3 and λ. Skipping tedious details of the solution procedure, the final result in the form of three different pairs of directions corresponding to local stationarity points of tangential stress value τ is

$$\begin{aligned} \left\{n_i^{(1)}\right\} &= \tfrac{1}{\sqrt{2}}\{0, \pm 1, \pm 1\}; & \tau^{(1)} &= \tfrac{1}{2}|\sigma_{\mathrm{II}} - \sigma_{\mathrm{III}}|, \\ \left\{n_i^{(2)}\right\} &= \tfrac{1}{\sqrt{2}}\{\pm 1, 0, \pm 1\}; & \tau^{(2)} &= \tfrac{1}{2}|\sigma_{\mathrm{III}} - \sigma_{\mathrm{I}}|, \\ \left\{n_i^{(3)}\right\} &= \tfrac{1}{\sqrt{2}}\{\pm 1, \pm 1, 0\}; & \tau^{(3)} &= \tfrac{1}{2}|\sigma_{\mathrm{I}} - \sigma_{\mathrm{II}}|. \end{aligned} \tag{5.13}$$

This means that the extreme values of tangential stress occur on cross-section planes forming the angles of $\pm 45°$ with pairs of principal directions of the stress tensor $\boldsymbol{\sigma}$

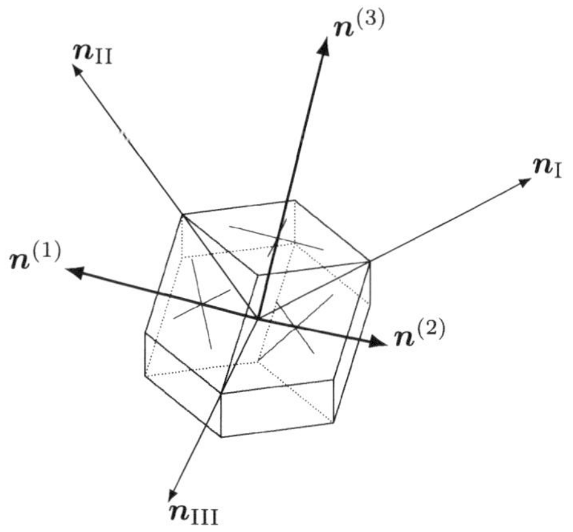

Fig. 5.7 Orientations of cross-sections corresponding to extreme tangential stress values, in relation to principal stress directions

(Fig. 5.7). Unit versors of the cross-section planes may also be expressed as

$$\begin{aligned} \boldsymbol{n}^{(1)} &= \pm\frac{1}{\sqrt{2}}\left(\boldsymbol{n}_{\mathrm{II}} \pm \boldsymbol{n}_{\mathrm{III}}\right), \\ \boldsymbol{n}^{(2)} &= \pm\frac{1}{\sqrt{2}}\left(\boldsymbol{n}_{\mathrm{III}} \pm \boldsymbol{n}_{\mathrm{I}}\right), \\ \boldsymbol{n}^{(3)} &= \pm\frac{1}{\sqrt{2}}\left(\boldsymbol{n}_{\mathrm{I}} \pm \boldsymbol{n}_{\mathrm{II}}\right). \end{aligned} \tag{5.14}$$

The highest of the three values $\tau^{(k)}$ equals the maximum tangential stress acting at the considered material point.

Exercises 5.3
Components of the stress tensor at a given point P, expressed in a selected Cartesian coordinate system, are given in a matrix form as

$$[\sigma_{ij}] = \begin{bmatrix} \frac{3}{2} & 1 & 1 \\ 1 & 2 & 0 \\ 1 & 0 & 1 \end{bmatrix} \text{ [MPa]}.$$

Compute (i) principal stresses and their corresponding principal directions, (ii) maximum tangential stress and its corresponding directions.

Solution. Let us start with the solution of the tensor eigenproblem using the method presented in Exercise 3.16. The characteristic equation for the zero matrix determinant $[\sigma_{ij} - \lambda\delta_{ij}]$ has the form

$$\lambda^3 - \tfrac{9}{2}\lambda^2 + \tfrac{9}{2}\lambda = 0,$$

which can also be written after simple transformations as

$$\lambda(\lambda - \tfrac{3}{2})(\lambda - 3) = 0.$$

The eigenvalues can thus be expressed as

$$\sigma_{\mathrm{I}} = \lambda_1 = 0\,\text{[MPa]}, \qquad \sigma_{\mathrm{II}} = \lambda_2 = \tfrac{3}{2}\,\text{[MPa]}, \qquad \sigma_{\mathrm{III}} = \lambda_3 = 3\,\text{[MPa]}.$$

Components of the principal direction versors $\boldsymbol{n}_{\mathrm{I}}, \boldsymbol{n}_{\mathrm{II}}, \boldsymbol{n}_{\mathrm{III}}$, i.e. stress tensor eigenvectors $\bar{\boldsymbol{v}}^{(k)}$, can be determined from homogeneous equation systems of the form $(\sigma_{ij} - \lambda_k\delta_{ij})\bar{v}_j^{(k)} = 0$, i.e.

$$\begin{bmatrix} \frac{3}{2} & 1 & 1 \\ 1 & 2 & 0 \\ 1 & 0 & 1 \end{bmatrix} \begin{Bmatrix} \bar{v}_1^{(1)} \\ \bar{v}_2^{(1)} \\ \bar{v}_3^{(1)} \end{Bmatrix} = \begin{Bmatrix} 0 \\ 0 \\ 0 \end{Bmatrix} \quad \Longrightarrow \quad \{n_{\mathrm{I}\,i}\} = \left\{\bar{v}_i^{(1)}\right\} = \left\{-\tfrac{2}{3}, \tfrac{1}{3}, \tfrac{2}{3}\right\},$$

$$\begin{bmatrix} 0 & 1 & 1 \\ 1 & \frac{1}{2} & 0 \\ 1 & 0 & -\frac{1}{2} \end{bmatrix} \begin{Bmatrix} \bar{v}_1^{(2)} \\ \bar{v}_2^{(2)} \\ \bar{v}_3^{(2)} \end{Bmatrix} = \begin{Bmatrix} 0 \\ 0 \\ 0 \end{Bmatrix} \quad \Longrightarrow \quad \{n_{\mathrm{II}\,i}\} = \left\{\bar{v}_i^{(2)}\right\} = \left\{\tfrac{1}{3}, -\tfrac{2}{3}, \tfrac{2}{3}\right\},$$

$$\begin{bmatrix} -\frac{3}{2} & 1 & 1 \\ 1 & -1 & 0 \\ 1 & 0 & -2 \end{bmatrix} \begin{Bmatrix} \bar{v}_1^{(3)} \\ \bar{v}_2^{(3)} \\ \bar{v}_3^{(3)} \end{Bmatrix} = \begin{Bmatrix} 0 \\ 0 \\ 0 \end{Bmatrix} \quad \Longrightarrow \quad \{n_{\mathrm{III}\,i}\} = \left\{\bar{v}_i^{(3)}\right\} = \left\{\tfrac{2}{3}, \tfrac{2}{3}, \tfrac{1}{3}\right\}.$$

It is easy to check that the resulting direction versors $\boldsymbol{n}_{\mathrm{I}}, \boldsymbol{n}_{\mathrm{II}}, \boldsymbol{n}_{\mathrm{III}}$ are mutually orthogonal.

Extreme tangential stresses can be determined from Eq. (5.13) which yield

$$\tau^{(1)} = \tfrac{3}{4}\,[\mathrm{MPa}], \qquad \tau^{(2)} = \tfrac{3}{2}\,[\mathrm{MPa}], \qquad \tau^{(3)} = \tfrac{3}{4}\,[\mathrm{MPa}].$$

The highest of the three values, $\tau^{(2)} = \frac{3}{2}$ [MPa], is the maximum tangential stress at the point P, revealed on the cross-section with the normal direction

$$\boldsymbol{n}^{(2)} = \pm\tfrac{1}{\sqrt{2}}(\boldsymbol{n}_{\mathrm{I}} \pm \boldsymbol{n}_{\mathrm{III}}),$$

see Eq. (5.14), i.e.

$$\left\{n_i^{(2)}\right\} = \pm\tfrac{1}{\sqrt{2}}\,\{0, 1, 1\} \quad \text{or} \quad \left\{n_i^{(2)}\right\} = \pm\tfrac{1}{3\sqrt{2}}\,\{4, 1, -1\}.$$

5.2 Description of Stress in Deformable Body

5.2.1 Cauchy and Piola–Kirchhoff Stress Tensors

Up to now in this chapter, we have been consequently neglecting the fact that the considered medium is deformable. We were considering just one and only geometric configuration of the body in which stress vectors and tensors were defined. In this section our discussion will be extended to include the case of a body changing its geometry, i.e. the motion of the body.

Consider a deformable body $\mathcal{B}$ and a certain reference configuration C^r in which material points are assigned Cartesian coordinates $\boldsymbol{x}^r \in \Omega^r$. The body is in motion

Fig. 5.8 Internal force vectors on a deformable body cross-section

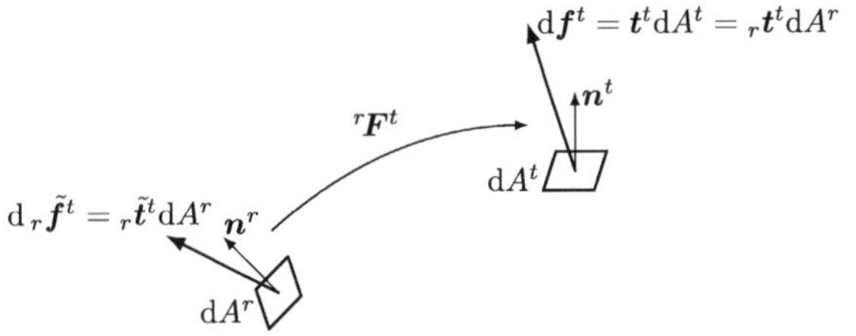

and its material points assume at each time instant τ a configuration C^τ. Moreover, at each time instant τ, the body is subject to a transient set of external forces which result in a corresponding transient field of internal forces.

Consider a material point $P \in \mathcal{B}$ and an infinitesimal element of a certain cross-section surface at this point. In the reference configuration C^r this element has the area $\mathrm{d}A^r$ and the normal direction versor $\boldsymbol{n}^r$. During the motion, the point's location $\boldsymbol{x}^\tau$, the element area $\mathrm{d}A^\tau$ and its orientation $\boldsymbol{n}^\tau$ are continuously changing. The internal forces $\mathrm{d}\boldsymbol{f}^\tau$ acting on this surface element are changing in time, too.

Let us select a time instant $\tau = t$, further referred to as current. All points of the body assume at this instant the configuration C^t and the transformation of their coordinates between C^r and C^t is described by the gradient ${}^r\boldsymbol{F}^t = \mathrm{d}\boldsymbol{x}^t/\mathrm{d}\boldsymbol{x}^r$ (see Fig. 5.8). Knowing the internal forces $\mathrm{d}\boldsymbol{f}^t$ acting on the surface element $\mathrm{d}A^t$ we can define the stress vector

$$\boldsymbol{t}^t = \frac{\mathrm{d}\boldsymbol{f}^t}{\mathrm{d}A^t} \tag{5.15}$$

and the stress tensor $\boldsymbol{\sigma}^t$ describing current internal forces acting at the point P on an arbitrary cross-section surface element:

$$\boldsymbol{t}^t = \boldsymbol{\sigma}^t\boldsymbol{n}^t. \tag{5.16}$$

This tensor is called the *Cauchy stress tensor*. As it can be seen, there is a full analogy between this tensor and the stress tensor introduced in the previous section. In particular, it is easy to show that the Cauchy stress is symmetric, too.

Although defined in a very straightforward, intuitive manner, the Cauchy stress tensor is not sufficient to describe the stress state in a deformable body. The reason is its uncomfortable feature: it is defined with respect to an instantaneous—i.e. variable in time—geometric configuration of the body. It is thus difficult to e.g. operate on the so-defined stress (say, time-differentiate it) in a manner conforming to the Lagrangian description concept adopted here. For this reason, two other tensorial stress measures will be introduced, related to a fixed configuration C^r.

Let us start by defining the nominal stress vector, i.e. the vector that expresses at each time t the relation between internal forces $\mathrm{d}\boldsymbol{f}^t$ acting on the surface element $\mathrm{d}A^t$ and the area of this element measured in the reference configuration (Fig. 5.8):

$$ {}_r\boldsymbol{t}^t = \frac{\mathrm{d}\boldsymbol{f}^t}{\mathrm{d}A^r} \,. \tag{5.17} $$

Next, let us accordingly define at the considered point P the *first Piola–Kirchhoff stress tensor* ${}_r\boldsymbol{\sigma}^t$, also called the *nominal stress tensor*, as

$$ {}_r\boldsymbol{t}^t = {}_r\boldsymbol{\sigma}^t\,\boldsymbol{n}^r \,. \tag{5.18} $$

Since

$$ \mathrm{d}\boldsymbol{f}^t = \boldsymbol{t}^t\,\mathrm{d}A^t = {}_r\boldsymbol{t}^t\,\mathrm{d}A^r \tag{5.19} $$

and, according to the so-called Nanson formula (4.41),

$$ \boldsymbol{n}^t\,\mathrm{d}A^t = {}^rJ^t\,({}^r\boldsymbol{F}^t)^{-\mathrm{T}}\,\boldsymbol{n}^r\,\mathrm{d}A^r , $$

the following relationship must hold true between the Cauchy and the first Piola–Kirchhoff stress tensors,

$$ {}_r\boldsymbol{\sigma}^t = {}^rJ^t\,\boldsymbol{\sigma}^t\,({}^r\boldsymbol{F}^t)^{-\mathrm{T}}, \qquad {}_r\sigma^t_{ij} = {}^rJ^t\,\sigma^t_{ik}\,\frac{\partial x^r_j}{\partial x^t_k} \,, \tag{5.20a} $$

or, in the reverse form,

$$ \boldsymbol{\sigma}^t = \frac{1}{{}^rJ^t}\,{}_r\boldsymbol{\sigma}^t\,({}^r\boldsymbol{F}^t)^{\mathrm{T}}, \qquad \sigma^t_{ij} = \frac{1}{{}^rJ^t}\,{}_r\sigma^t_{ik}\,\frac{\partial x^t_j}{\partial x^r_k} . \tag{5.20b} $$

The first Piola–Kirchhoff tensor, defined with respect to the fixed configuration C^r, is much more convenient for description of stress according to the Lagrangian concept. It is still not free from disadvantages, though, among which the most significant one is the lack of symmetry, directly visible in Eq. (5.20a). Since symmetry of tensors is as a rule a desired feature in derivation of equations of mechanics, let us introduce another Lagrangian measure of stress: the *second Piola–Kirchhoff stress tensor*. To this aim we define the following modified internal force vector (Fig. 5.8),

$$ \mathrm{d}\tilde{\boldsymbol{f}}^t = ({}^r\boldsymbol{F}^t)^{-1}\,\mathrm{d}\boldsymbol{f}^t , \tag{5.21} $$

and the modified stress vector

$$ {}_r\tilde{\boldsymbol{t}}^t = \frac{\mathrm{d}\tilde{\boldsymbol{f}}^t}{\mathrm{d}A^r} = ({}^r\boldsymbol{F}^t)^{-1}\,{}_r\boldsymbol{t}^t \,. \tag{5.22} $$

Each of these vectors may be interpreted as a result of a fictitious inverse transformation of the actual internal stress vector in which force vectors are transformed according to the same rules as linear material elements $\mathrm{d}\boldsymbol{x}$. The second Piola–Kirchhoff stress tensor is then defined as

$$ {}_r\tilde{\boldsymbol{t}}^t = {}_r\tilde{\boldsymbol{\sigma}}^t \, \boldsymbol{n}^r . \tag{5.23} $$

In view of Eq. (5.22), the following relationships between the two Piola–Kirchhoff tensors hold true,

$$ {}_r\tilde{\boldsymbol{\sigma}}^t = ({}^r\boldsymbol{F}^t)^{-1} \, {}_r\boldsymbol{\sigma}^t , \qquad {}_r\tilde{\sigma}^t_{ij} = \frac{\partial x^r_i}{\partial x^t_k} \, {}_r\sigma^t_{kj} , \tag{5.24a} $$

$$ {}_r\boldsymbol{\sigma}^t = {}^r\boldsymbol{F}^t \, {}_r\tilde{\boldsymbol{\sigma}}^t , \qquad {}_r\sigma^t_{ij} = \frac{\partial x^t_i}{\partial x^r_k} \, {}_r\tilde{\sigma}^t_{kj} , \tag{5.24b} $$

and the relationships between the second Piola–Kirchhoff and the Cauchy tensors assume the form

$$ {}_r\tilde{\boldsymbol{\sigma}}^t = {}^rJ^t \, ({}^r\boldsymbol{F}^t)^{-1} \, \boldsymbol{\sigma}^t \, ({}^r\boldsymbol{F}^t)^{-\mathrm{T}} , \qquad {}_r\tilde{\sigma}^t_{ij} = {}^rJ^t \, \frac{\partial x^r_i}{\partial x^t_k} \, \sigma^t_{kl} \, \frac{\partial x^r_j}{\partial x^t_l} , \tag{5.25a} $$

$$ \boldsymbol{\sigma}^t = \frac{1}{{}^rJ^t} \, {}^r\boldsymbol{F}^t \, {}_r\tilde{\boldsymbol{\sigma}}^t \, ({}^r\boldsymbol{F}^t)^{\mathrm{T}} , \qquad \sigma^t_{ij} = \frac{1}{{}^rJ^t} \, \frac{\partial x^t_i}{\partial x^r_k} \, {}_r\tilde{\sigma}^t_{kl} \, \frac{\partial x^t_j}{\partial x^r_l} . \tag{5.25b} $$

It is clear from Eq. (5.25a) and from symmetry of the Cauchy stress tensor that the second Piola–Kirchhoff tensor is symmetric.

In summary, the following three tensorial measures of stress have been introduced at a material point of a deformable body:

- Cauchy tensor $\boldsymbol{\sigma}^t$—symmetric, describing actual internal forces related to surface area in the current configuration,
- first Piola–Kirchhoff tensor ${}_r\boldsymbol{\sigma}^t$—non-symmetric, describing actual internal forces related to surface area in the reference configuration,
- second Piola–Kirchhoff tensor ${}_r\tilde{\boldsymbol{\sigma}}^t$—symmetric, describing fictitious, i.e. transformed internal forces related to surface area in the reference configuration.

The mutual relationships between these measures are given by Eqs. (5.20), (5.24) and (5.25).

If the reference configuration coincides with the current configuration, $C^r = C^t$, then all the three stress measures become equal,

$$ \boldsymbol{\sigma}^t = {}_t\boldsymbol{\sigma}^t = {}_t\tilde{\boldsymbol{\sigma}}^t . \tag{5.26} $$

However, even in this case we have to distinguish between the three tensors, since—as we are going to see soon—their increments and time derivatives are different.

Let us now reconsider the example of axially stretched bar from Sect. 2.1. We select the reference configuration as the initial bar configuration, i.e. the one in which the bar length and cross-section area are l^0 and A^0, respectively. At $\tau = t$ these geometrical parameters are $l^t = \lambda l^0$ and $A^t = \gamma^2 A^0$ (where γ is a certain coefficient, generally different from unity) which corresponds to the following deformation gradient components,

$$[{}^0F^t_{ij}] = \left[\frac{\partial x^t_i}{\partial x^r_j}\right] = \begin{bmatrix} \lambda & 0 & 0 \\ 0 & \gamma & 0 \\ 0 & 0 & \gamma \end{bmatrix}.$$

If the resultant internal axial force acting on a bar cross-section is F^t, then it is easy to show that the Cauchy stress tensor has only one non-zero component

$$\sigma^t_{11} = \frac{F^t}{A^t}.$$

Similarly, the first Piola–Kirchhoff stress tensor, computed from Eq. (5.20), has only one non-zero component

$${}_0\sigma^t_{11} = {}^0J^t \frac{\sigma^t_{11}}{{}^0F^t_{11}} = \gamma^2 \sigma^t_{11} = \frac{F^t}{A^0}.$$

Both the measures have thus a simple one-dimensional interpretation, cf. Definition (2.3); they describe internal forces acting on a bar cross-section related to the cross-section area in the initial (reference) and current configuration, respectively. Unfortunately, the axial component of the second Piola–Kirchhoff stress, ${}_0\tilde{\sigma}^t_{11} = \frac{1}{\lambda}\frac{F^t}{A^0}$, lacks any simple intuitive physical interpretation.

5.2.2 *Objectivity and Invariance of Stress Measures*

Careful readers have surely learned from all previous considerations of this section that values of stress in a deformable body depend on both the level of internal forces, i.e. mechanical interactions between material particles, and the body's motion resulting in changes of the cross section surface on which the internal forces act. The latter occur both at any rigid motion and at changes of the body's deformation (it is easy to figure out rigid motion of a rotating body in which constant interactions between material particles exist and yet the stress field changes in time due to changes in the cross-section surface orientation).

A question may thus arise whether, and to which extent, it is possible to distinguish between changes in stress tensor related to the two effects. It is particularly interesting how the rigid motion of the body affects the particular tensorial stress measures.

Consider the surface element $\mathrm{d}A$ of the body $\mathcal{B}$ (see Fig. 5.8) on which at the time instant $\tau = t$ internal forces $\mathrm{d}\boldsymbol{f}^t$ act. Let us now imagine a small modification of the geometric configuration C^t, further denoted by C', being a result of a small rotation

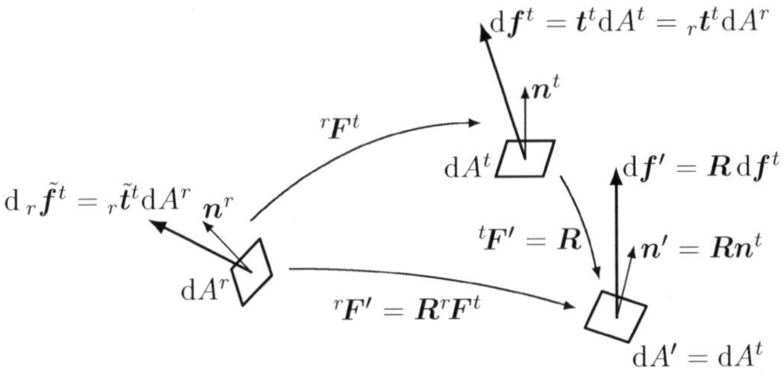

Fig. 5.9 Stress vectors at rigid rotation of the body

of C^t, defined by an orthogonal tensor $\boldsymbol{R}$ (see Fig. 5.9). The deformation gradient tensor from C^r to C' has thus the form

$$ {}^r\boldsymbol{F}' = \boldsymbol{R}\,{}^r\boldsymbol{F}^t . $$

Let us also assume that all external and internal forces in material points of $\mathcal{B}$ in the configuration C' are equal to the corresponding forces in the configuration C^t but rotated according to the same rotation operator $\boldsymbol{R}$. Briefly speaking, we have rotated in a rigid manner the entire body together with all forces acting at its points.

Let us now derive relations between the three stress measures defined at a material point in the two configurations C' and C^t. Since

$$ \mathrm{d}\boldsymbol{f}' = \boldsymbol{R}\,\mathrm{d}\boldsymbol{f}^t = \boldsymbol{R}\boldsymbol{t}^t\mathrm{d}A^t = \boldsymbol{R}\,{}_r\boldsymbol{t}^t\mathrm{d}A^r $$

and $\mathrm{d}A^t = \mathrm{d}A'$, we can write

$$ \boldsymbol{t}' = \boldsymbol{R}\,\boldsymbol{t}^t , \qquad {}_r\boldsymbol{t}' = \boldsymbol{R}\,{}_r\boldsymbol{t}^t \tag{5.27} $$

and

$$ {}_r\tilde{\boldsymbol{t}}' = ({}^r\boldsymbol{F}')^{-1}\,{}_r\boldsymbol{t}' = ({}^r\boldsymbol{F}^t)^{-1}\,{}_r\boldsymbol{t}^t = {}_r\tilde{\boldsymbol{t}}^t . \tag{5.28} $$

Substituting into Eqs. (5.27)–(5.28) definitions of stress tensors (5.16), (5.18) and (5.23) expressed in the configurations C' and C^t, we can get after simple transformation

$$ \boldsymbol{\sigma}' = \boldsymbol{R}\boldsymbol{\sigma}^t\boldsymbol{R}^{\mathrm{T}} , \qquad {}_r\boldsymbol{\sigma}' = \boldsymbol{R}\,{}_r\boldsymbol{\sigma}^t , \qquad {}_r\tilde{\boldsymbol{\sigma}}' = {}_r\tilde{\boldsymbol{\sigma}}^t . \tag{5.29} $$

The above relationships imply that among the three tensorial stress measures only the second Piola–Kirchhoff stress tensor is insensitive to rigid rotation of the body. We can thus call it an *invariant* stress measure (with respect to rigid rotation), that describes actual mechanical interactions between material particles at the body's material points.

Another interesting property at rigid rotation is featured by the Cauchy stress tensor. Equation $(5.29)_1$ may evoke in a careful reader an association with the formula (3.44) that describes transformation of the rank-2 tensor components at rotation of the Cartesian coordinate system. Indeed, comparing the two formulae, it is easy to conclude that if the coordinate system gets rotated along with the body according to the rotation operator $\boldsymbol{R}$, i.e. if the orthogonal transformation matrix $[Q_{ij}]$ in Eq. (3.44) equals the transposed component matrix $[R_{ji}]$ of the body's rotation tensor, then the Cauchy stress components σ'_{ij} expressed in the rotated coordinate system remain unchanged, i.e. equal to the components σ^t_{ij} in the original (non-rotated) coordinate system. This does not mean that the stress tensor remains unchanged—to the contrary, it is emphasized that $\boldsymbol{\sigma}'$ and $\boldsymbol{\sigma}^t$ are different tensors—but the components of the two different tensors expressed in the so-defined two different Cartesian coordinate systems appear to be equal. A tensor that meets the condition $(5.29)_1$ at rigid rotation is called *objective tensor.*[1]

5.3 Increments and Rates of Stress Tensors

Consider the time interval $\tau \in [t, t+\Delta t]$ in which material points of the body $\mathcal{B}$ have changed their location by $\Delta \boldsymbol{x}^t$ and forces acting on different cross-sections at these points have changed their values by $\Delta(\mathrm{d}\boldsymbol{f}^t)$. These changes obviously imply changes in the values of stress tensor components. Let us define the tensor increments as

$$\Delta\, {}_r\tilde{\boldsymbol{\sigma}}^t = {}_r\tilde{\boldsymbol{\sigma}}^{t+\Delta t} - {}_r\tilde{\boldsymbol{\sigma}}^t, \tag{5.30a}$$

$$\Delta\, {}_r\boldsymbol{\sigma}^t = {}_r\boldsymbol{\sigma}^{t+\Delta t} - {}_r\boldsymbol{\sigma}^t, \tag{5.30b}$$

$$\Delta\boldsymbol{\sigma}^t = \boldsymbol{\sigma}^{t+\Delta t} - \boldsymbol{\sigma}^t. \tag{5.30c}$$

Note that only the first two of the above definitions—those regarding the two Piola–Kirchhoff tensors—describe stress increments according to the Lagrangian concept, i.e. increments defined with respect to the same, fixed reference configuration C^r. The Cauchy stress tensor increment (5.30c), in fact equal to ${}_{t+\Delta t}\boldsymbol{\sigma}^{t+\Delta t} - {}_t\boldsymbol{\sigma}^t$, is an Eulerian increment, including the change of the reference configuration.

This fact is worth particular mentioning in view of the issue of experimental stress measuring. Note that the quantity that is actually measured in experiments is force, not stress. It may thus be argued that the increment $\Delta\, {}_r\boldsymbol{\sigma}^t$ is more "physical" than $\Delta\boldsymbol{\sigma}^t$ because its value describes the increment of a physical and not an artificial

[1] Different definitions of tensor objectivity can be found in literature. In our study, the definition of Truesdell and Noll [2] is adopted. According to another definition, given by Hill [1], the second Piola–Kirchhoff tensor is considered the objective stress measure.

quantity. In the latter case, the changed force is related to the changed cross-section area while in the Piola–Kirchhoff tensors—to a fixed (e.g. initial) cross-section area.

Even if the reference configuration C^r is identified with the current configuration C^t, i.e.—in view of Eq. (5.26)—all the three tensor measures have the same values, the stress increments defined by Eq. (5.30) are different:

$$\Delta\,{}_t\tilde{\boldsymbol{\sigma}}^t = {}_t\tilde{\boldsymbol{\sigma}}^{t+\Delta t} - {}_t\tilde{\boldsymbol{\sigma}}^t = {}_t\tilde{\boldsymbol{\sigma}}^{t+\Delta t} - \boldsymbol{\sigma}^t, \tag{5.31a}$$

$$\Delta\,{}_t\boldsymbol{\sigma}^t = {}_t\boldsymbol{\sigma}^{t+\Delta t} - {}_t\boldsymbol{\sigma}^t = {}_t\boldsymbol{\sigma}^{t+\Delta t} - \boldsymbol{\sigma}^t, \tag{5.31b}$$

$$\Delta\boldsymbol{\sigma}^t = \boldsymbol{\sigma}^{t+\Delta t} - \boldsymbol{\sigma}^t = {}_{t+\Delta t}\boldsymbol{\sigma}^{t+\Delta t} - \boldsymbol{\sigma}^t. \tag{5.31c}$$

The first one only describes changes of forces acting between material points, the second one also includes changes of their directions resulting from the body's motion while the third one additionally includes changes of cross-section areas on which the forces act and to which they are related.

Stress rates, i.e. their time derivatives, can be defined by passing in Eq. (5.30) to the limit as $\Delta t \to 0$. Stress rates play an important role in solid mechanics formulations since they appear in constitutive equations for a wide class of materials. Since the constitutive equations describe relations between deformation of material continuum and physical internal forces acting in this medium, the issues of invariance and objectivity of the rates appear to have crucial importance. Without coming into detail, let us only observe that not every stress rate is appropriate for use in mathematically correct constitutive formulations.

Consider the second Piola–Kirchhoff stress rate ${}_r\dot{\tilde{\boldsymbol{\sigma}}}^t = \mathrm{d}_r\tilde{\boldsymbol{\sigma}}^\tau/\mathrm{d}\tau|_{\tau=t}$. Invariance of the tensor ${}_r\tilde{\boldsymbol{\sigma}}^t$ itself (discussed in Sect. 5.2.2) implies that its rate is also invariant, i.e. insensitive to any rigid rotation: if ${}^r\boldsymbol{F}' = \boldsymbol{R}\,{}^r\boldsymbol{F}^t$ then

$${}_r\dot{\tilde{\boldsymbol{\sigma}}}' = {}_r\dot{\tilde{\boldsymbol{\sigma}}}^t. \tag{5.32}$$

Hence, the tensor ${}_r\dot{\tilde{\boldsymbol{\sigma}}}^t$ describes the rate of the actual forces between material particles at a selected point of the body.

One could now ask a question whether the rate of an objective stress tensor (like the Cauchy tensor) is an objective tensor, too. To discuss this issue in more detail, let us consider Eq. (5.25a) and rewrite it in a simplified manner, skipping the configuration-related indices, as

$$\boldsymbol{\sigma} = \frac{1}{J}\,\boldsymbol{F}\tilde{\boldsymbol{\sigma}}\boldsymbol{F}^{\mathrm{T}}. \tag{5.33}$$

We recall that the reference configuration C^r remains by definition fixed during the motion in the surrounding of the time instant t. Differentiation of Eq. (5.33) with respect to time followed by a series of simple transformations (in which the relation (4.72) is used) leads to the following formula for the Cauchy stress rate:

$$\begin{aligned}\dot{\boldsymbol{\sigma}} &= -\frac{\dot{J}}{J^2}\,\boldsymbol{F}\tilde{\boldsymbol{\sigma}}\boldsymbol{F}^{\mathrm{T}} + \frac{1}{J}\,\dot{\boldsymbol{F}}\tilde{\boldsymbol{\sigma}}\boldsymbol{F}^{\mathrm{T}} + \frac{1}{J}\,\boldsymbol{F}\dot{\tilde{\boldsymbol{\sigma}}}\boldsymbol{F}^{\mathrm{T}} + \frac{1}{J}\,\boldsymbol{F}\tilde{\boldsymbol{\sigma}}\dot{\boldsymbol{F}}^{\mathrm{T}} \\ &= -\operatorname{tr}(\dot{\boldsymbol{F}}\boldsymbol{F}^{-1})\,\boldsymbol{\sigma} + \frac{1}{J}\,\dot{\boldsymbol{F}}\boldsymbol{F}^{-1}\boldsymbol{F}\tilde{\boldsymbol{\sigma}}\boldsymbol{F}^{\mathrm{T}} + \frac{1}{J}\,\boldsymbol{F}\dot{\tilde{\boldsymbol{\sigma}}}\boldsymbol{F}^{\mathrm{T}} + \frac{1}{J}\,\boldsymbol{F}\tilde{\boldsymbol{\sigma}}\boldsymbol{F}^{\mathrm{T}}\boldsymbol{F}^{-\mathrm{T}}\dot{\boldsymbol{F}}^{\mathrm{T}} \\ &= -(\operatorname{tr}\boldsymbol{L})\,\boldsymbol{\sigma} + \boldsymbol{L}\boldsymbol{\sigma} + \frac{1}{J}\,\boldsymbol{F}\dot{\tilde{\boldsymbol{\sigma}}}\boldsymbol{F}^{\mathrm{T}} + \boldsymbol{\sigma}\boldsymbol{L}^{\mathrm{T}},\end{aligned} \tag{5.34}$$

where $\boldsymbol{L}$ denotes the gradient of velocity $\boldsymbol{v}^t = \dot{\boldsymbol{x}}^t$ with respect to the current configuration coordinates, cf. Eq. (4.67),

$$\boldsymbol{L} = \dot{\boldsymbol{F}}\boldsymbol{F}^{-1}, \qquad L_{ij} = \frac{\partial \dot{x}^t_i}{\partial x^r_k}\,\frac{\partial x^r_k}{\partial x^t_j} = \frac{\partial v^t_i}{\partial x^t_j}\,, \tag{5.35}$$

and its trace equals the divergence of the velocity field in this configuration,

$$\operatorname{tr}\boldsymbol{L} = \frac{\partial v^t_i}{\partial x^t_i}\,. \tag{5.36}$$

Let us now rewrite the relation (5.34) for a modified deformation gradient combined with rigid rotation, $\boldsymbol{F}' = \boldsymbol{R}\boldsymbol{F}$. In such a case, $\dot{\boldsymbol{F}}' = \dot{\boldsymbol{R}}\boldsymbol{F} + \boldsymbol{R}\dot{\boldsymbol{F}}$ and $\boldsymbol{L}' = \dot{\boldsymbol{R}}\boldsymbol{R}^{\mathrm{T}} + \boldsymbol{R}\boldsymbol{L}\boldsymbol{R}^{\mathrm{T}}$. It is easy to see that the tensor $\dot{\boldsymbol{\sigma}}$ does not meet the condition of objectivity, i.e. $\dot{\boldsymbol{\sigma}}' \neq \boldsymbol{R}\dot{\boldsymbol{\sigma}}\boldsymbol{R}^{\mathrm{T}}$, and thus it cannot be used to objectively describe the stress rate.

Objective time derivatives of the Cauchy stress tensor do exist, though, but they have to be defined in a more elaborated manner. An example of such a derivative is the *Truesdell rate* defined as

$$\overset{\triangledown}{\boldsymbol{\sigma}} = \dot{\boldsymbol{\sigma}} + (\operatorname{tr}\boldsymbol{L})\,\boldsymbol{\sigma} - \boldsymbol{L}\boldsymbol{\sigma} - \boldsymbol{\sigma}\boldsymbol{L}^{\mathrm{T}}, \tag{5.37}$$

i.e.

$$\overset{\triangledown}{\boldsymbol{\sigma}} = \frac{1}{J}\boldsymbol{F}\dot{\tilde{\boldsymbol{\sigma}}}\boldsymbol{F}^{\mathrm{T}}. \tag{5.38}$$

Modifying the deformation gradient by combining it with the rigid rotation, $\boldsymbol{F}' = \boldsymbol{R}\boldsymbol{F}$, $J' = J$, and considering the relations $(5.29)_1$, $(5.29)_3$ and (5.32), one can rewrite the Truesdell rate of the Cauchy tensor as

$$\overset{\triangledown}{\boldsymbol{\sigma}}{}' = \frac{1}{J'}\,\boldsymbol{F}'\,\dot{\tilde{\boldsymbol{\sigma}}}{}'\,\boldsymbol{F}'^{\mathrm{T}} = \frac{1}{J}\,\boldsymbol{R}\boldsymbol{F}\,\dot{\tilde{\boldsymbol{\sigma}}}\,\boldsymbol{F}^{\mathrm{T}}\boldsymbol{R}^{\mathrm{T}} = \boldsymbol{R}\,\overset{\triangledown}{\boldsymbol{\sigma}}\,\boldsymbol{R}^{\mathrm{T}}. \tag{5.39}$$

The Truesdell rate $\overset{\triangledown}{\boldsymbol{\sigma}}$ is thus an objective tensor. Note that if the current configuration is taken as the reference configuration, $C^t = C^r$, then the Truesdell rate of the Cauchy tensor is exactly equal to the second Piola–Kirchhoff tensor rate:

$$\overset{\nabla}{\sigma} = \dot{\tilde{\sigma}}\,. \tag{5.40}$$

The Truesdell rate is not the only objective rate that can be defined for the Cauchy stress tensor. Other formulations that may be found in the literature, like the *Zaremba–Jaumann rate*, will not be discussed here, though.

5.4 Work of Internal Forces. Conjugate Stress–Strain Pairs

In Chap. 4, a number of tensorial measures of deformation and strain in a material point have been introduced. In this chapter, a number of stress tensor definitions have been described. Such a large diversity of notions used to describe similar phenomena is probably perceived by careful readers as a factor greatly complicating the analysis. The following discussion in this section is intended to introduce some order into the various strain and stress measures introduced so far. The tool we are going to use for this purpose is the notion of the internal force work and energy accumulated due to it.

If a material point subjected to an acting force $\boldsymbol{f}$ is moved by an infinitesimal displacement $\mathrm{d}\boldsymbol{u}$ then one can say that the force has performed an infinitesimal work expressed as

$$\mathrm{d}W = \boldsymbol{f} \cdot \mathrm{d}\boldsymbol{u} = f_i\, \mathrm{d}u_i\,. \tag{5.41}$$

$\mathrm{d}W$ is a scalar quantity. If the point has been moved over a finite distance along a certain trajectory S then the total work performed by the force is expressed by the integral

$$W = \int_S \mathrm{d}W = \int_S \boldsymbol{f} \cdot \mathrm{d}\boldsymbol{u}, \tag{5.42}$$

in which the force $\boldsymbol{f}$ may generally change in time. This work gets converted to various forms of energy—it may be accumulated as a potential energy, converted to a kinetic energy, dissipated as heat, etc. It is frequently convenient to express the formula (5.41) in the rate form as

$$\dot{W} = \boldsymbol{f} \cdot \dot{\boldsymbol{u}} = f_i\, \dot{u}_i\,, \tag{5.43}$$

in which the rate $\dot{W}$ denotes the power of the force $\boldsymbol{f}$.

In particular, in a deformable body subject to loads, internal and external forces perform work on displacements of material points at which they are applied.

Consider a material particle P with coordinates x_i in the deformable body $\mathcal{B}$. Assume its deformations to be sufficiently small so that any differences between geometric configurations in motion may be disregarded and the linearized Cauchy tensor $\varepsilon_{ij} = \frac{1}{2}(u_{i,j} + u_{j,i})$ is an acceptable approximate of strain. Consider an

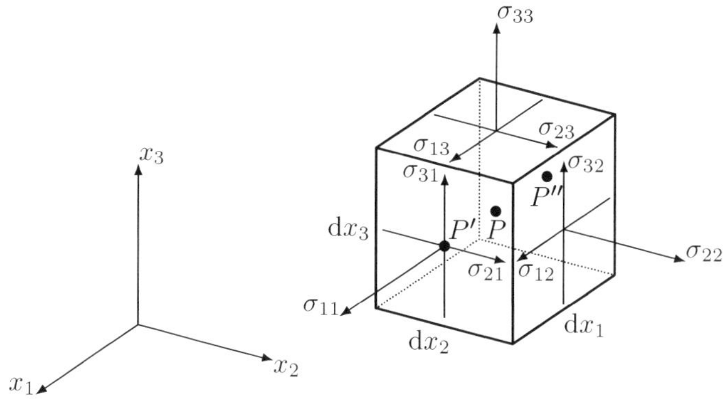

Fig. 5.10 Work of stress tensor components

infinitesimal volumetric element $\mathrm{d}V$ around this particle in the form of a cuboid with edges aligned with the coordinate system axes and with linear dimensions $\mathrm{d}x_1, \mathrm{d}x_2, \mathrm{d}x_3$ (see Fig. 5.10—the particle P is located in the cuboid's geometrical centre). Assume uniform stress state described by the tensor σ_{ij} and a possibly non-uniform velocity field $\dot{u}_i(x_k)$ within the element $\mathrm{d}V$.

Let us now compute the power of the particular internal force components, acting on the cuboid's walls and performing work on their displacements. Consider the three stress components $\sigma_{11}, \sigma_{21}, \sigma_{31}$, i.e. those acting on the walls orthogonal to the axis x_1. Since the forces are uniformly distributed on the walls, they can be replaced by their resultant forces applied at the wall centres (P' and P''; in the latter point the forces act in the opposite directions). and equal to $\sigma_{11}\mathrm{d}x_2\mathrm{d}x_3$, $\sigma_{21}\mathrm{d}x_2\mathrm{d}x_3$, $\sigma_{31}\mathrm{d}x_2\mathrm{d}x_3$, respectively. The forces perform work on displacements of the points P' and P'', equal to $\dot{u}_i + \frac{1}{2}\dot{u}_{i,1}\mathrm{d}x_1$ and $\dot{u}_i - \frac{1}{2}\dot{u}_{i,1}\mathrm{d}x_1$, respectively ($\dot{u}_i$ denotes the displacement of P). Hence, the total power values of the particular three stress components are

$$\begin{aligned}
\dot{W}(\sigma_{11}) &= \sigma_{11}\mathrm{d}x_2\mathrm{d}x_3\left(\dot{u}_1 + \tfrac{1}{2}\dot{u}_{1,1}\mathrm{d}x_1\right) - \sigma_{11}\mathrm{d}x_2\mathrm{d}x_3\left(\dot{u}_1 - \tfrac{1}{2}\dot{u}_{1,1}\mathrm{d}x_1\right)\\
&= \sigma_{11}\dot{u}_{1,1}\,\mathrm{d}x_1\mathrm{d}x_2\mathrm{d}x_3\,,\\
\dot{W}(\sigma_{21}) &= \sigma_{21}\mathrm{d}x_2\mathrm{d}x_3\left(\dot{u}_2 + \tfrac{1}{2}\dot{u}_{2,1}\mathrm{d}x_1\right) - \sigma_{21}\mathrm{d}x_2\mathrm{d}x_3\left(\dot{u}_2 - \tfrac{1}{2}\dot{u}_{2,1}\mathrm{d}x_1\right)\\
&= \sigma_{21}\dot{u}_{2,1}\,\mathrm{d}x_1\mathrm{d}x_2\mathrm{d}x_3\,,\\
\dot{W}(\sigma_{31}) &= \sigma_{31}\mathrm{d}x_2\mathrm{d}x_3\left(\dot{u}_3 + \tfrac{1}{2}\dot{u}_{3,1}\mathrm{d}x_1\right) - \sigma_{31}\mathrm{d}x_2\mathrm{d}x_3\left(\dot{u}_3 - \tfrac{1}{2}\dot{u}_{3,1}\mathrm{d}x_1\right)\\
&= \sigma_{31}\dot{u}_{3,1}\,\mathrm{d}x_1\mathrm{d}x_2\mathrm{d}x_3\,.
\end{aligned}$$

Carrying out the same reasoning for the remaining 6 stress components acting on the remaining cuboid's walls and using the equality $\mathrm{d}x_1\mathrm{d}x_2\mathrm{d}x_3 = \mathrm{d}V$, one concludes that the total power of internal forces in the continuum element (being the sum of the powers related to the particular components) is expressed by the full product of the stress and velocity gradient tensors multiplied by the element volume, i.e.

$$\dot{W} = \sigma_{ij}\dot{u}_{i,j}\,\mathrm{d}V = \boldsymbol{\sigma}\cdot\nabla\dot{\boldsymbol{u}}\,\mathrm{d}V. \tag{5.44}$$

In view of the stress tensor symmetry and the relation (3.65), the above formula may be rewritten as

$$\dot{W} = \sigma_{ij}\dot{\varepsilon}_{ij}\,\mathrm{d}V = \boldsymbol{\sigma}\cdot\dot{\boldsymbol{\varepsilon}}\,\mathrm{d}V \tag{5.45}$$

or, in a differential form, as

$$\mathrm{d}W = \sigma_{ij}\mathrm{d}\varepsilon_{ij}\,\mathrm{d}V = \boldsymbol{\sigma}\cdot\mathrm{d}\boldsymbol{\varepsilon}\,\mathrm{d}V. \tag{5.46}$$

Remark. While deriving the above relations we have made an assumption that the stress state is uniform, i.e. that the gradients $\sigma_{ij,k} = 0$. This may make the careful readers anxious because—if the velocity field is assumed non-uniform—it is difficult to justify such a restrictive assumption about the stress field. Looking at Fig. 5.10 and the formulae derived above, it is easy to show that if non-uniformly distributed stresses were allowed, the stress components acting on the "front" and "back" walls would be expressed as $\sigma_{ij} \pm \frac{1}{2}\sigma_{ij,j}\mathrm{d}x_j$, respectively (no summation on j, for nine values of the pairs i, j). Simple transformations lead to the conclusion that the right-hand side of Eq. (5.45) would have to be extended in this case by the expression $\sigma_{ij,j}\dot{u}_i\,\mathrm{d}V = \mathrm{div}\,\boldsymbol{\sigma}\cdot\dot{\boldsymbol{u}}\,\mathrm{d}V$. Anticipating a discussion of Chap. 6, we only mention here that—at no other loads acting on the material particle—the stress divergence $\sigma_{ij,j}$ equals zero which is a consequence of the momentum conservation law as will be shown in Sect. 6.2.

The work of internal forces $\mathrm{d}W$ defined in Eq. (5.46) is accumulated in the material as its internal energy. Its form (the way it is accumulated) depends on the material properties. If the material is elastic, the work $\mathrm{d}W$ is equal to the internal elastic energy accumulated in the material element's volume $\mathrm{d}V$. Otherwise, if there are any inelastic (plastic, viscous, etc.) phenomena accompanying deformation processes in the material, then at least a part of the work is converted to heat. This kind of energy, still being a form of internal energy of the body, is no longer a potential energy, though, and is usually irreversibly dissipated. This issue will be discussed in more detail in subsequent chapters of the book.

The above derivations and definitions were based on the assumption that the body does not significantly change its geometric configuration. Let us now generalize our discussion to the case of large deformations.

Consider two configurations of the body: the current C^t and the reference C^r. Looking at the formulae derived above, it is easy to notice that they are still valid upon our new assumptions provided that the stress $\boldsymbol{\sigma}$ is understood as the Cauchy stress (5.16) and the strain $\boldsymbol{\varepsilon}$—as the tensor ${}^0_t\boldsymbol{\varepsilon}^t$, cf. Eq. (4.65) defining its rate, i.e.

if the reference configuration is identified with the current one, $C^r = C^t$. We have then:

$$\dot{W}^t = \boldsymbol{\sigma}^t \cdot {}_t\dot{\boldsymbol{\varepsilon}}^t \, \mathrm{d}V^t = \sigma^t_{ij} \, {}_t\dot{\varepsilon}^t_{ij} \, \mathrm{d}V^t. \tag{5.47}$$

Passing now to another, arbitrary reference configuration and making use of the transformation formulae (4.45), (5.25b) and (4.69), we may transform Eq. (5.47) to the form (in which—for clarity—all upper indices at ${}^r\boldsymbol{F}^t$ and ${}^rJ^t$ are skipped):

$$\begin{aligned}\dot{W}^t &= \left(\frac{1}{J} F_{ik} \, {}_r\tilde{\sigma}^t_{kl} \, F_{jl}\right) \left(F^{-1}_{mi} \, {}_r\dot{\varepsilon}^t_{mn} \, F^{-1}_{nj}\right) \mathrm{d}V^t = \frac{1}{J} \delta_{mk} \delta_{ln} \, {}_r\tilde{\sigma}^t_{kl} \, {}_r\dot{\varepsilon}^t_{mn} \, \mathrm{d}V^t \\ &= {}_r\tilde{\sigma}^t_{kl} \, {}_r\dot{\varepsilon}^t_{kl} \, \mathrm{d}V^r = {}_r\tilde{\boldsymbol{\sigma}}^t \cdot {}_r\dot{\boldsymbol{\varepsilon}}^t \, \mathrm{d}V^r. \end{aligned} \tag{5.48}$$

In this equation, the second Piola–Kirchhoff stress tensor (5.23) and the rate of the strain defined by Eq. (4.15) appear instead of the Cauchy stress and the corresponding strain defined in C^t. If $C^r = C^t$, Eq. (5.48) becomes identical as Eq. (5.47) because the tensors $\boldsymbol{\sigma}^t$ and ${}_t\tilde{\boldsymbol{\sigma}}^t$ are equal in this case.

The quantity $\dot{W}^t$ can also be expressed in terms of the first Piola–Kirchhoff stress tensor. Making subsequently use of Eq. (5.24a), the relation

$${}_r\dot{\boldsymbol{\varepsilon}}^t = \frac{1}{2}\left(({}^r\dot{\boldsymbol{F}}^t)^{\mathrm{T}} \, {}^r\boldsymbol{F}^t + ({}^r\boldsymbol{F}^t)^{\mathrm{T}} \, {}^r\dot{\boldsymbol{F}}^t\right) = \operatorname{sym}\left(({}^r\boldsymbol{F}^t)^{\mathrm{T}} \, {}^r\dot{\boldsymbol{F}}^t\right)$$

and the property (3.65), implying for the symmetric tensor ${}_r\tilde{\boldsymbol{\sigma}}^t$ the following relation,

$${}_r\tilde{\boldsymbol{\sigma}}^t \cdot {}_r\dot{\boldsymbol{\varepsilon}}^t = {}_r\tilde{\boldsymbol{\sigma}}^t \cdot \left(({}^r\boldsymbol{F}^t)^{\mathrm{T}} \, {}^r\dot{\boldsymbol{F}}^t\right),$$

one may transform Eq. (5.48) as follows

$$\begin{aligned}\dot{W}^t &= {}_r\tilde{\sigma}^t_{kl} \, {}_r\dot{\varepsilon}^t_{kl} \, \mathrm{d}V^r = \left(F^{-1}_{ki} \, {}_r\sigma^t_{il}\right)\left(F_{jk} \dot{F}_{jl}\right) \mathrm{d}V^r = \frac{1}{\rho^r} \delta_{ij} \, {}_r\sigma^t_{il} \, \dot{F}_{jl} \, \mathrm{d}V^r \\ &= {}_r\sigma^t_{il} \, \dot{F}_{il} \, \mathrm{d}V^r = {}_r\boldsymbol{\sigma}^t \cdot {}^r\dot{\boldsymbol{F}}^t \, \mathrm{d}V^r. \end{aligned} \tag{5.49}$$

The above considerations allow to conclude that—from the point of view of internal force energy—there are certain pairs of internal force and deformation measures such that the full product of the stress tensor and the corresponding tensor rate of strain or deformation gradient yields as a result the internal force power related to the volume unit. Such tensor pairs are called *work-conjugate* or *energy-conjugate*. In view of the above derivations, examples of work-conjugate pairs can be listed as follows:

- the Cauchy stress tensor $\boldsymbol{\sigma}^t$ and the strain tensor in the current configuration ${}^0_t\varepsilon^t$, cf. Eq. (5.47),
- the first Piola–Kirchhoff stress tensor ${}_r\boldsymbol{\sigma}^t$ and the deformation gradient ${}^r\boldsymbol{F}^t$, cf. Eq. (5.49),
- the second Piola–Kirchhoff stress tensor ${}_r\tilde{\boldsymbol{\sigma}}^t$ and the strain tensor ${}^0_r\varepsilon^t$, both expressed in the same reference configuration, cf. Eq. (5.48). In particular, if this is the initial configuration, then the conjugate pair consists of the stress ${}_0\tilde{\boldsymbol{\sigma}}^t$ and the Green strain ε^G.

In summary, the notion of work-conjugate variables introduces a kind of ordering within the variety of stress and deformation measures defined up to now.

Denote by ${}_rU^t$ the internal force energy density W^t related to the volume unit measured in the configuration C^r. We have then

$$\dot{W}^t = {}_r\dot{U}^t \, \mathrm{d}V^r, \tag{5.50}$$

and, in the current configuration, $\dot{W}^t = {}_t\dot{U}^t \, \mathrm{d}V^t$. The rate of this quantity can thus be expressed as the full product of two work-conjugate stress and strain rate tensors as

$${}_t\dot{U}^t = \boldsymbol{\sigma}^t \cdot {}_t\dot{\varepsilon}^t , \qquad {}_r\dot{U}^t = {}_r\tilde{\boldsymbol{\sigma}}^t \cdot {}_r\dot{\varepsilon}^t = {}_r\boldsymbol{\sigma}^t \cdot {}^r\dot{\boldsymbol{F}}^t . \tag{5.51}$$

To make the notion of energy density independent of the choice of the reference configuration, let us relate the energy to the mass rather than the volume unit. If ρ denotes the scalar field of local density of material mass m, i.e. $\mathrm{d}m = \rho \mathrm{d}V$, one can define the quantity

$$\dot{e}^t = \frac{1}{\rho^t} \, {}_t\dot{U}^t \tag{5.52}$$

as the internal force energy density with respect to the mass unit. Anticipating discussion in the next chapter (in particular, the mass conservation law discussed there), let us postulate the equality $\rho^t \mathrm{d}V^t = \rho^r \mathrm{d}V^r$ which implies $\dot{e}^t = \frac{1}{\rho^r} \, {}_r\dot{U}^t$. Therefore, in view of Eqs. (5.50)–(5.51),

$$\dot{e}^t = \frac{1}{\rho^t} \boldsymbol{\sigma}^t \cdot {}_t\dot{\varepsilon}^t = \frac{1}{\rho^r} \, {}_r\tilde{\boldsymbol{\sigma}}^t \cdot {}_r\dot{\varepsilon}^t = \frac{1}{\rho^r} \, {}_r\boldsymbol{\sigma}^t \cdot {}^r\dot{\boldsymbol{F}}^t . \tag{5.53}$$

The rate equations for strain energy density (5.51)–(5.53) may be expressed in the incremental form, too. Assuming stress to remain constant within the interval $[t, t+\Delta t]$, the following result can be obtained:

$$\Delta e^t = \frac{1}{\rho^t} \boldsymbol{\sigma}^t \cdot \Delta_t\varepsilon^t = \frac{1}{\rho^r} \, {}_r\tilde{\boldsymbol{\sigma}}^t \cdot \Delta_r\varepsilon^t = \frac{1}{\rho^r} \, {}_r\boldsymbol{\sigma}^t \cdot \Delta^r\boldsymbol{F}^t , \tag{5.54}$$

In the above formula, all strain/deformation increments are understood in the Lagrangian sense, i.e. e.g. $\Delta_t\varepsilon^t = {}_t\varepsilon^{t+\Delta t} - {}_t\varepsilon^t$. If stress varies during the increment, the more precise formula for the energy increment reads:

$$\Delta e^t = \frac{1}{\rho^{t+\alpha\Delta t}}\,{}_t\tilde{\boldsymbol{\sigma}}^{t+\alpha\Delta t}\cdot\Delta_t\varepsilon^t = \frac{1}{\rho^r}\,{}_r\tilde{\boldsymbol{\sigma}}^{t+\alpha\Delta t}\cdot\Delta_r\varepsilon^t = \frac{1}{\rho^r}\,{}_r\boldsymbol{\sigma}^{t+\alpha\Delta t}\cdot\Delta^r\boldsymbol{F}^t, \quad (5.55)$$

where $\alpha \in (0, 1)$.

References

1. Hill R., 1978. *Aspects of invariance in solid mechanics*. Adv. Appl. Mech., 18:1–75.
2. Truesdell C., Noll W., 1960. *The Non-linear Field Theories of Mechanics*, volume III, Part 3 of *Encyclopaedia of Physics*. Springer-Verlag.

Chapter 6
Conservation Laws in Continuum Mechanics

The process of motion and deformation of a continuum must remain in accordance with certain universal (independent of the continuum's properties) conditions called conservation laws (or conservation rules). In this chapter, the following four such laws will be introduced:

- mass conservation law,
- momentum conservation law,
- angular momentum conservation law,
- mechanical energy conservation law.

The laws governing momentum and angular momentum conservation are a generalization of the mechanical equilibrium rules, discussed in the context of forces and moments acting on material elements. The mechanical energy conservation law results in fact from the other three laws. However, it deserves separate treatment here since—due to its simple scalar form—it has a great practical significance in addressing the problems on hand.

6.1 Mass Conservation Law

Mass is one of fundamental physical properties of matter constituting deformable bodies. In material continuum, mass is by definition continuously distributed and the notion of mass density ρ (further simply referred to as "density") may be defined at each material point as

$$\mathrm{d}m = \rho\,\mathrm{d}V, \tag{6.1}$$

where $\mathrm{d}m$ is the mass of an infinitesimal volume $\mathrm{d}V$. Hence, the total mass of the body $\mathcal{B}$ occupying in a selected physical configuration the domain $\Omega \subset E^3$ equals

M. Kleiber and P. Kowalczyk, *Introduction to Nonlinear Thermomechanics of Solids*, Lecture Notes on Numerical Methods in Engineering and Sciences,
DOI 10.1007/978-3-319-33455-4_6

$$m = \int_{\Omega} \rho(\boldsymbol{x})\, \mathrm{d}V. \tag{6.2}$$

The mass conservation law says that mass does not change during the motion and deformation, i.e. the entire body as well as its parts preserve the initial mass. This means that at each time instant $\tau = t \in [0, \bar{t}]$

$$\int_{\Omega^t} \rho^t(\boldsymbol{x}^t)\, \mathrm{d}V^t = \int_{\Omega^r} \rho^r(\boldsymbol{x}^r)\, \mathrm{d}V^r, \tag{6.3}$$

where $\rho^t(\boldsymbol{x}^t) = \rho(\boldsymbol{x}(\boldsymbol{x}^r, t), t)$ denotes instantaneous mass density at the material point at $\tau = t$ while $\rho^r(\boldsymbol{x}^r)$ is density at the point in the reference configuration. The relation (6.3) must also hold true for any part of the body which implies the local condition in the form

$$\rho^t \mathrm{d}V^t = \rho^r \mathrm{d}V^r. \tag{6.4}$$

In other words, the mass conservation law says that changes in local density at material points cannot be arbitrary—they are strictly related to local deformation changes and particularly to local volume changes. Employing Eq. (4.45) to determine the instantaneous volume of the element $\mathrm{d}V$

$$\mathrm{d}V^t = {}^rJ^t\, \mathrm{d}V^r, \tag{6.5}$$

where ${}^rJ^t = \det\, {}^r\boldsymbol{F}^t$, one can rewrite Eq. (6.3) as

$$\int_{\Omega^r} \left[{}^rJ^t(\boldsymbol{x}^r)\, \rho^t(\boldsymbol{x}^t(\boldsymbol{x}^r)) - \rho^r(\boldsymbol{x}^r) \right] \mathrm{d}V^r = 0. \tag{6.6}$$

Since this equality is valid for any part of the body $\mathcal{B}$, the relation

$$\rho^r = {}^rJ^t\, \rho^t \tag{6.7}$$

must hold true between the instantaneous density ρ^t at a material point in the deformable body and the relative density ρ^r at this point.

Generally, the same relation holds true for two arbitrary time instants $\tau = t_1$ and $\tau = t_2$,

$$\rho^{t_1} = {}^{t_1}J^{t_2}\, \rho^{t_2}. \tag{6.8}$$

If the function $\rho(\boldsymbol{x}(\boldsymbol{x}^r, \tau), \tau)$ is smooth with respect to its arguments, Eq. (6.7) can be differentiated with respect to time. Employing the formula (4.73) for the rate of the determinant ${}^rJ^t$ and skipping upper indices for clarity, one obtains

$$0 = J\dot{\rho} + \dot{J}\rho = J\,(\dot{\rho} + v_{i,i}), \tag{6.9}$$

where v_i denotes velocity while the notation $(\cdot)_{,i}$ stands for the gradient components with respect to the current coordinates x_i^t. Dividing the above formula side-wise by (always positive) J one comes at the well known differential form of the mass conservation law, also called the *continuity equation*:

$$\dot{\rho} + \rho v_{i,i} = 0. \tag{6.10}$$

An interesting consequence of the mass conservation law (6.4) is the following relation, that holds true for an arbitrary tensor field $\boldsymbol{a}(\boldsymbol{x}^r, t)$ of any order:

$$\int_{\Omega^t} \rho^t \boldsymbol{a}^t \, \mathrm{d}V^t = \int_{\Omega^r} \rho^r \boldsymbol{a}^t \, \mathrm{d}V^r. \tag{6.11}$$

If $\boldsymbol{a}$ denotes density of a certain physical quantity in a mass unit of material continuum, then the above formula shows how to compute the total content of this quantity accumulated in the deformable body $\mathcal{B}$ by integration of the density $\boldsymbol{a}$ either in the current configuration C^t or the reference configuration C^r. Equation (6.11) may be very useful to compute the rate of this content: instead differentiating the left-hand side, which is an integral of a variable-in-time expression over a variable-in-time domain, one may pass to the reference configuration and take the right-hand side instead, which is just an integral over a fixed domain Ω^r, i.e.

$$\frac{\mathrm{d}}{\mathrm{d}\tau} \int_{\Omega^\tau} \rho^\tau \boldsymbol{a}^\tau \, \mathrm{d}V^\tau \bigg|_{\tau=t} = \frac{\mathrm{d}}{\mathrm{d}\tau} \int_{\Omega^r} \rho^r \boldsymbol{a}^\tau \, \mathrm{d}V^r \bigg|_{\tau=t} = \int_{\Omega^r} \rho^r \dot{\boldsymbol{a}}^t \, \mathrm{d}V^r, \tag{6.12}$$

Applying Eq. (6.11) to $\dot{\boldsymbol{a}}^t$, one can finally write

$$\frac{\mathrm{d}}{\mathrm{d}\tau} \int_{\Omega^\tau} \rho^\tau \boldsymbol{a}^\tau \, \mathrm{d}V^\tau \bigg|_{\tau=t} = \int_{\Omega^t} \rho^t \dot{\boldsymbol{a}}^t \, \mathrm{d}V^t. \tag{6.13}$$

Considering a finite time interval between $\tau = t$ and $\tau = t + \Delta t$, on can conclude from Eq. (6.7) that

$$0 = {}^rJ^{t+\Delta t} \, \rho^{t+\Delta t} - {}^rJ^t \, \rho^t, \tag{6.14}$$

i.e. denoting $\Delta\, {}^rJ^t = {}^rJ^{t+\Delta t} - {}^rJ^t$ and $\Delta\rho^t = \rho^{t+\Delta t} - \rho^t$,

$$\Delta\, {}^rJ^t \, \rho^t + {}^rJ^t \, \Delta\rho^t + \Delta\, {}^rJ^t \, \Delta\rho^t = 0. \tag{6.15}$$

This equation expresses the mass conservation law in an incremental form. For sufficiently small increments Δt the increment $\Delta\, {}^rJ^t$ is an incremental extrapolation of the derivative (4.72):

$$\Delta\, {}^rJ^t \approx {}^rJ^t \, \Delta\, {}^rF^t_{ik} \, {}^tF^r_{ki} = {}^rJ^t \, \frac{\partial \Delta u^t_i}{\partial x^t_i}.$$

Therefore, dividing Eq. (6.15) side-wise by ${}^{r}J^{t}$ and referring again the notation $(\cdot)_{,i}$ to differentiation with respect to $\boldsymbol{x}^{t}$, we can write the approximate relation

$$(\rho^{t} + \Delta\rho^{t})\,\Delta u^{t}_{i,i} + \Delta\rho^{t} = 0. \tag{6.16}$$

6.2 Momentum Conservation Law

The momentum conservation law discussed below is in fact a generalization of the Newton's second law of motion to deformable continuum. According to the Newton's law, any change in momentum of a body (or its part) equals the sum of all forces acting on the body (or its part). If the sum is zero then the law is reduced to the force equilibrium rule mentioned in the previous chapters.

Before we pass to the general form of the momentum conservation law, let us go back to our discussion about stress equilibrium of Sect. 5.1 and the example shown in Fig. 5.3. The uniform stress state defined by the symmetric tensor $\boldsymbol{\sigma}$, cf. Eq. (5.3), has been analysed there and our conclusion was that the so-defined stress ensures equilibrium of internal forces. Let us now try to examine the equilibrium conditions in the case of a non-uniform stress in the form of an arbitrary continuous and differentiable field $\boldsymbol{\sigma}(\boldsymbol{x})$.

Figure 6.1 presents values of stress components on the walls of a cuboidal element $\mathrm{d}V$ in this general case. It is assumed that the tensor components $[\sigma_{ij}]$ describe the stress state in the element's geometrical centre. On the walls, the stress component values are modified by the first terms in the corresponding Taylor series, containing appropriate stress gradient components $\sigma_{ij,k}$. For clarity, stress components acting on the front (visible) and the back (invisible) cuboid walls have been depicted separately.

The resultant internal force components acting on particular walls are products of the stress components times the areas of the corresponding walls. Let us examine the internal force equilibrium in the direction x_1. The sum of the forces is expressed as

$$\begin{aligned}
&\left(\sigma_{11} + \tfrac{1}{2}\sigma_{11,1}\mathrm{d}x_1\right)\mathrm{d}x_2\mathrm{d}x_3 - \left(\sigma_{11} - \tfrac{1}{2}\sigma_{11,1}\mathrm{d}x_1\right)\mathrm{d}x_2\mathrm{d}x_3\\
&\quad + \left(\sigma_{12} + \tfrac{1}{2}\sigma_{12,2}\mathrm{d}x_2\right)\mathrm{d}x_3\mathrm{d}x_1 - \left(\sigma_{12} - \tfrac{1}{2}\sigma_{12,2}\mathrm{d}x_2\right)\mathrm{d}x_3\mathrm{d}x_1\\
&\quad + \left(\sigma_{13} + \tfrac{1}{2}\sigma_{13,3}\mathrm{d}x_3\right)\mathrm{d}x_1\mathrm{d}x_2 - \left(\sigma_{13} - \tfrac{1}{2}\sigma_{13,3}\mathrm{d}x_3\right)\mathrm{d}x_1\mathrm{d}x_2\\
&\quad = \sigma_{11,1}\,\mathrm{d}x_1\mathrm{d}x_2\mathrm{d}x_3 + \sigma_{12,2}\,\mathrm{d}x_2\mathrm{d}x_3\mathrm{d}x_1 + \sigma_{13,3}\,\mathrm{d}x_3\mathrm{d}x_1\mathrm{d}x_2\\
&\quad = (\sigma_{11,1} + \sigma_{12,2} + \sigma_{13,3})\,\mathrm{d}V.
\end{aligned}$$

According to the equilibrium rule, this quantity should vanish, similarly as analogous quantities computed for the two remaining directions x_2 and x_3, i.e. $(\sigma_{21,1} + \sigma_{22,2} + \sigma_{23,3})\,\mathrm{d}V$ and $(\sigma_{31,1} + \sigma_{32,2} + \sigma_{33,3})\,\mathrm{d}V$, respectively. Since $\mathrm{d}V > 0$, the resulting vector equation on the internal force equilibrium can be written as

$$\sigma_{ij,j} = 0 \qquad \text{or} \qquad \operatorname{div}\boldsymbol{\sigma} = 0. \tag{6.17}$$

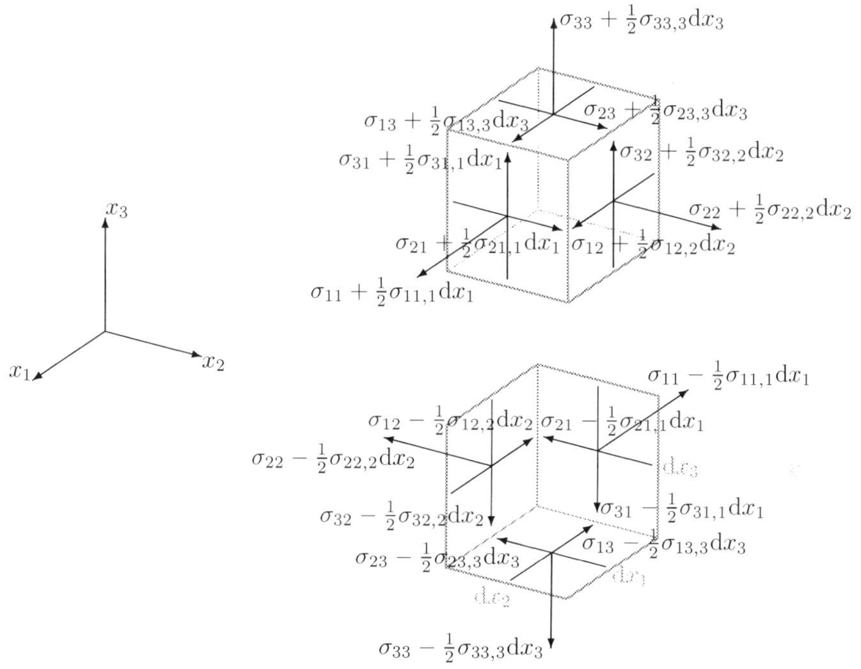

Fig. 6.1 Stress tensor components in the non-uniform case

This is a system of partial differential equation for the stress tensor components. The equations are local, i.e. must hold true at each point of the body.

If there are other forces acting on the considered element, like mass forces (gravity, magnetism, etc.), Eq. (6.17) must be extended to include these forces as they affect the overall equilibrium. If the mass forces (related to the mass unit of continuum) are described by the field $\boldsymbol{f}(\boldsymbol{x})$ and the mass of our element is $\mathrm{d}m=\rho\mathrm{d}V$, then the additional force to complete the equilibrium equations is[1] $\rho f_i \mathrm{d}V$ and the equations assume the form

$$\sigma_{ij,j}+\rho f_i=0 \qquad \text{or} \qquad \operatorname{div}\boldsymbol{\sigma}+\rho\boldsymbol{f}=0. \tag{6.18}$$

Equations (6.17) and (6.18) express the momentum conservation law. In the derivation, we have so far neglected the fact that the volume element $\mathrm{d}V$ remains in general in motion and is subject to deformation. We shall now derive equations appropriate for such a case.

Consider again a deformable body $\mathcal{B}$ occupying the domain Ω^r in the reference configuration and assuming the domain Ω^t in the current configuration at the time instant $\tau=t$ during its motion (Fig. 6.2). There is a system of external load forces

[1] The mass forces f_i have been averaged within the element's volume and applied at its geometrical centre.

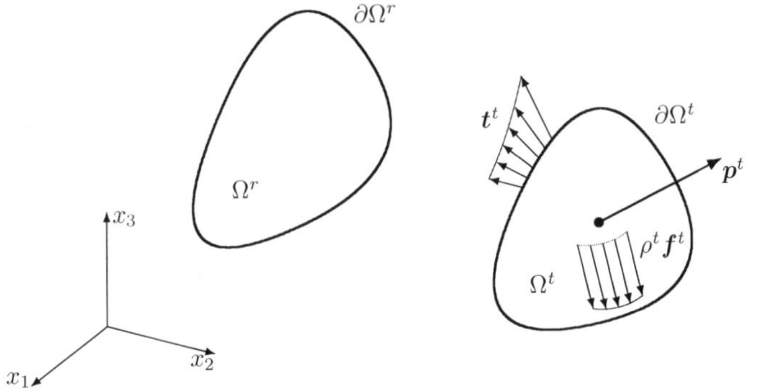

Fig. 6.2 Forces acting on a body; momentum conservation law

acting on the body consisting of mass forces $\boldsymbol{f}^t$ distributed within the body's volume and surface forces distributed on its boundary.[2] The resultant force is

$$\boldsymbol{P}^t = \int_{\Omega^t} \rho^t \boldsymbol{f}^t \, \mathrm{d}V^t + \int_{\partial\Omega^t} \boldsymbol{t}^t \, \mathrm{d}A^t. \tag{6.19}$$

The momentum $\boldsymbol{p}$ of the body is defined in the current configuration as

$$\boldsymbol{p}^t = \int_{\Omega^t} \rho^t \boldsymbol{v}^t \, \mathrm{d}V^t = \int_{\Omega^t} \rho^t \dot{\boldsymbol{x}}^t \, \mathrm{d}V^t. \tag{6.20}$$

The momentum conservation law says that the change of momentum in time or its rate $\dot{\boldsymbol{p}}^t$ equals the sum of all external loads $\boldsymbol{P}^t$ acting on the body, i.e. in our case

$$\frac{\mathrm{d}}{\mathrm{d}\tau} \int_{\Omega^\tau} \rho^\tau \boldsymbol{v}^\tau \, \mathrm{d}V^\tau \Big|_{\tau=t} = \int_{\Omega^t} \rho^t \boldsymbol{f}^t \, \mathrm{d}V^t + \int_{\partial\Omega^t} \boldsymbol{t}^t \, \mathrm{d}A^t. \tag{6.21}$$

For convenience of our further derivations it is desirable to express the integrals in Eq. (6.21) in the reference configuration C^r. Making use of Eqs. (6.4) and (6.12) resulting from the mass conservation law and of Eq. (5.19) resulting from the nominal stress definition, one can transform Eq. (6.21) as

$$\int_{\Omega^r} \rho^r \dot{\boldsymbol{v}}^t \, \mathrm{d}V^r = \int_{\Omega^r} \rho^r \boldsymbol{f}^t \, \mathrm{d}V^r + \int_{\partial\Omega^r} {}_r\boldsymbol{t}^t \, \mathrm{d}A^r. \tag{6.22}$$

---

[2] Concentrated forces are not distinguished as a separate class of load—they are treated as a limit case of distributed forces acting on a very small area.

The last integral on the right-hand side may further be transformed using the first Piola–Kirchhoff stress tensor (5.18) and the Gauss–Ostrogradski theorem as

$$\int_{\partial\Omega^r} {}_r\boldsymbol{t}^t \,\mathrm{d}A^r = \int_{\partial\Omega^r} {}_r\boldsymbol{\sigma}^t \boldsymbol{n}^t \,\mathrm{d}A^r = \int_{\Omega^r} \operatorname{div} {}_r\boldsymbol{\sigma}^t \,\mathrm{d}V^r,$$

$$\int_{\partial\Omega^r} {}_r t_i^t \,\mathrm{d}A^r = \int_{\partial\Omega^r} {}_r\sigma_{ij}^t n_j^t \,\mathrm{d}A^r = \int_{\Omega^r} {}_r\sigma_{ij,j}^t \,\mathrm{d}V^r.$$

This allows to express the vector equation (6.22) in a form containing only integrals over Ω^r. Grouping them together, we arrive at

$$\int_{\Omega^r} \left[\operatorname{div} {}_r\boldsymbol{\sigma}^t + \rho^r \boldsymbol{f}^t - \rho^r \dot{\boldsymbol{v}}^t\right] \mathrm{d}V^r = 0, \tag{6.23a}$$

$$\int_{\Omega^r} \left[{}_r\sigma_{ij,j}^t + \rho^r f_i^t - \rho^r \dot{v}_i^t\right] \mathrm{d}V^r = 0. \tag{6.23b}$$

Since the above equality must in fact hold true for any part of the body, (the forces $\boldsymbol{t}^t$ acting at a cross-section surface of $\mathcal{B}$ bounding such a part are then understood as internal forces acting on this surface), it is concluded that the integrand in the above equation vanishes at each material point of the body. Thus, the momentum conservation law reads:

$$\operatorname{div} {}_r\boldsymbol{\sigma}^t + \rho^r \boldsymbol{f}^t - \rho^r \dot{\boldsymbol{v}}^t = 0, \qquad {}_r\sigma_{ij,j}^t + \rho^r f_i^t - \rho^r \dot{v}_i^t = 0. \tag{6.24}$$

The above vector equation (also called the *equation of motion*) is a generalization of Eq. (6.18) derived earlier for the case of a linear deformable body in motion. It is a system of partial differential equations with respect to the first Piola–Kirchhoff stress tensor components in the reference configuration. The last term on the left-hand side, $\rho^r \dot{\boldsymbol{v}}^t$, denotes inertial forces acting on the material particle related to the unit volume. If the motion is quasi-stationary (i.e. accelerations are negligible), the equation of motion assumes the form

$$\operatorname{div} {}_r\boldsymbol{\sigma}^t + \rho^r \boldsymbol{f}^t = 0 \tag{6.25}$$

while in the case of no mass forces it simplifies to

$$\operatorname{div} {}_r\boldsymbol{\sigma}^t = 0. \tag{6.26}$$

Similarly as in the case of the mass conservation law, one may present the just derived law in an incremental form. According to (6.24), at $\tau = t + \Delta t$ we have

$${}_r\sigma_{ij,j}^{t+\Delta t} + \rho^r f_i^{t+\Delta t} - \rho^r \dot{v}_i^{t+\Delta t} = 0. \tag{6.27}$$

Subtracting side-wise Eq. (6.24) from the above equation, we arrive at the following relation:

$$\Delta\sigma^t_{ij,j} + \rho^r \Delta f^t_i - \rho^r \Delta \dot{v}^t_i = 0. \tag{6.28}$$

6.3 Angular Momentum Conservation Law

The angular momentum conservation law is again a generalization of the Newton's second law of motion. It says that any change in angular momentum (also called moment of momentum) of a body (or its part) equals the sum of all moments of forces acting on the body (or its part). All the moments mentioned above must be defined with respect to the same arbitrary point in space, say, the origin of the coordinate system. According to the definition (3.29), the moment of an arbitrary vector $\boldsymbol{a}$ with respect to the coordinate system origin may be expressed as the vector product $\boldsymbol{x} \times \boldsymbol{a}$. Hence, the conservation law discussed here requires that the following relation holds true:

$$\frac{\mathrm{d}}{\mathrm{d}\tau} \int_{\Omega^\tau} \rho^\tau \boldsymbol{x}^\tau \times \boldsymbol{v}^\tau \,\mathrm{d}V^\tau \bigg|_{\tau=t} = \int_{\Omega^t} \rho^t \boldsymbol{x}^t \times \boldsymbol{f}^t \,\mathrm{d}V^t + \int_{\partial\Omega^t} \boldsymbol{x}^t \times \boldsymbol{t}^t \,\mathrm{d}A^t. \tag{6.29}$$

Analogy to Eq. (6.21) is visible and our derivation may thus follow the sequence of transformations employed in the previous section. Let us first transform the integrals to express them in the reference configuration C^r, making use of the relations (6.4), (6.12) and (5.19):

$$\int_{\Omega^r} \rho^r \left(\dot{\boldsymbol{x}}^t \times \boldsymbol{v}^t + \boldsymbol{x}^t \times \dot{\boldsymbol{v}}^t\right) \mathrm{d}V^r = \int_{\Omega^r} \rho^r \boldsymbol{x}^t \times \boldsymbol{f}^t \,\mathrm{d}V^r + \int_{\partial\Omega^r} \boldsymbol{x}^t \times {}_r\boldsymbol{t}^t \,\mathrm{d}A^r.$$

Since $\dot{\boldsymbol{x}}^t = \boldsymbol{v}^t$ (and thus $\dot{\boldsymbol{x}}^t \times \boldsymbol{v}^t \equiv 0$) and ${}_r\boldsymbol{t}^t = {}_r\boldsymbol{\sigma}^t \boldsymbol{n}^r$, this equation may be rewritten in the following form:

$$\int_{\Omega^r} \rho^r \boldsymbol{x}^t \times \dot{\boldsymbol{v}}^t \,\mathrm{d}\Omega^r = \int_{\Omega^r} \rho^r \boldsymbol{x}^t \times \boldsymbol{f}^t \,\mathrm{d}V^r + \int_{\partial\Omega^r} \boldsymbol{x}^t \times {}_r\boldsymbol{\sigma}^t \boldsymbol{n}^r \,\mathrm{d}A^r, \tag{6.30a}$$

$$\int_{\Omega^r} \rho^r x^t_i \dot{v}^t_j \epsilon_{ijk} \,\mathrm{d}V^r = \int_{\Omega^r} \rho^r x^t_i f^t_j \epsilon_{ijk} \,\mathrm{d}V^r + \int_{\partial\Omega^r} x^t_i \,{}_r\sigma^t_{jl} n^r_l \epsilon_{ijk} \,\mathrm{d}A^r. \tag{6.30b}$$

The last integral may be further transformed according to the Gauss–Ostrogradski theorem and the relation (5.20b) between the Cauchy and the first Piola–Kirchhoff stress tensors:

$$\begin{aligned}\int_{\partial\Omega^r} x^t_i \,{}_r\sigma^t_{jl} n^r_l \epsilon_{ijk} \,\mathrm{d}A^r &= \int_{\Omega^r} \left(x^t_{i,l} \,{}_r\sigma^t_{jl} + x^t_i \,{}_r\sigma^t_{jl,l}\right) \epsilon_{ijk} \,\mathrm{d}V^r \\ &= \int_{\Omega^r} \left({}^rJ^t \sigma^t_{ji} + x^t_i \,{}_r\sigma^t_{jl,l}\right) \epsilon_{ijk} \,\mathrm{d}V^r.\end{aligned}$$

The vector equation (6.30) has thus been expressed in the form containing only integrals over Ω^r. Grouping the integrands and taking into account the arbitrary choice of the integration domain (the conservation rule must hold true for every part of the body), we arrive at the following system of equations ($i = 1, 2, 3$):

$$\left[{}^rJ^t\,\sigma^t_{ji} + x^t_i\left({}_r\sigma^t_{jl,l} + \rho^r f^t_j - \rho^r \dot{v}^t_j\right)\right]\epsilon_{ijk} = 0, \tag{6.31}$$

which in view of the momentum conservation law (6.24) is equivalent to

$$\sigma^t_{ji}\epsilon_{ijk} = 0, \tag{6.32}$$

i.e. simply

$$\sigma^t_{ij} = \sigma^t_{ji} \quad \text{or} \quad \boldsymbol{\sigma}^t = (\boldsymbol{\sigma}^t)^{\mathrm{T}}. \tag{6.33}$$

In summary, the only conclusion resulting from the angular momentum conservation law is symmetry of the Cauchy stress. This condition has been already proven in Sect. 5.1 under the assumption of a uniform stress field; here we have generalized it to the case of arbitrary smooth stress field in a deformable body.

6.4 Mechanical Energy Conservation Law

The energy conservation law plays in physics a fundamental role. There exist several forms of energy—mechanical, thermal, chemical, electrical, nuclear, etc. In this study, only the first two of these are of interest. Let us consider the mechanical energy and consequently assume that the body $\mathcal{B}$ is subject to only mechanical interactions. Other effects, like those related to heat, will be subject of our discussion in subsequent chapters.

Let us start by forming a scalar product of the vector equation of motion (6.24) and the velocity vector $\boldsymbol{v}^t$,

$${}_r\sigma^t_{ij,j}v^t_i + \rho^r f^t_i v^t_i - \rho^r \dot{v}^t_i v^t_i = 0 \tag{6.34}$$

which can be integrated over the domain Ω^r to yield

$$\int_{\Omega^r} {}_r\sigma^t_{ij,j}v^t_i\,\mathrm{d}V^r + \int_{\Omega^r} \rho^r f^t_i v^t_i\,\mathrm{d}V^r - \int_{\Omega^r} \rho^r \dot{v}^t_i v^t_i\,\mathrm{d}V^r = 0. \tag{6.35}$$

Let us now transform the integrals in the above equation. For the first of them, making use of the equation

$$\left({}_r\sigma^t_{ij}v^t_i\right)_{,j} = {}_r\sigma^t_{ij,j}v^t_i + {}_r\sigma^t_{ij}v^t_{i,j}$$

and, subsequently, of the Gauss–Ostrogradski theorem and the relations (5.49), (5.19), (6.4) and (6.11) we obtain

$$\begin{aligned}
\int_{\Omega^r} {}_r\sigma^t_{ij,j} v^t_i \,\mathrm{d}V^r &= \int_{\Omega^r} \left({}_r\sigma^t_{ij} v^t_i\right)_{,j} \mathrm{d}V^r - \int_{\Omega^r} {}_r\sigma^t_{ij} v^t_{i,j} \,\mathrm{d}V^r \\
&= \int_{\partial\Omega^r} {}_r\sigma^t_{ij}\, v^t_i\, n^r_j \,\mathrm{d}A^r - \int_{\Omega^r} {}_r\sigma^t_{ij}\, {}^r\dot{F}^t_{ij} \,\mathrm{d}V^r \\
&= \int_{\partial\Omega^r} {}_r t^t_i\, v^t_i \,\mathrm{d}A^r - \int_{\Omega^r} \rho^r \dot{e}^t \,\mathrm{d}V^r \\
&= \int_{\partial\Omega^t} t^t_i\, v^t_i \,\mathrm{d}A^t - \int_{\Omega^t} \rho^t \dot{e}^t \,\mathrm{d}V^t .
\end{aligned} \tag{6.36}$$

In view of the mass conservation law (6.4), the second integral may be rewritten as

$$\int_{\Omega^r} \rho^r f^t_i v^t_i \,\mathrm{d}V^r = \int_{\Omega^t} \rho^t f^t_i v^t_i \,\mathrm{d}V^t . \tag{6.37}$$

The third integral, in view of Eq. (6.13), takes the form

$$\int_{\Omega^r} \rho^r\, \dot{v}^t_i v^t_i \,\mathrm{d}V^r = \int_{\Omega^r} \rho^r \left.\frac{\mathrm{d}}{\mathrm{d}\tau}\left(\frac{1}{2}\, v^\tau_i v^\tau_i\right)\right|_{\tau=t} \mathrm{d}V^r = \left.\frac{\mathrm{d}}{\mathrm{d}\tau}\int_{\Omega^\tau} \frac{1}{2}\,\rho^\tau v^\tau_i v^\tau_i \,\mathrm{d}V^\tau\right|_{\tau=t} . \tag{6.38}$$

Denoting by $\dot{L}$ the power of external forces acting on the body $\mathcal{B}$,

$$\dot{L} = \int_{\partial\Omega} t_i v_i \,\mathrm{d}A + \int_{\Omega} \rho f_i v_i \,\mathrm{d}V = \int_{\partial\Omega} \boldsymbol{t}\cdot\boldsymbol{v} \,\mathrm{d}A + \int_{\Omega} \rho \boldsymbol{f}\cdot\boldsymbol{v} \,\mathrm{d}V, \tag{6.39}$$

by $\dot{W}$ the power of internal forces,

$$\dot{W} = \int_{\Omega} \rho\dot{e} \,\mathrm{d}V, \tag{6.40}$$

and by K the kinetic energy of the body,

$$K = \int_{\Omega} \frac{1}{2}\rho v_i v_i \,\mathrm{d}V = \int_{\Omega} \frac{1}{2}\rho \boldsymbol{v}^2 \,\mathrm{d}V, \tag{6.41}$$

one can rewrite Eq. (6.35) at an arbitrary time instant $\tau = t$ in the following concise form:

$$\dot{L}^t = \dot{K}^t + \dot{W}^t . \tag{6.42}$$

Equation (6.42) expresses the mechanical energy conservation law. It says that the energy supplied to the body (in a time unit) due to the work performed by external forces (the left-hand side of the equation) is equal to the corresponding increase in the body's internal energy (the right-hand side of the equation) consisting of the kinetic energy and the energy accumulated in material due to the work of internal forces.

Remark. The internal force energy and the way it is treated here require a comment. As we have mentioned in Sect. 5.4, this energy has partially a potential character (this refers to materials featuring elastic properties) while partially it is accumulated as thermal energy (this refers to materials subject to inelastic deformations which are accompanied by irreversible energy conversion processes). This subject will be addressed in more detail in Chap. 7, when constitutive properties of inelastic materials are discussed.

Since it has been assumed at the beginning of this section that all other than mechanical effects in the body $\mathcal{B}$ are neglected, we have to consequently disregard in the above derivation the effects related to the internal dissipation in the form of heat generation. This means that Eq. (6.42) should be regarded as valid for limited class of elastic materials only. In Chap. 9, an extended equation of energy conservation (the first law of thermodynamics) will be discussed in which terms related to thermal effects will also be present and the above-mentioned internal energy forms appropriately taken into account.

Chapter 7
Constitutive Equations

7.1 Introductory Remarks

In the previous sections mathematical description of displacements, strains and stresses in a deformable body as well as the laws governing motion of the body's particles and of the entire body have been discussed. Let us now resume our considerations by collecting the fundamental equations formulated up to now and pointing out the unknowns appearing there. Let us assume for clarity that the body's configuration at $\tau = t$ is analysed in its motion from $\tau = 0$ and the reference configuration is the current one, $C^r = C^t$. Upon these assumptions, the first Piola–Kirchhoff stress tensor is replaced by the Cauchy tensor and the general strain tensor by the Almansi tensor. Moreover, all the burdensome configuration-related indices are skipped in the discussion to follow.

We thus have the following equations:

$$\text{continuity equation} \quad (1) \qquad \dot{\rho} + \rho u_{i,i} = 0, \tag{7.1a}$$

$$\text{equation of motion} \quad (3) \qquad \sigma_{ij,j} + \rho f_i = \rho \ddot{u}_i\,, \tag{7.1b}$$

$$\text{geometric equations} \quad (6) \qquad \varepsilon_{ij} = \tfrac{1}{2}(u_{i,j} + u_{j,i} - u_{k,i}u_{k,j}), \tag{7.1c}$$

where the numbers of component equations are given in parentheses. The number 6 at the geometric equations indicates that the symmetry conditions have been included for the strain tensor.

The following unknown fields appear in the equations (the numbers of components are again given in parentheses, with tensor symmetries taken into account):

$$\text{density} \quad (1) \qquad \rho, \tag{7.2a}$$

$$\text{displacement} \quad (3) \qquad u_i, \tag{7.2b}$$

$$\text{strain} \quad (6) \qquad \varepsilon_{ij}\,, \tag{7.2c}$$

$$\text{stress} \quad (6) \qquad \sigma_{ij}\,. \tag{7.2d}$$

M. Kleiber and P. Kowalczyk, *Introduction to Nonlinear Thermomechanics of Solids*, Lecture Notes on Numerical Methods in Engineering and Sciences, DOI 10.1007/978-3-319-33455-4_7

There are thus 10 equations and 16 unknowns altogether. These numbers may be reduced in the case of incompressible bodies (ρ = const), as well as in several geometrically linearized models in which density changes are negligible—the continuity equation is then assumed fulfilled by definition and we end up with 9 equations and 15 unknowns. Note that if equations of the system (7.1) are replaced by their incremental forms and the unknown fields with their increments, the above balance remains unchanged—in every case 6 equations are missing to complete the system.

The lack of a certain number of equations should not be surprising to the readers as no equations characterizing physical properties of materials of which the deformable bodies are constituted have appeared in our discussion yet. These properties are described by relations between stress and strain in each of the material particles. Since both the quantities are described by 6 independent symmetric tensor components, the missing equations are relations between the 6 components of stress and strain tensors. The equations are called *constitutive equations* as they describe properties resulting from the material's internal structure (constitution).

Introduction to the constitutive equations in classical continuum mechanics handbooks typically follows the way "from general to specific". First, general requirements are formulated for the equations regarding such properties as objectivity of description or material symmetries. Then, particular models of material behaviour are derived and discussed according to the required postulates. However, in our computationally oriented approach, we concentrate our attention only on mathematical fundamentals of selected constitutive models most frequently used in engineering applications.

When talking about linear or nonlinear material behaviour, we usually mean the character of the stress–strain relationship. It should be noted, however, that this kind of nonlinearity (the so-called material nonlinearity) is not the only reason for nonlinearity in the equations of mechanics. Other sources of it may come from nonlinear terms appearing in the strain–displacement relations (the so-called geometrical nonlinearity) and the nonlinearity of boundary conditions (inequality bounds imposed on boundary displacements for instance).

Furthermore, the notion of material nonlinearity is not limited to nonlinearities in constitutive equations relating the total stress to the total strain in the material. Properties of real structural materials are frequently much more complex and require description that also includes stress and strain rates or their time integrals. Constitutive equations may additionally contain several other parameters among which particularly important are the so-called internal state parameters, which is a crucial notion in the theory of constitutive equations of inelastic materials.

7.2 Elastic Materials

A material is called elastic if a body made from this material returns to its original shape after removal of loads that made it deform and there is a unique relation between the strain and stress states at each material particle. These conditions are fulfilled with satisfactory accuracy (although only for a specified range of deformation) for a wide class of structural materials of practical application, like metals, wood, glass, ceramics, rocks, etc.

7.2.1 Linear Elasticity

Let us start with the simplest case—the linear elasticity, described by the constitutive equation in which stress components are linear functions of strain components. The most general, classical form of this equation, relating the Cauchy stress tensor components σ_{ij} to the linearized Cauchy strain components ε_{ij} and preserving transformation rules under changes of the coordinate system, has the form:

$$\sigma_{ij} = C_{ijkl}\,\varepsilon_{kl}\,, \qquad \boldsymbol{\sigma} = \boldsymbol{C} \cdot \boldsymbol{\varepsilon}. \tag{7.3}$$

Equation (7.3) is called the *generalized Hooke law*. The tensors appearing in it are generally fields on $\boldsymbol{x} \in \Omega$. The fourth order tensor $\boldsymbol{C}$ is called the elastic stiffness tensor—its $3^4 = 81$ components C_{ijkl} depend only on properties of the material considered. In the one-dimensional case discussed in Sect. 2.1 Eq. (7.3) takes the simplified form (2.4).

Since strain is a dimensionless quantity, the elastic constants C_{ijkl} have the dimension of stress and are typically expressed in pascals.

The fact that the "simplest case" of linear elasticity involves 81 elastic stiffness coefficients may raise some surprise. Fortunately, in many situations of practical importance, the coefficients turn out strongly interdependent and their number can be reduced to just 2.

The need to employ the constitutive equation (7.3) in the computation-oriented context suggests the use of matrix notation. While vectors and second order tensors may easily be expressed by 3×1 and 3×3 matrices, respectively, an analogous notation is no longer applicable to fourth order tensors. Therefore, another convention has been adopted in computational practice in which second order tensors are expressed as 9×1 column matrices while fourth order tensors as 9×9 square matrices. Equation (7.3) can thus be expressed in the matrix form as

$$\begin{Bmatrix} \sigma_{11} \\ \sigma_{22} \\ \sigma_{33} \\ \sigma_{12} \\ \sigma_{23} \\ \sigma_{31} \\ \hline \sigma_{21} \\ \sigma_{32} \\ \sigma_{13} \end{Bmatrix} = \left[\begin{array}{cccccc|ccc} C_{1111} & C_{1122} & C_{1133} & C_{1112} & C_{1123} & C_{1131} & C_{1121} & C_{1132} & C_{1113} \\ C_{2211} & C_{2222} & C_{2233} & C_{2212} & C_{2223} & C_{2231} & C_{2221} & C_{2232} & C_{2213} \\ C_{3311} & C_{3322} & C_{3333} & C_{3312} & C_{3323} & C_{3331} & C_{3321} & C_{3332} & C_{3313} \\ C_{1211} & C_{1222} & C_{1233} & C_{1212} & C_{1223} & C_{1231} & C_{1221} & C_{1232} & C_{1213} \\ C_{2311} & C_{2322} & C_{2333} & C_{2312} & C_{2323} & C_{2331} & C_{2321} & C_{2332} & C_{2313} \\ C_{3111} & C_{3122} & C_{3133} & C_{3112} & C_{3123} & C_{3131} & C_{3121} & C_{3132} & C_{3113} \\ \hline C_{2111} & C_{2122} & C_{2133} & C_{2112} & C_{2123} & C_{2131} & C_{2121} & C_{2132} & C_{2113} \\ C_{3211} & C_{3222} & C_{3233} & C_{3212} & C_{3223} & C_{3231} & C_{3221} & C_{3232} & C_{3213} \\ C_{1311} & C_{1322} & C_{1333} & C_{1312} & C_{1323} & C_{1331} & C_{1321} & C_{1332} & C_{1313} \end{array}\right] \begin{Bmatrix} \varepsilon_{11} \\ \varepsilon_{22} \\ \varepsilon_{33} \\ \varepsilon_{12} \\ \varepsilon_{23} \\ \varepsilon_{31} \\ \hline \varepsilon_{21} \\ \varepsilon_{32} \\ \varepsilon_{13} \end{Bmatrix}. \tag{7.4}$$

Symmetry of the strain and stress tensors ($\sigma_{ij} = \sigma_{ji}$, $\varepsilon_{kl} = \varepsilon_{lk}$) implies symmetries $C_{ijkl} = C_{jikl} = C_{ijlk}$ of the stiffness tensor. Looking at Eq. (7.4) one can thus conclude that (i) the last three of 9 equations are identical with the previous three ones and may thus be omitted and (ii) the last three elements of the strain matrix (ε_{21}, ε_{32}, ε_{13}) may also be removed along with the last three columns of the stiffness matrix. The latter operation obviously requires additional multiplication of the remaining shear strain components (ε_{12}, ε_{23}, ε_{31}) by 2. Upon these modifications, Eq. (7.4) is reduced to the form

$$\begin{Bmatrix} \sigma_{11} \\ \sigma_{22} \\ \sigma_{33} \\ \sigma_{12} \\ \sigma_{23} \\ \sigma_{31} \end{Bmatrix} = \begin{bmatrix} C_{1111} & C_{1122} & C_{1133} & C_{1112} & C_{1123} & C_{1131} \\ C_{2211} & C_{2222} & C_{2233} & C_{2212} & C_{2223} & C_{2231} \\ C_{3311} & C_{3322} & C_{3333} & C_{3312} & C_{3323} & C_{3331} \\ C_{1211} & C_{1222} & C_{1233} & C_{1212} & C_{1223} & C_{1231} \\ C_{2311} & C_{2322} & C_{2333} & C_{2312} & C_{2323} & C_{2331} \\ C_{3111} & C_{3122} & C_{3133} & C_{3112} & C_{3123} & C_{3131} \end{bmatrix} \begin{Bmatrix} \varepsilon_{11} \\ \varepsilon_{22} \\ \varepsilon_{33} \\ 2\varepsilon_{12} \\ 2\varepsilon_{23} \\ 2\varepsilon_{31} \end{Bmatrix}, \tag{7.5}$$

in which second order symmetric tensors are represented by 6×1 column matrices of their components. This equation contains only 36 independent material coefficients. It can be shown that this number is even smaller—the additional symmetry $C_{ijkl} = C_{klij}$, which will be justified below, implies symmetry of the stiffness matrix in Eq. (7.5) and reduces the number of material coefficients to 21.

In the deformation process of an elastic medium energy is accumulated in the material. The energy equals the work performed by stresses on strains in the material. This energy, related to the volume of a material particle, is expressed by Eq. (5.51) rewritten here in the differential form as

$$\mathrm{d}U = \sigma_{ij}\,\mathrm{d}\varepsilon_{ij}\,. \tag{7.6}$$

Observations indicate that this energy has a potential character, i.e. its total value $U = \int \mathrm{d}U$ is independent of the integration path (the history of the particle's deformation) and is just a scalar function of the current strain state,

$$U = U(\boldsymbol{\varepsilon}). \tag{7.7}$$

Substituting constitutive relation (7.3) into Eq. (7.6) one can see that integration of the latter to obtain the closed form (7.7) is only possible if the above-mentioned symmetry of the stiffness tensor, $C_{ijkl} = C_{klij}$, is assumed. We have then

$$U = \frac{1}{2} C_{ijkl}\varepsilon_{ij}\varepsilon_{kl} = \frac{1}{2}\boldsymbol{\varepsilon} \cdot \boldsymbol{C} \cdot \boldsymbol{\varepsilon}, \tag{7.8}$$

i.e.

$$\sigma_{ij} = \frac{\partial U}{\partial \varepsilon_{ij}}, \qquad C_{ijkl} = \frac{\partial^2 U}{\partial \varepsilon_{ij}\partial \varepsilon_{kl}}\,. \tag{7.9}$$

Symmetry of the second partial derivative of U with respect to its arguments (here ε_{ij} and ε_{kl}) results from the Schwarz theorem.

Equation (7.8) expresses thus the *potential strain energy density* in linearly elastic material. Laws of thermodynamics require this energy to be positive for each non-zero strain tensor (and zero for $\varepsilon_{ij} \equiv 0$) which imposes certain limitations on values of C_{ijkl} components.

Upon certain algebraic conditions Eq. (7.3) may be transformed to its inverse form

$$\varepsilon_{ij} = D_{ijkl}\,\sigma_{kl}\,, \qquad \boldsymbol{\varepsilon} = \boldsymbol{D}\cdot\boldsymbol{\sigma}, \tag{7.10}$$

in which the elastic compliance tensor $\boldsymbol{D}$ is the inverse of the stiffness tensor $\boldsymbol{C}$ and features the same symmetries, i.e. $D_{ijkl} = D_{jikl} = D_{ijlk} = D_{klij}$. This equation may also be written in the matrix notation as

$$\begin{Bmatrix} \varepsilon_{11} \\ \varepsilon_{22} \\ \varepsilon_{33} \\ 2\varepsilon_{12} \\ 2\varepsilon_{23} \\ 2\varepsilon_{31} \end{Bmatrix} = \begin{bmatrix} D_{1111} & D_{1122} & D_{1133} & D_{1112} & D_{1123} & D_{1131} \\ & D_{2222} & D_{2233} & D_{2212} & D_{2223} & D_{2231} \\ & & D_{3333} & D_{3312} & D_{3323} & D_{3331} \\ & & & D_{1212} & D_{1223} & D_{1231} \\ & \text{sym.} & & & D_{2323} & D_{2331} \\ & & & & & D_{3131} \end{bmatrix} \begin{Bmatrix} \sigma_{11} \\ \sigma_{22} \\ \sigma_{33} \\ \sigma_{12} \\ \sigma_{23} \\ \sigma_{31} \end{Bmatrix}. \tag{7.11}$$

In summary, only 21 of 81 material coefficients in Eq. (7.3) or (7.10) are independent. This is still a large number. However, the equation considered has still a very general form and does not include certain internal symmetries of the material. Note that the relation between the particular components σ_{11} and ε_{11}, described by the stiffness coefficient C_{1111}, is in this equation different than an analogous relation between the components σ_{22} and ε_{22}, described by the stiffness coefficient C_{2222}. This means that material properties generally depend on spatial directions. However, in many materials (like metals), the properties observed are the same in all directions. In others (like wood), certain characteristic directions may be shown in which properties are different, with properties in all the remaining directions with them To sum up: in most elastic materials there are several additional relations between elastic constants that result from internal symmetries in material microstructure and significantly reduce the number of independent material constants in the constitutive equation.

7.2.1.1 Isotropy

Consider the simplest case of a linearly elastic material, i.e. material with the same properties in all spatial directions. Such material is called *isotropic*. In such a case the stiffness tensor $\boldsymbol{C}$ must have the same representation C_{ijkl} in each Cartesian coordinate system, i.e. $\boldsymbol{C}$ must be an isotropic tensor. Such tensors have been discussed in Sect. 3.2.3. The general form of a fourth order tensor featuring this property is given by Eq. (3.62). In view of the required symmetry $C_{ijkl} = C_{klij}$, this form has to be additionally modified for our needs as follows,

$$C_{ijkl} = \lambda\delta_{ij}\delta_{kl} + \mu\,(\delta_{ik}\delta_{jl} + \delta_{il}\delta_{jk}). \tag{7.12}$$

Substituting (7.12) into Eq. (7.3) one obtains the relation

$$\sigma_{ij} = \lambda\,\delta_{ij}\varepsilon_{kk} + 2\mu\,\varepsilon_{ij} \tag{7.13}$$

in which only two elastic coefficients λ and μ appear and are called the *Lamé constants*. Written in the matrix form (7.5) this relation may be expressed as

$$\begin{Bmatrix} \sigma_{11} \\ \sigma_{22} \\ \sigma_{33} \\ \sigma_{12} \\ \sigma_{23} \\ \sigma_{31} \end{Bmatrix} = \begin{bmatrix} \lambda{+}2\mu & \lambda & \lambda & & & \\ \lambda & \lambda{+}2\mu & \lambda & & & \\ \lambda & \lambda & \lambda{+}2\mu & & & \\ & & & \mu & & \\ & & & & \mu & \\ & & & & & \mu \end{bmatrix} \begin{Bmatrix} \varepsilon_{11} \\ \varepsilon_{22} \\ \varepsilon_{33} \\ 2\varepsilon_{12} \\ 2\varepsilon_{23} \\ 2\varepsilon_{31} \end{Bmatrix} \tag{7.14}$$

(zeros have been skipped for clarity). Note that this form of the constitutive equation remains the same in all rotated coordinate systems.

The above isotropic form of the constitutive equation has several interesting implications. The first is coaxiality of the principal directions of strain and stress tensors. Since Eq. (7.13) is valid in each coordinate system, it is also valid in the system related to the strain principal directions (in which $\varepsilon_{12} = \varepsilon_{23} = \varepsilon_{31} = 0$). In this case it is clear from Eq. (7.14) that the tangential stress components σ_{12}, σ_{23}, σ_{31} must also vanish, i.e. the coordinate system axes are also the principal directions of stress.

It is also interesting to note that the stress tensor trace computed from Eq. (7.13) is a linear function of the strain tensor trace:

$$\sigma_{ii} = (3\lambda + 2\mu)\,\varepsilon_{ii}\,. \tag{7.15}$$

Let us now recall the theorem on decomposition of a second order tensor into a sum of its spherical and deviatoric parts, Eqs. (3.67)–(3.68). For the strain and stress tensors we obtain

$$\begin{aligned} \sigma^{\mathrm{S}}_{ij} &= \frac{1}{3}\,\sigma_{kk}\delta_{ij}\,,, \qquad & \sigma^{\mathrm{D}}_{ij} &= \sigma_{ij} - \sigma^{\mathrm{S}}_{ij}\,, \\ \varepsilon^{\mathrm{S}}_{ij} &= \frac{1}{3}\,\varepsilon_{kk}\delta_{ij}, \qquad & \varepsilon^{\mathrm{D}}_{ij} &= \varepsilon_{ij} - \varepsilon^{\mathrm{S}}_{ij}\,. \end{aligned} \tag{7.16}$$

It is easy to deduce from Eq. (7.13) that the deviatoric and spherical parts of these tensors are related to each other,

$$\sigma^{\mathrm{D}}_{ij} = 2\mu\,\varepsilon^{\mathrm{D}}_{ij}, \qquad \sigma^{\mathrm{S}}_{ij} = (3\lambda + 2\mu)\,\varepsilon^{\mathrm{S}}_{ij}, \tag{7.17}$$

and that the relations are much more simple in their forms than the matrix equation (7.14) relating the non-decomposed tensor components. The above relations are not interdependent, i.e. spherical stress is not related to deviatoric strain for instance.

The trace of the Cauchy strain tensor describes the volumetric strain i.e. the relative change of the material particle's volume, $e = \varepsilon_{kk}$, cf. Eq. (4.47). The spherical part of this tensor, $\varepsilon^{\mathrm{S}}_{ij} = \frac{1}{3}e\delta_{ij}$, describes the averaged proportional change of all linear particle dimensions. The spherical part of the stress tensor corresponds to an averaged normal stress acting in all directions on the particle's boundary, like hydrostatic pressure, $\sigma^{\mathrm{S}}_{ij} = -p\delta_{ij}$ (in which case the stress is negative). The second of Eq. (7.17) may thus be rewritten as

$$p = -\left(\lambda + \tfrac{2}{3}\mu\right) e. \tag{7.18}$$

This equation says that simultaneous change of all the particle's linear dimensions, i.e. the volume change at no shape change, is accompanied in the material by appearance of spherical stress, i.e. the state of uniform multiaxial stretch or compression. On the other hand, if a particle is subject to load by spherical stress (submerged in a liquid under hydrostatic pressure, for instance), then its strain is also spherical, i.e. all its linear dimensions are changed proportionally.

Inverting the relation (7.15), $\varepsilon_{kk} = \sigma_{kk}/(3\lambda + 2\mu)$, and substituting it into Eq. (7.13) it is easy to express ε_{ij} as a function of σ_{ij} and write the inverse form of the constitutive equation of isotropic linear elasticity as

$$\varepsilon_{ij} = \frac{1}{2\mu}\sigma_{ij} - \frac{\lambda}{2\mu(3\lambda + 2\mu)}\sigma_{kk}\delta_{ij}\,. \tag{7.19}$$

In engineering practice the Lamé constants are frequently replaced by other parameters called engineering elastic constants. These are the Young modulus E, the Poisson ratio ν, the shear modulus G and the bulk modulus K. They are related with each other and with the Lamé constants λ, μ by the following unique relations:

$$\begin{aligned}
\lambda = K - \frac{2}{3}G \quad &= \frac{E\nu}{(1+\nu)(1-2\nu)}\,, \\
\mu = G \quad &= \frac{E}{2(1+\nu)}\,, \\
K = \frac{E}{3(1-2\nu)} \quad &= \lambda + \frac{2}{3}\mu\,, \\
\nu = \frac{3K - 2G}{6K + 2G} \quad &= \frac{\lambda}{2(\lambda+\mu)}\,, \\
E = \frac{9KG}{3K + G} \quad &= 2\mu(1+\nu) = \frac{\mu(3\lambda + 2\mu)}{\lambda + \mu}\,.
\end{aligned} \tag{7.20}$$

All these constants, except for the dimensionless Poisson ratio, are expressed in the units of stress. Let us underline that—no matter which of the elastic constants are used in the constitutive formulation—only two of them are independent while the remaining ones can be expressed as their functions by Eq. (7.20).

Equations (7.17)–(7.18) may thus be written in an alternative simple form as

$$\sigma^{\mathrm{D}}_{ij} = 2G\,\varepsilon^{\mathrm{D}}_{ij}\,, \qquad \sigma^{\mathrm{S}}_{ij} = 3K\,\varepsilon^{\mathrm{S}}_{ij}\,, \qquad p = -Ke\,, \tag{7.21}$$

from which the physical interpretation of the engineering constants $K,\ G$ may be deduced as the material stiffnesses with respect to volumetric and shear strains, respectively. Equation (7.19) may be transformed to a similar simple form as

$$\varepsilon_{ij} = \frac{1+\nu}{E}\,\sigma_{ij} - \frac{\nu}{E}\,\sigma_{kk}\,\delta_{ij}\,, \tag{7.22}$$

which also reads

$$\begin{Bmatrix} \varepsilon_{11} \\ \varepsilon_{22} \\ \varepsilon_{33} \\ 2\varepsilon_{12} \\ 2\varepsilon_{23} \\ 2\varepsilon_{31} \end{Bmatrix} = \begin{bmatrix} 1/E & -\nu/E & -\nu/E & & & \\ -\nu/E & 1/E & -\nu/E & & & \\ -\nu/E & -\nu/E & 1/E & & & \\ & & & 1/G & & \\ & & & & 1/G & \\ & & & & & 1/G \end{bmatrix} \begin{Bmatrix} \sigma_{11} \\ \sigma_{22} \\ \sigma_{33} \\ \sigma_{12} \\ \sigma_{23} \\ \sigma_{31} \end{Bmatrix}. \tag{7.23}$$

Physical meaning of the constants E and ν can be easily deduced from Eq. (7.23). Consider a stress state in which $\sigma_{11} \neq 0$ and all the remaining stress components $\sigma_{ij} = 0$. This is the state corresponding to a uniaxial stretch (or compression) of the material particle. Equation (7.23) is then reduced to

$$\varepsilon_{11} = \sigma_{11}/E\,, \qquad \varepsilon_{22} = \varepsilon_{33} = -\nu\varepsilon_{11}\,. \tag{7.24}$$

Thus, the Young modulus E appears to be the material stiffness in the direction of the uniaxial load while the Poisson ratio ν indicates how the particle dimensions change in the transversal directions. If, for instance, the stress $\sigma_{11} = 0.01E$ and the Poisson ratio $\nu = 0.3$, then according to Eq. (7.23) the material particle is stretched along the x_1 direction by 1 % ($\varepsilon_{11} = 0.01$) and its transversal dimensions decrease by 0.3 % ($\varepsilon_{22} = \varepsilon_{33} = -0.003$).

The stress state considered here is a 3D counterpart of the 1D example with uniaxially stretched bar discussed in Sect. 2.1. If the stress σ and strain ε from that example are identified with the tensor components σ_{11} and ε_{11}, respectively, and if we realize that the bar in the example was not loaded by any transversal loads (i.e. the remaining stress components were zero) then Eq. (2.4) postulated there appears to be identical with Eq. $(7.24)_1$. The additional formulae $(7.24)_2$ for the transversal strain components did not appear in the discussion in Sect. 2.1 as the analysis of the bar's cross-section deformation was not of our interest at that moment.

Remark. The notation used here, in which the Greek letters σ and ε denote stress and strain, respectively, is commonly accepted in the literature of the subject. It is noteworthy, however, that—especially in the classical handbooks on the linear theory of elasticity—the tangential stress components are frequently denoted by τ_{ij} while the doubled shear strain components by γ_{ij} (we have then γ_{12} instead of $2\varepsilon_{12}$ and τ_{12} instead of σ_{12}). The reason for this duality in the notation can be deduced from the form of Eq. (7.14) which can be naturally decoupled into two separate systems of equations—one relating axial stress and strain components and the other (of a quite different form) relating tangential stresses with shear strains:

$$\begin{Bmatrix} \sigma_{11} \\ \sigma_{22} \\ \sigma_{33} \end{Bmatrix} = \begin{bmatrix} \lambda+2\mu & \lambda & \lambda \\ \lambda & \lambda+2\mu & \lambda \\ \lambda & \lambda & \lambda+2\mu \end{bmatrix} \begin{Bmatrix} \varepsilon_{11} \\ \varepsilon_{22} \\ \varepsilon_{33} \end{Bmatrix}, \qquad \begin{Bmatrix} \tau_{12} \\ \tau_{23} \\ \tau_{31} \end{Bmatrix} = \mu \begin{Bmatrix} \gamma_{12} \\ \gamma_{23} \\ \gamma_{31} \end{Bmatrix}.$$

Equation (7.23) may be rewritten in analogy as:

$$\begin{Bmatrix} \varepsilon_{11} \\ \varepsilon_{22} \\ \varepsilon_{33} \end{Bmatrix} = \frac{1}{E} \begin{bmatrix} 1 & -\nu & -\nu \\ -\nu & 1 & -\nu \\ -\nu & -\nu & 1 \end{bmatrix} \begin{Bmatrix} \sigma_{11} \\ \sigma_{22} \\ \sigma_{33} \end{Bmatrix}, \qquad \begin{Bmatrix} \gamma_{12} \\ \gamma_{23} \\ \gamma_{31} \end{Bmatrix} = \frac{1}{G} \begin{Bmatrix} \tau_{12} \\ \tau_{23} \\ \tau_{31} \end{Bmatrix}.$$

Substitution of the relations (7.20) into the matrix equation (7.14) allows to express it in terms of the engineering material constants. It takes then the following form:

$$\begin{Bmatrix} \sigma_{11} \\ \sigma_{22} \\ \sigma_{33} \\ \sigma_{12} \\ \sigma_{23} \\ \sigma_{31} \end{Bmatrix} = \frac{E}{(1+\nu)(1-2\nu)} \begin{bmatrix} 1-\nu & \nu & \nu & & & \\ \nu & 1-\nu & \nu & & & \\ \nu & \nu & 1-\nu & & & \\ & & & \frac{1-2\nu}{2} & & \\ & & & & \frac{1-2\nu}{2} & \\ & & & & & \frac{1-2\nu}{2} \end{bmatrix} \begin{Bmatrix} \varepsilon_{11} \\ \varepsilon_{22} \\ \varepsilon_{33} \\ 2\varepsilon_{12} \\ 2\varepsilon_{23} \\ 2\varepsilon_{31} \end{Bmatrix}. \tag{7.25}$$

Values of material constants are subject to certain limitations resulting from thermodynamical restrictions. Let us recall the general form of the strain energy function (7.8). For an isotropic material, this function is expressed as

$$U(\varepsilon_{mn}) = \mu\varepsilon_{ij}\varepsilon_{ij} + \frac{\lambda}{2}\varepsilon_{ii}\varepsilon_{jj} \tag{7.26}$$

or, after utilizing the decomposition $(7.16)_2$ and substitution of the engineering constants K and G, as

$$U(\varepsilon_{mn}) = G\,\varepsilon^{\mathrm{D}}_{ij}\varepsilon^{\mathrm{D}}_{ij} + \frac{K}{2}\,\varepsilon_{ii}\varepsilon_{jj}\,. \tag{7.27}$$

The energy U must be positive for each non-zero strain tensor $\boldsymbol{\varepsilon}$. Since the trace ε_{kk} (volumetric strain) and $\varepsilon^{\rm D}_{ij}$ (five independent deviatoric strain components) do not depend on each other, the necessary and sufficient condition for the positive definition of U is

$$K > 0 \quad \text{i} \quad G > 0. \tag{7.28}$$

Besides, Eq. (7.20) imply that

$$E > 0, \qquad -1 < \nu < \tfrac{1}{2}\,. \tag{7.29}$$

The conditions $K > 0$, $G > 0$ and $E > 0$ are intuitively obvious—since no deformation (of the volumetric, shear or stretch type) can result from a load acting in the direction opposite to it. Besides, if ν exceeded $\frac{1}{2}$ at $E > 0$ then the material stiffness with respect to volumetric deformation K would be negative. The limit case of $\nu = \frac{1}{2}$ corresponds to the so-called incompressible material, i.e. retaining its constant local volume at each physical deformation and at each stress state. Such material implies the infinite value of the bulk modulus K.

Since materials with the negative value of the Poisson ratio ν hardly ever appear in practical engineering applications (its typical values for metals are about 0.3 while for plastics are even higher), we will further restrict the condition $(7.29)_2$ to the form

$$0 < \nu < \tfrac{1}{2}\,. \tag{7.30}$$

7.2.1.2 Two-Dimensional Formulations

Many solid mechanics problems of engineering significance can be reduced within acceptable accuracy to two-dimensions. Two classes of such problems will be discussed here known as the plane stress and the plane strain problems. In the *plane stress* formulation, it is assumed that in a selected coordinate system there are only three non-zero stress components: σ_{11}, σ_{22} and $\sigma_{12} = \sigma_{21}$. An example of plane stress problem is a planar plate loaded by stretching or compressing forces acting solely in its plane (Fig. 7.1). No assumptions about strains are made, however, a short glance on the matrix constitutive equation (7.23) allows to conclude that

- the components $\varepsilon_{13} = \varepsilon_{23} = 0$,
- the components ε_{11}, ε_{22} and ε_{12} are related to the stress components by the matrix formula

$$\begin{Bmatrix} \varepsilon_{11} \\ \varepsilon_{22} \\ 2\varepsilon_{12} \end{Bmatrix} = \begin{bmatrix} 1/E & -\nu/E & \\ -\nu/E & 1/E & \\ & & 1/G \end{bmatrix} \begin{Bmatrix} \sigma_{11} \\ \sigma_{22} \\ \sigma_{12} \end{Bmatrix}, \tag{7.31}$$

- the component ε_{33} describing local relative change of the plate's thickness is not independent—comparison of Eqs. (7.23) and (7.31) implies that

$$\varepsilon_{33} = \frac{-\nu}{1-\nu}(\varepsilon_{11} + \varepsilon_{22}). \tag{7.32}$$

Inversion of the matrix in Eq. (7.31) leads to the following inverse equation:

$$\begin{Bmatrix} \sigma_{11} \\ \sigma_{22} \\ \sigma_{12} \end{Bmatrix} = \frac{E}{1-\nu^2} \begin{bmatrix} 1 & \nu & \\ \nu & 1 & \\ & & \frac{1-\nu}{2} \end{bmatrix} \begin{Bmatrix} \varepsilon_{11} \\ \varepsilon_{22} \\ 2\varepsilon_{12} \end{Bmatrix} \tag{7.33}$$

in which we observe that the material coefficient matrix is not a result of a simple removal of some rows and columns from the full matrix appearing in the 3D equation (7.25). It is the additional relation (7.32) between the strain components that implies this matrix to be quite different.

In the *plane strain* formulations it is assumed that in a properly selected coordinate system there are only three non-zero strain components: ε_{11}, ε_{22} and $\varepsilon_{12} = \varepsilon_{21}$. An example of a plane strain problem is the analysis of a thick wall or a dam with a constant cross-section and fixed ends, loaded by forces acting only transversally to its length and uniformly distributed along the length. Cross-sections of this structure (or more precisely its finite-thickness "slices") are only deformed in their plane and retain their thickness unchanged (Fig. 7.1).

Based on Eq. (7.25) for a 3D linear elastic isotropy and following analogous sequence of transformations to that carried out in the plane stress analysis, we can derive the following equations describing the plane strain problem:

$$\begin{Bmatrix} \sigma_{11} \\ \sigma_{22} \\ \sigma_{12} \end{Bmatrix} = \frac{E}{(1+\nu)(1-2\nu)} \begin{bmatrix} 1-\nu & \nu & \\ \nu & 1-\nu & \\ & & \frac{1-2\nu}{2} \end{bmatrix} \begin{Bmatrix} \varepsilon_{11} \\ \varepsilon_{22} \\ 2\varepsilon_{12} \end{Bmatrix}, \tag{7.34}$$

$$\sigma_{33} = \nu(\sigma_{11} + \sigma_{22}) \tag{7.35}$$

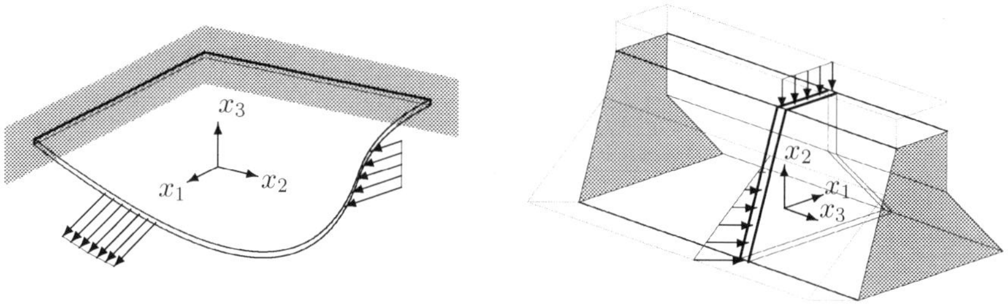

Fig. 7.1 Examples of plane stress (*left*) and plane strain (*right*) problem

and

$$\begin{Bmatrix} \varepsilon_{11} \\ \varepsilon_{22} \\ 2\varepsilon_{12} \end{Bmatrix} = \frac{1+\nu}{E} \begin{bmatrix} 1-\nu & -\nu & \\ -\nu & 1-\nu & \\ & & 2 \end{bmatrix} \begin{Bmatrix} \sigma_{11} \\ \sigma_{22} \\ \sigma_{12} \end{Bmatrix}. \tag{7.36}$$

Exercise 7.1
For a material with the properties $E = 80\,000\,\text{MPa}$, $\nu = \frac{1}{3}$, at a strain state with the components $\varepsilon_{11} = 0.002$, $\varepsilon_{22} = -0.001$ and $\varepsilon_{12} = 0.001$, find:

- under the assumption of plane stress—the stress components σ_{ij} and the transversal strain ε_{33},
- under the assumption of plane strain—the stress components σ_{ij}.

Solution. In the case of plane stress, the three nonzero stress components can be determined from Eq. (7.33) that upon substitution of the material constant values assumes the form

$$\begin{Bmatrix} \sigma_{11} \\ \sigma_{22} \\ \sigma_{12} \end{Bmatrix} = \begin{bmatrix} 90\,000 & 30\,000 & 0 \\ 30\,000 & 30\,000 & 0 \\ 0 & 0 & 30\,000 \end{bmatrix} \begin{Bmatrix} \varepsilon_{11} \\ \varepsilon_{22} \\ 2\varepsilon_{12} \end{Bmatrix}.$$

For the given values of strain components the result is

$$\sigma_{11} = 150\,\text{MPa}, \qquad \sigma_{22} = -30\,\text{MPa}, \qquad \sigma_{12} = \sigma_{21} = 60\,\text{MPa}$$

(with all remaining stress components being zero). The transversal strain can be computed from Eq. (7.32) as

$$\varepsilon_{33} = -0.0005.$$

In the case of plane strain $\varepsilon_{33} = 0$ and the stress components are given by Eqs. (7.34) and (7.35). For the given material constants the first of them has the form

$$\begin{Bmatrix} \sigma_{11} \\ \sigma_{22} \\ \sigma_{12} \end{Bmatrix} = \begin{bmatrix} 120\,000 & 60\,000 & 0 \\ 60\,000 & 120\,000 & 0 \\ 0 & 0 & 30\,000 \end{bmatrix} \begin{Bmatrix} \varepsilon_{11} \\ \varepsilon_{22} \\ 2\varepsilon_{12} \end{Bmatrix},$$

i.e., for the given values of strain components, one obtains

$$\sigma_{11} = 180\,\text{MPa}, \qquad \sigma_{22} = 0\,\text{MPa}, \qquad \sigma_{12} = \sigma_{21} = 60\,\text{MPa}.$$

The remaining nonzero transversal stress component is $\sigma_{33} = 60\,\text{MPa}$.

7.2.1.3 Orthotropy

Isotropy is the simplest but not the only type of material symmetry. There are several other kinds of symmetry that imply additional relations between elastic material constants appearing in Eq. (7.5) or (7.11). One of them, typical of many structural materials known in engineering practice is called the *orthotropic symmetry*.

An orthotropic elastic material has three specific orthogonal directions in space, called principal directions of orthotropy. If the coordinate system axes are co-linear with these directions, then the constitutive equation can be written in the form

$$\begin{Bmatrix} \sigma_{11} \\ \sigma_{22} \\ \sigma_{33} \\ \sigma_{12} \\ \sigma_{23} \\ \sigma_{31} \end{Bmatrix} = \begin{bmatrix} C_{1111} & C_{1122} & C_{1133} & & & \\ C_{1122} & C_{2222} & C_{2233} & & & \\ C_{1133} & C_{2233} & C_{3333} & & & \\ & & & C_{1212} & & \\ & & & & C_{2323} & \\ & & & & & C_{3131} \end{bmatrix} \begin{Bmatrix} \varepsilon_{11} \\ \varepsilon_{22} \\ \varepsilon_{33} \\ 2\varepsilon_{12} \\ 2\varepsilon_{23} \\ 2\varepsilon_{31} \end{Bmatrix} \tag{7.37}$$

(for more clarity, zeros have been skipped and the matrix symmetry taken into account). The elastic stiffness matrix contains thus only 9 independent coefficients. The form of this equation resembles to some extent that of isotropic elasticity. Note for example that the axial stress components depend only on axial strain components while the tangential stresses—on shear strains only. This property is true in the selected coordinate system. Since the stiffness matrix does not have the isotropic form, one can conclude that in another coordinate system, when components of $\boldsymbol{\sigma}$, ε and $\boldsymbol{C}$ are transformed according to some appropriate rules, the constitutive equation may generally contain the full 6×6 stiffness matrix (being functions of the above 9 independent coefficients and the directional cosines of the rotation matrix). Thus, simple stretching of an orthotropic material in an arbitrary direction oblique to the principal directions will generally induce both axial and shear strains, for instance.

An example of an orthotropic material is wood. Its principal directions conform to the natural direction of the trunk axis and the two orthogonal directions—circumferential (tangential to the rings) and axial (transversal to the rings). The axial stiffness in wood is usually much higher than that in the other directions. Orthotropic properties are also featured by several fibrous composites, i.e. materials consisting of a matrix reinforced by a mesh of fibres with specified directionality, as well as by various cellular materials whose cells have specifically ordered structures.

Particular cases of orthotropy are e.g. *transversal isotropy* (the material exhibits isotropic properties in a specified plane defined by two principal directions of orthotropy and anisotropic ones in other planes) and *cubic symmetry* (material has the same stiffness in all principal directions but—contrary to isotropic materials—the shear stiffness is not related to the longitudinal stiffness). The stiffness matrix of the transversally isotropic material has 5 independent coefficients while for a material with the cubic symmetry—only 3 independent coefficients.

7.2.2 Nonlinear Elasticity

The linear equation (7.3) is the simplest example of the constitutive equation of an elastic material. Its specific feature is the constant stiffness tensor $\boldsymbol{C}$ being the proportionality coefficient between the stress and strain tensors. In many materials this tensor is subject to changes during the deformation process and may be expressed as an explicit function of strain or stress. In such cases we have to do with nonlinear formulations of elastic constitutive equations.

Consider the following general form of such an equation:

$$\sigma_{ij} = f_{ij}(\varepsilon_{kl}). \tag{7.38}$$

A material model described by this equation is called elastic in the Cauchy sense. The form of the tensor function f_{ij} cannot be arbitrary—it must fulfill appropriate component transformation rules under changes of the coordinate system.

A particular form of such a material is obviously the linear elastic material. The linear equation (7.3) may be easily generalized by introduction of nonlinearity in the form

$$\sigma_{ij} = C_{ijkl}(\varepsilon_{mn})\,\varepsilon_{kl}\,, \qquad \boldsymbol{\sigma} = \boldsymbol{C}(\boldsymbol{\varepsilon})\cdot\boldsymbol{\varepsilon}. \tag{7.39}$$

The above equation is again a particular case of the Cauchy equation of elasticity (7.38) in which the stiffness tensor is an explicit function of the strain state.

Let us compute from Eq. (7.39) the infinitesimal stress increment $\mathrm{d}\sigma_{ij}$ accompanying the given strain increment $\mathrm{d}\varepsilon_{kl}$:

$$\begin{aligned} \mathrm{d}\sigma_{ij} &= C_{ijkl}\,\mathrm{d}\varepsilon_{kl} + \frac{\partial C_{ijkl}}{\partial \varepsilon_{mn}}\,\mathrm{d}\varepsilon_{mn}\,\varepsilon_{kl} = C_{ijkl}\,\mathrm{d}\varepsilon_{kl} + \frac{\partial C_{ijmn}}{\partial \varepsilon_{kl}}\,\varepsilon_{mn}\,\mathrm{d}\varepsilon_{kl} \\ &= \left(C_{ijkl} + \frac{\partial C_{ijmn}}{\partial \varepsilon_{kl}}\,\varepsilon_{mn}\right)\mathrm{d}\varepsilon_{kl} \end{aligned} \tag{7.40}$$

(in the transformations, the pairs of dummy indices kl and mn have been switched). The resulting relation can be concisely written as

$$\mathrm{d}\sigma_{ij} = \hat{C}_{ijkl}\,\mathrm{d}\varepsilon_{kl}\,, \qquad \mathrm{d}\boldsymbol{\sigma} = \hat{\boldsymbol{C}}\cdot\mathrm{d}\boldsymbol{\varepsilon}, \tag{7.41}$$

where the new tensor $\hat{\boldsymbol{C}}$ describes instantaneous (at a particular strain state $\boldsymbol{\varepsilon}$) incremental elastic stiffness of the material. Its components are expressed as

$$\hat{C}_{ijkl}(\varepsilon_{mn}) = C_{ijkl}(\varepsilon_{mn}) + \frac{\partial C_{ijpr}}{\partial \varepsilon_{kl}}\,\varepsilon_{pr}\,. \tag{7.42}$$

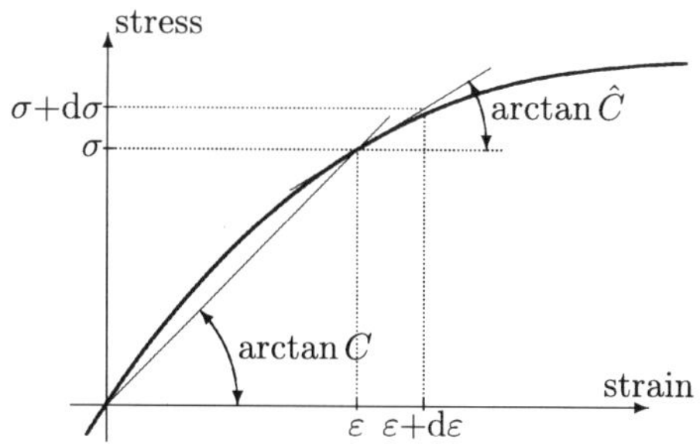

Fig. 7.2 Elastic moduli: secant and tangent

The tensor $\hat{\boldsymbol{C}}(\varepsilon)$, called *tangent stiffness tensor* (or tangent stiffness modulus), is generally different from the tensor $\boldsymbol{C}(\varepsilon)$ appearing in Eq. (7.39), called *secant stiffness tensor*. Figure 7.2 illustrates the difference between the two stiffness moduli in the one-dimensional case.

In the case of linear elasticity (and only then), the tangent and the secant stiffness tensors are equal, $\hat{\boldsymbol{C}} = \boldsymbol{C}$. The necessity of distinguishing between the two kinds of elastic stiffness is an inherent feature of nonlinear elastic material models.

Equation (7.41) is obviously valid for stress and strain rates, assuming then the form

$$\dot{\sigma}_{ij} = \hat{C}_{ijkl}\,\dot{\varepsilon}_{kl}\,, \qquad \dot{\boldsymbol{\sigma}} = \hat{\boldsymbol{C}} \cdot \dot{\boldsymbol{\varepsilon}}. \tag{7.43}$$

7.2.2.1 Hyperelasticity

One of the ways to formulate nonlinear elastic equations used in a wide class of material models is defining the explicit form of the strain energy density U as a function of the strain state. For the linear elastic material, this function has the form (7.8). Given the function $U(\varepsilon)$ in an arbitrary form and defining the stress as, cf. Eq. (7.9),

$$\sigma_{ij} = \frac{\partial U}{\partial \varepsilon_{ij}}\,, \qquad \boldsymbol{\sigma} = \frac{\mathrm{d}U}{\mathrm{d}\boldsymbol{\varepsilon}}\,, \tag{7.44}$$

we obtain the constitutive equation of a *hyperelastic* material.

Equation (7.44) is a particular case of Eq. (7.38). In many cases it can also be transformed to the form (7.39) while in all cases it can be expressed in the rate form (7.43) with the tangent stiffness tensor defined as

$$\hat{C}_{ijkl} = \frac{\partial \sigma_{ij}}{\partial \varepsilon_{ij}} = \frac{\partial^2 U}{\partial \varepsilon_{ij} \partial \varepsilon_{kl}}\,. \tag{7.45}$$

Similarly as in linear elasticity, the nonlinear constitutive equations may be either isotropic or anisotropic. It is easy to show that isotropy of the hyperelastic formulation is guaranteed if the strain energy function is expressed in the basic invariants of strain,

$$U(\varepsilon) = U(I_1, I_2, I_3), \tag{7.46}$$

where, cf. Eq. (3.81)

$$\begin{aligned} I_1 &= \operatorname{tr}\varepsilon = \varepsilon_{ii}\,, \\ I_2 &= \tfrac{1}{2}\left[\operatorname{tr}\varepsilon^2 - (\operatorname{tr}\varepsilon)^2\right] = \tfrac{1}{2}(\varepsilon_{ij}\varepsilon_{ji} - \varepsilon_{ii}\varepsilon_{jj}), \\ I_3 &= \det\varepsilon = \tfrac{1}{6}\epsilon_{ijk}\epsilon_{lmn}\,\varepsilon_{il}\varepsilon_{jm}\varepsilon_{kn}\,. \end{aligned} \tag{7.47}$$

Since the basic invariant values do not depend on the coordinate system, the form of Eq. (7.44) is the same in each of the systems.

In the discussion regarding elastic constitutive models influence of geometric configuration changes on the form of the constitutive equations has not so far been considered. To the contrary, small deformation and linear Cauchy strain approximation were rather assumed. When talking about nonlinear elasticity this issue can be no longer disregarded, though. At linear strains—which are very small by the definition—each nonlinear elastic constitutive equation can be replaced by its linear approximate with a practically acceptable error. Considering large deformations and nonlinear strains in formulations of hyperelasticity leads to several fundamental difficulties—note, for instance, that the function U describes the strain energy density related to unit volume and thus the question immediately arises in which configuration of the body this volume is measured.

Let C^r be the reference configuration for the body $\mathcal{B}$ in a motion. Let ${}_rU$ denote the elastic strain energy accumulated in a unit volume of material measured in this configuration. Thus, the energy accumulated at the time instant t in a material particle with the reference volume $\mathrm{d}V^r$ is $W^t = {}_rU^t\,\mathrm{d}V^r$. Besides, the energy density related to the mass unit is defined as $e^t = \rho^r\,{}_rU^t$.

The discussion in Sect. 5.4 allows to conclude that if the energy density ${}_rU^t$ is expressed as a function of the strain tensor ${}^0_r\varepsilon^t$ then the stress tensor appearing on the left-hand side of the constitutive equation (7.44) is the second Piola–Kirchhoff tensor ${}_r\tilde{\sigma}^t$, i.e. the one work-conjugate to the tensor ${}^0_r\varepsilon^t$. We have then

$${}_r\tilde{\sigma}^t_{ij} = \frac{\partial\,{}_rU^t}{\partial\,{}^0_r\varepsilon^t_{ij}}\,. \tag{7.48}$$

In particular, if the reference configuration is identified with the initial configuration, $C^r = C^0$, then the strain energy density is expressed as a function of the Green strain tensor, ${}_0U^t = {}_0U^t(\varepsilon^G)$, and the hyperelastic constitutive equation has the form

$${}_0\tilde{\sigma}^t_{ij} = \frac{\partial\,{}_0U^t}{\partial\varepsilon^G_{ij}}\,. \tag{7.49}$$

On the other hand, if the current configuration is taken as the reference one, $C^r = C^t$, then the strain energy density is expressed as a function of the Almansi strain, ${}_tU^t = {}_tU^t(\varepsilon^A)$, and the hyperelastic constitutive equation assumes the form:

$$\sigma^t_{ij} = {}_t\tilde{\sigma}^t_{ij} = \frac{\partial\, {}_tU^t}{\partial \varepsilon^A_{ij}} \,. \tag{7.50}$$

We can thus conclude that for materials subject to large deformations the hyperelastic constitutive equations can be consistently formulated in various reference configurations.

Exercise 7.2
Derive the constitutive equation and the tangent stiffness tensor for the following strain energy density function (proposed by Fung to describe elastic properties of soft biological tissues [1]):

$$U(\varepsilon) = C\,\exp(aI_1^2 + bI_2),$$

where I_1, I_2 denote the basic invariants of ε (7.47); a, b, C—material constants.

Solution. Let us start from derivation of the partial derivatives of I_1 and I_2 with respect to strain tensor components. Making use of the obvious equality $\partial\varepsilon_{ij}/\partial\varepsilon_{kl} = \delta_{ik}\delta_{jl}$ (the derivative equals 1 if and only if $ij = kl$, otherwise it is zero) and the symmetry of the tensor ε, we obtain:

$$\frac{\partial I_1}{\partial \varepsilon_{kl}} = \delta_{ik}\delta_{il} = \delta_{kl}\,,$$
$$\frac{\partial I_2}{\partial \varepsilon_{kl}} = \frac{1}{2}(\delta_{ik}\delta_{jl}\varepsilon_{ji} + \varepsilon_{ij}\delta_{jk}\delta_{il} + 2\delta_{kl}\varepsilon_{ii}) = \varepsilon_{kl} + \delta_{kl}\varepsilon_{ii}\,.$$

Substituting this into Eq. (7.44) we can derive the constitutive relation

$$\begin{aligned}\sigma_{kl} &= \frac{\partial U}{\partial \varepsilon_{kl}} = C\,\exp(aI_1^2 + bI_2)\left(2aI_1\frac{\partial I_1}{\partial \varepsilon_{kl}} + b\frac{\partial I_2}{\partial \varepsilon_{kl}}\right)\\ &= U(\varepsilon)\,[(2a-b)\delta_{kl}\varepsilon_{ii} + b\varepsilon_{kl}]\,.\end{aligned}$$

Comparing the above equation with Eq. (7.13), one can see the similarity: substituting $\lambda(\varepsilon) = (2a-b)\,U(\varepsilon)$ and $2\mu(\varepsilon) = b\,U(\varepsilon)$ we come at nonlinear, strain-dependent expressions on isotropic Lamé constants.

The tangent stiffness components $\hat{C}_{klmn}$ can be determined by differentiation of the constitutive equation with respect to ε_{mn}:

$$\hat{C}_{klmn} = \frac{\partial \sigma_{kl}}{\partial \varepsilon_{mn}} = \frac{\partial U}{\partial \varepsilon_{mn}} \left[(2a-b)\delta_{kl}\varepsilon_{ii} + b\varepsilon_{kl}\right] + U \left[(2a-b)\delta_{kl} \frac{\partial \varepsilon_{ii}}{\partial \varepsilon_{mn}} + b \frac{\partial \varepsilon_{kl}}{\partial \varepsilon_{mn}}\right]$$
$$= U(\varepsilon) \left\{ \left[(2a-b)\delta_{kl}\varepsilon_{ii} + b\varepsilon_{kl}\right]\left[(2a-b)\delta_{mn}\varepsilon_{jj} + b\varepsilon_{mn}\right] + (2a-b)\delta_{kl}\delta_{mn} + b\delta_{km}\delta_{ln} \right\}.$$

Careful readers will surely notice that the form of the resulting components $\hat{C}_{klmn}$ is not symmetric with respect to the pairs kl and mn. This is because the symmetry $\varepsilon_{mn} = \varepsilon_{nm}$ was not fully taken into account in differentiation of the strain components (to do it consistently, we should have rather assumed that $\partial \varepsilon_{ij}/\partial \varepsilon_{kl} = \frac{1}{2}(\delta_{ik}\delta_{jl} + \delta_{jk}\delta_{il})$). Note, however, that the full product of C_{klmn} and the symmetric tensor $\dot{\varepsilon}_{mn}$ is equivalent to the full product of the symmetric part of C_{klmn} (with respect to the indices mn) and $\dot{\varepsilon}_{mn}$. Hence, the term $b\delta_{km}\delta_{ln}$ in the above result may be replaced by its symmetric part $\frac{b}{2}(\delta_{km}\delta_{ln} + \delta_{kn}\delta_{lm})$ which yields the final, symmetric form of the tangent stiffness tensor $\hat{C}_{klmn}$.

7.2.2.2 Hypoelasticity

Yet another nonlinear elastic material is defined by equations of *hypoelasticity*. These are rate relations in the following general form:

$$\dot{\sigma}_{ij} = \tilde{C}_{ijkl}(\sigma_{mn})\, \dot{\varepsilon}_{kl}\,, \tag{7.51}$$

where the tangent modulus $\tilde{C}_{ijkl}$ has the same meaning as $\hat{C}_{ijkl}$ but is expressed as an explicit function of stress. Equation (7.51) is not equivalent to Eq. (7.39), although it is frequently possible to derive it from the corresponding hyperelastic form.[1] One may define the following potential quantity expressed in terms of strain rate:

$$\tilde{U} = \frac{1}{2}\tilde{C}_{ijkl}(\sigma_{mn})\, \dot{\varepsilon}_{ij}\dot{\varepsilon}_{kl}\,, \tag{7.52}$$

which allows to express the stress rate as

$$\dot{\sigma}_{ij} = \frac{\partial \tilde{U}}{\partial \dot{\varepsilon}_{ij}}\,. \tag{7.53}$$

[1] The definitions of Cauchy elasticity, hyperelasticity and hypoelasticity are generally not equivalent; conditions of their equivalence is an interesting problem of the theory of elasticity.

Let us briefly discuss possible forms of the tensor $\tilde{C}_{ijkl}$ in Eq. (7.51) for isotropic materials. Isotropy requires that under an arbitrary orthogonal transformation of the coordinate system $x'_i = Q_{ij}x_j$, the tensor components $\tilde{C}'_{ijkl}$ expressed as functions of σ'_{mn} have the same form as $\tilde{C}_{ijkl}$ expressed in σ_{mn}. According to Eqs. (3.44) and (3.51),

$$\tilde{C}'_{pqrs} = Q_{pi}Q_{qj}Q_{rk}Q_{sl}\tilde{C}_{ijkl}, \qquad \sigma'_{ab} = Q_{am}Q_{bn}\sigma_{mn}. \tag{7.54}$$

Moreover, symmetry of $\dot{\sigma}_{ij}$ and $\dot{\varepsilon}_{ij}$ implies the following symmetries of $\tilde{C}_{ijkl}$:

$$\tilde{C}_{ijkl} = \tilde{C}_{jikl} = \tilde{C}_{ijlk} = \tilde{C}_{jilk}. \tag{7.55}$$

The most general form of $\tilde{C}_{ijkl}$ that meets the above conditions appears to be

$$\begin{aligned} \tilde{C}_{ijkl}(\sigma_{mn}) = {} & A_1\delta_{ij}\delta_{kl} + A_2(\delta_{ik}\delta_{jl} + \delta_{jk}\delta_{il}) + A_3\sigma_{ij}\delta_{kl} + A_4\delta_{ij}\sigma_{kl} \\ & + A_5(\delta_{ik}\sigma_{jl} + \delta_{il}\sigma_{jk} + \delta_{jk}\sigma_{il} + \delta_{jl}\sigma_{ik}) \\ & + A_6\delta_{ij}\sigma_{km}\sigma_{ml} + A_7\delta_{kl}\sigma_{im}\sigma_{mj} \\ & + A_8(\delta_{ik}\sigma_{jm}\sigma_{ml} + \delta_{il}a_{jm}\sigma_{mk} + \delta_{jk}\sigma_{im}\sigma_{ml} + \delta_{jl}\sigma_{im}\sigma_{mk}) \\ & + A_9\sigma_{ij}\sigma_{kl} + A_{10}\sigma_{ij}\sigma_{km}\sigma_{ml} + A_{11}\sigma_{im}\sigma_{mj}\sigma_{kl} \\ & + A_{12}\sigma_{im}\sigma_{mj}\sigma_{kn}\sigma_{nl}, \end{aligned} \tag{7.56}$$

where the 12 coefficients $A_1, A_2, \ldots, A_{12}$ may only depend on invariants of the stress tensor σ_{ij}.

Assuming $A_3 = A_4 = \cdots = A_{12} = 0$ and $A_1 = \text{const}$, $A_2 = \text{const}$, we arrive at the so-called zero order hypoelastic material, equivalent to the linear elastic material (7.3) discussed earlier in this section. In the first order hypoelastic material the stiffness tensor is by definition a linear function of stress—in this case A_1 and A_2 are only functions of the first basic invariant of stress, A_3, A_4 and A_5 are constant while $A_6 = \cdots = A_{12} = 0$.

7.3 Viscoelastic Materials

A large group of materials of significant practical importance (like gums, plastics, biological tissues) have *viscoelastic* properties. Description of their constitutive behaviour is more complex than in the case of elasticity because of the time variable appearing in the constitutive equations. An elastic material "remembers" only the undeformed initial configuration to which the strain is related at each deformation stage. Viscoelastic materials have the memory of the entire deformation process, though. A relatively simple description of their constitutive properties is only possible for those of them that exhibit linear relations between the cause (e.g. force) and the effect (e.g. displacement). This class of materials will be discussed below.

For the sake of clarity, we will limit ourselves to small linearized deformations only, neglecting configuration changes of the body and its particles.

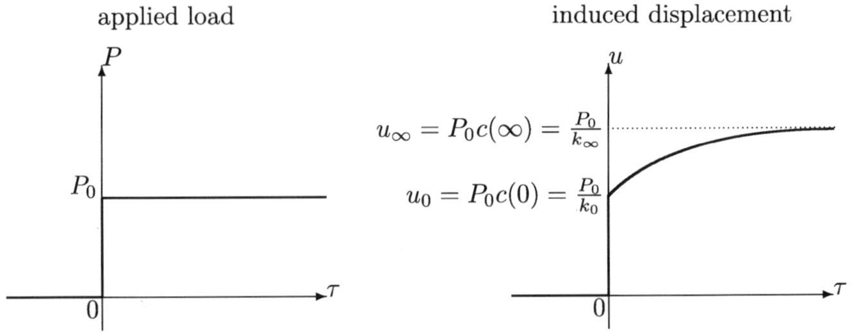

Fig. 7.3 Creep

7.3.1 One-Dimensional Models

Consider the 1D problem of a bar stretched by the constant force P_0 that has been instantaneously applied at $\tau = 0$. Assume the history of the load be described by the *Heaviside function* $H(\tau)$

$$P(\tau) = P_0 H(\tau), \qquad H(\tau) = \begin{cases} 0 & \tau < 0 \\ \frac{1}{2} & \tau = 0 \\ 1 & \tau > 0. \end{cases} \tag{7.57}$$

Denote by u the end point displacement, i.e. the change of the bar's length. If the bar is elastic, the displacement has also a step function form, $u(\tau) = u_0 H(\tau)$ where $u_0 = P_0/k$, cf. Eq. (2.1), and the stiffness coefficient $k = EA/l$ depends on the bar's parameters, cf. Eq. (2.5). However, if the bar material is viscoelastic, such a load will result in a displacement $u(\tau)$, $\tau \geq 0$, that is changing in time under this load and usually converges asymptotically at $\tau \to \infty$ to a certain value. This phenomenon is called *creep*. The time history of the displacement may thus be expressed as

$$u(\tau) = P_0\, c(\tau), \tag{7.58}$$

where $c(\tau)$ is the so-called *creep function*, which is a specific material property.

Creep is illustrated in Fig. 7.3. The initial displacement value $u(0) = P_0/k_0$ ($k_0 = E_0 A/l$) corresponds to the instantaneous elastic behaviour of the bar. As the time passes, the bar gets further stretched under the constant load as a result of "weakening" material properties—the Young modulus E is decreasing in time and tends to a certain limit value E_∞ corresponding to the limit bar stiffness $k_\infty = E_\infty A/l$. The creep function assumes non-zero values only for $\tau \geq 0$ and is monotonically non-decreasing.

The above discussion may be generalized to the case of arbitrary load history $P(\tau)$ initialized at $\tau = 0$ and acting up to $\tau = t$. Treating the load history as a series of

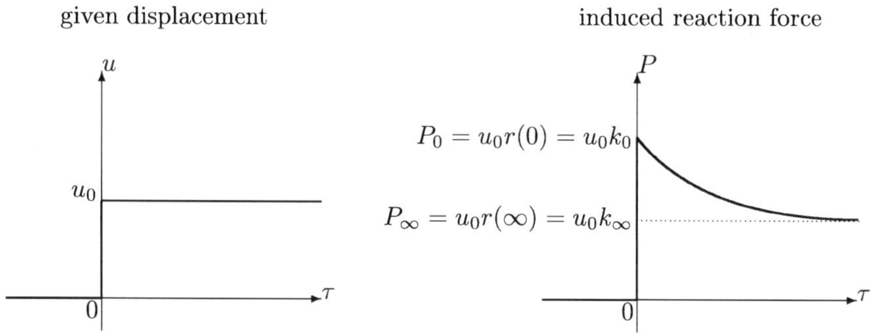

Fig. 7.4 Relaxation

small increments $\mathrm{d}P(\tau) = \dot{P}\mathrm{d}\tau$ we postulate the displacement u at the time instant t to be a sum of displacements induced by all the subsequent step loads preceding this time instant. Passing to the limit as $\mathrm{d}\tau \to 0$ we may write the result in the integral form as

$$u(t) = \int_0^t \dot{P}(\tau)\, c(t-\tau)\, \mathrm{d}\tau. \tag{7.59}$$

Another phenomenon, typical of viscoelastic materials and closely related to creep, is *stress relaxation*. Staying with our 1D example of a stretched bar let us figure out a kinematical excitation by a displacement defined by the step Heaviside function

$$u(\tau) = u_0 H(\tau).$$

The reaction force measured at the bar end is a non-zero time function given for $\tau \geq 0$ as

$$P(\tau) = u_0\, r(\tau), \tag{7.60}$$

where $r(\tau)$ denotes the *relaxation function*. At $\tau = 0$ this function equals $k_0 = E_0 A/l$ while for $\tau \geq 0$ it is monotonically non-increasing and convergent to $k_\infty = E_\infty A/l$ as $\tau \to \infty$, see Fig. 7.4. The relaxation function describes the transient force necessary to keep the bar stretched with constant unit displacement. The forms of the creep and relaxation functions are related to each other.

Similarly as in the case of creep, an analogous integral formula may be derived for the transient reaction force at a given arbitrary displacement course $u(\tau)$,

$$P(t) = \int_0^t \dot{u}(\tau)\, r(t-\tau)\, \mathrm{d}\tau. \tag{7.61}$$

Let us illustrate the above considerations with three simple examples of one-dimensional viscoelastic material models. They are built up of two types of mechanical elements: an elastic spring and a hydraulic damper (Fig. 7.5). The spring is assumed to obey the linear force–elongation relation,

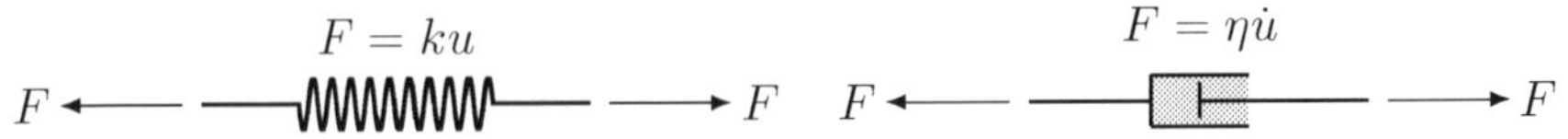

Fig. 7.5 Simple mechanical elements in viscoelastic material models

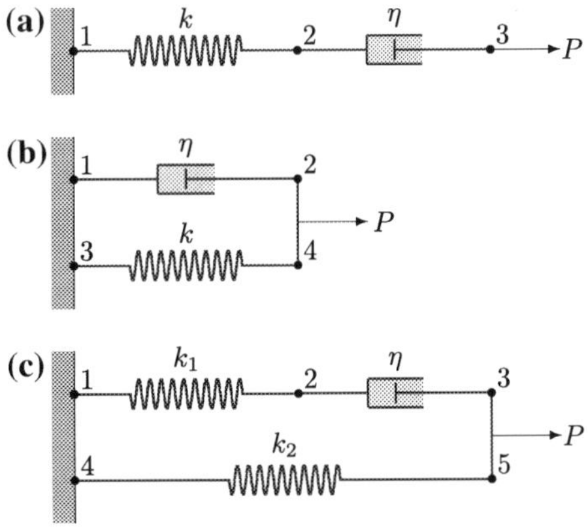

Fig. 7.6 One-dimensional viscoelastic material models: **a** Maxwell model, **b** Voigt model, **c** standard linear model

$$F = ku, \tag{7.62}$$

while the damper the linear relation between force and velocity (elongation rate),

$$F = \eta \dot{u}. \tag{7.63}$$

No physical limitations on the elements' displacements (elongations) are imposed.

The three models are depicted in Fig. 7.6 with appropriate derivations of the creep and relaxation functions. P denotes the load force stretching the system, u is the displacement of the point where the force is applied and the symbols F_{i-j} and u_{i-j} denote stretching forces and elongations in elements located between the nodes i and j.

7.3.1.1 Maxwell Model (Fig. 7.6a)

- Continuity and equilibrium equations:

$$u = u_{1-2} + u_{2-3}, \qquad P = F_{1-2} = F_{2-3}.$$

- Constitutive equations for elements:

$$F_{1-2} = k u_{1-2}, \qquad F_{2-3} = \eta \dot{u}_{2-3}.$$

- Differential equation of motion:

$$\dot{u} = \tfrac{1}{k}\,\dot{P} + \tfrac{1}{\eta}\,P. \tag{7.64}$$

- Creep function—we solve Eq. (7.64) with respect to u under the assumption $P(\tau) = H(\tau)$; the creep function $c(\tau)$ is then equal to the solution $u(\tau)$. For $\tau < 0$ ($P \equiv 0$) the solution is zero, for $\tau > 0$ ($P \equiv 1$) we obtain $c(\tau) = \frac{1}{k} + \frac{\tau}{\eta}$, while at $\tau = 0$ the solution does not exist. Let us write generally

$$c(\tau) = \left(\frac{1}{k} + \frac{1}{\eta}\tau\right) H(\tau). \tag{7.65}$$

- Relaxation function—we solve Eq. (7.64) with respect to P under the assumption $u(\tau) = H(\tau)$; the relaxation function $r(\tau)$ equals then the solution $P(\tau)$:

$$r(\tau) = k\,e^{-\frac{k}{\eta}\tau}\,H(\tau). \tag{7.66}$$

- Limit values of the stiffness coefficient:

$$k_0 = r(0) = k, \qquad k_\infty = r(\infty) = 0.$$

7.3.1.2 Voigt Model (Fig. 7.6b)

- Continuity and equilibrium equations:

$$u = u_{1-2} = u_{3-4}, \qquad P = F_{1-2} + F_{3-4}\,.$$

- Constitutive equations for elements:

$$F_{1-2} = \eta \dot{u}_{1-2}, \qquad F_{3-4} = k u_{3-4}\,.$$

- Differential equation of motion:

$$P = \eta \dot{u} + k u. \tag{7.67}$$

- Creep function—we solve Eq. (7.67) with respect to u under the assumption $P(\tau) = H(\tau)$:

$$c(\tau) = \frac{1}{k}\left(1 - e^{-\frac{k}{\eta}\tau}\right) H(\tau). \tag{7.68}$$

- Relaxation function—we solve Eq. (7.67) with respect to P under the assumption $u(\tau) = H(\tau)$:

$$r(\tau) = \eta\,\delta(\tau) + k\,H(\tau), \tag{7.69}$$

where $\delta(\tau)$ is the Dirac function, being the derivative of the Heaviside function:

$$\delta(\tau) = \dot{H}(\tau) = \begin{cases} 0 & \tau \neq 0, \\ \infty & \tau = 0, \end{cases} \tag{7.70}$$

$$\forall\, f(\tau)\ \forall\, \varepsilon{>}0 \quad \int_{-\varepsilon}^{\varepsilon} f(\tau)\, \delta(\tau)\, \mathrm{d}\tau = f(0).$$

- Limit values of the stiffness coefficient:

$$k_0 = r(0) = \infty, \qquad k_\infty = r(\infty) = k.$$

7.3.1.3 Standard Linear Model (Fig. 7.6c)

- Continuity and equilibrium equations:

$$u = u_{1-2} + u_{2-3} = u_{4-5}, \quad P = F_{1-3} + F_{4-5}, \quad F_{1-3} = F_{1-2} = F_{2-3}\,.$$

- Constitutive equations for elements:

$$F_{1-2} = k_1 u_{1-2}, \qquad F_{2-3} = \eta \dot{u}_{2-3}\,, \qquad F_{4-5} = k_2 u_{4-5}.$$

- Differential equations of motion:

$$P = k_1 u_{1-2} + k_2 u, \qquad k_1 u_{1-2} = \eta(\dot{u} - \dot{u}_{1-2}). \tag{7.71}$$

- Creep function—we solve Eq. (7.71) with respect to u under the assumption $P(\tau) = H(\tau)$:

$$c(\tau) = \frac{1}{k_2}\left(1 - \frac{k}{k_2}\, e^{-\frac{k}{\eta}\tau}\right) H(\tau), \qquad \frac{1}{k} = \frac{1}{k_1} + \frac{1}{k_2}\,. \tag{7.72}$$

- Relaxation function—we solve Eq. (7.71) with respect to P under the assumption $u(\tau) = H(\tau)$:

$$r(\tau) = \left(k_2 + k_1\, e^{-\frac{k_1}{\eta}\tau}\right) H(\tau). \tag{7.73}$$

- Limit values of the stiffness coefficient:

$$k_0 = r(0) = k_1 + k_2, \qquad k_\infty = r(\infty) = k_2.$$

7.3.2 *Continuum Formulation*

Let us generalize the above discussion to the case of 3D continuum. A linear viscoelastic material is defined as a material whose constitutive equation at an arbitrary time instant t can be expressed in the form

$$\sigma_{ij}(\boldsymbol{x}, t) = \int_0^t C_{ijkl}(\boldsymbol{x}, t-\tau) \frac{\partial \varepsilon_{kl}(\boldsymbol{x}, \tau)}{\partial \tau} \, \mathrm{d}\tau, \tag{7.74}$$

where $\tau = 0$ has been assumed as the initial time instant of the deformation process. Skipping for clarity the argument $\boldsymbol{x}$ and denoting by dot the material time derivative, we may write Eq. (7.74) as

$$\sigma_{ij}(t) = \int_0^t C_{ijkl}(t-\tau)\, \dot{\varepsilon}_{kl}(\tau) \, \mathrm{d}\tau, \qquad \boldsymbol{\sigma}(t) = \int_0^t \boldsymbol{C}(t-\tau) \cdot \dot{\boldsymbol{\varepsilon}}(\tau) \, \mathrm{d}\tau. \tag{7.75}$$

In the above relations, $\varepsilon(\tau)$ is a given history of strain for $\tau < t$. The fourth order tensor $\boldsymbol{C}(\tau)$ is a tensor (precisely: tensor function) of relaxation for the material.

The constitutive equation (7.74) may be expressed in the inverse form as

$$\varepsilon_{ij}(t) = \int_0^t D_{ijkl}(t-\tau)\, \dot{\sigma}_{kl}(\tau) \, \mathrm{d}\tau, \qquad \varepsilon(t) = \int_0^t \boldsymbol{D}(t-\tau) \cdot \dot{\boldsymbol{\sigma}}(\tau) \, \mathrm{d}\tau, \tag{7.76}$$

where $\boldsymbol{D}(\tau)$ is the material tensor (function) of creep. Existence of the inverse relation (7.76) can be proved if $\boldsymbol{C}(\tau)$ is twice differentiable with respect to τ and $C_{ijkl}(0) \neq 0$.

Remark. In the above relations the time instant $\tau = 0$ has been assumed as the beginning of the deformation process. Note that—upon this assumption—any other negative value could be taken as the lower integration limit. Thus, in many handbooks, the equations of viscoelasticity are presented in an alternative form, with integration from $-\infty$ to t.

Note that the linear elasticity theory discussed in Sect. 7.2.1 is in fact a limit case of the linear viscoelasticity theory where the relaxation function $C_{ijkl}(\tau)$ is a constant (precisely: constant multiplied by the Heaviside function):

$$\begin{aligned} \sigma_{ij}(t) &= \int_0^t C_{ijkl} H(t-\tau)\, \dot{\varepsilon}_{kl}(\tau) \, \mathrm{d}\tau = C_{ijkl} \int_0^t \dot{\varepsilon}_{kl}(\tau) \, \mathrm{d}\tau \\ &= C_{ijkl}\, \varepsilon_{kl}(t). \end{aligned} \tag{7.77}$$

The integral form (7.75) or (7.76) of the viscoelastic constitutive equation has a serious disadvantage from the point of view of numerical applications. For instance, in order to compute the stress state in a material particle at a given time instant from Eq. (7.75), one has to know the whole history of its strain from the beginning of the deformation process up to this time instant. Making use of a natural in this case discrete time integration procedure in which the time axis is divided into finite time intervals separated by a series of instants $t_0 = 0, t_1, t_2, \ldots,$ computation of the stress σ_{ij} at $\tau = t_{n+1}$ requires the strain values at all the previous instants from t_0 to t_n. It is thus necessary to accumulate in the computer memory a huge amount of data whose size is growing with each time step. To avoid this extremely undesired feature, formulations of constitutive equations are sought in which, to compute stress at $\tau = t_{n+1}$, it is only necessary to know the state of the material particle at the previous time instant, $\tau = t_n$. Such formulations appear to exist upon certain assumptions about the form of the relaxation function $C_{ijkl}(\tau)$.

A simple way of mathematical representation of material relaxation properties is an exponential function that asymptotically decreases from a certain initial value $\boldsymbol{C}^0$ (called instantaneous stiffness tensor) to the value $\boldsymbol{C}^\infty$ (called "equilibrium" stiffness). Such a function can be expressed as

$$\boldsymbol{C}(\tau) = \boldsymbol{C}^\infty + \tilde{\boldsymbol{C}}\, e^{-\frac{\tau}{t_r}} \tag{7.78}$$

or equivalently

$$\boldsymbol{C}(\tau) = \boldsymbol{C}^0 - \tilde{\boldsymbol{C}}\left(1 - e^{-\frac{\tau}{t_r}}\right), \tag{7.79}$$

where $\tilde{\boldsymbol{C}} = \boldsymbol{C}^0 - \boldsymbol{C}^\infty$. Making use of Eq. (7.79) one may rewrite Eq. (7.75) as

$$\begin{aligned} \sigma_{ij}(t) &= \int_0^t \left[C^0_{ijkl} - \tilde{C}_{ijkl}\left(1 - e^{-\frac{t-\tau}{t_r}}\right)\right]\dot{\varepsilon}_{kl}(\tau)\,\mathrm{d}\tau \\ &= C^0_{ijkl}\varepsilon_{kl}(t) - \alpha_{ij}(t), \end{aligned} \tag{7.80}$$

where

$$\alpha_{ij}(t) = \int_0^t \tilde{C}_{ijkl}\left(1 - e^{-\frac{t-\tau}{t_r}}\right)\dot{\varepsilon}_{kl}(\tau)\,\mathrm{d}\tau \tag{7.81}$$

is the difference between the purely elastic instantaneous stress $C^0_{ijkl}\varepsilon_{kl}(t)$ and the actual stress $\sigma_{ij}(t)$.

Let us now recall a known property of an arbitrary function of two arguments $g(t, u)$: if one defines $f(t)$ as

$$f(t) = \int_0^t g(t, u)\,\mathrm{d}u \tag{7.82}$$

then, upon fulfillment of certain smoothness requirements for $f(t)$, the following relation holds true:

$$\dot{f}(t) \equiv \left.\frac{\mathrm{d}f(\tau)}{\mathrm{d}\tau}\right|_{\tau=t} = g(t, t) + \int_0^t \frac{\mathrm{d}g(t, u)}{\mathrm{d}t}\,\mathrm{d}u. \tag{7.83}$$

With this property Eq. (7.81) can be transformed to the form

$$\dot{\alpha}_{ij}(t) = \frac{1}{t_r}\int_0^t \left(\tilde{C}_{ijkl}\, e^{-\frac{t-\tau}{t_r}}\right)\dot{\varepsilon}_{kl}(\tau)\,\mathrm{d}\tau = \frac{1}{t_r}\left[\tilde{C}_{ijkl}\varepsilon_{kl}(t) - \alpha_{ij}(t)\right], \tag{7.84}$$

which will be called the evolution equation for α_{ij}. Note that the above relation no longer includes the time integral and that the current rate $\dot{\alpha}_{ij}(t)$ only depends on the particle state at $\tau = t$ (described by the current values of the tensors ε_{ij} and α_{ij}). Denoting by $\boldsymbol{D}^0$ the inverse tensor to $\boldsymbol{C}^0$ (i.e., cf. Eq. (7.80), $\varepsilon_{ij} = D^0_{ijkl}(\sigma_{kl} + \alpha_{kl})$) one can transform the evolution equation so as to express its right-hand side in terms of stress instead of strain

$$\dot{\alpha}_{ij}(t) = \frac{1}{t_r}\left[\tilde{C}_{ijkl} D^0_{klmn}\left[\sigma_{mn}(t) + \alpha_{mn}(t)\right] - \alpha_{ij}(t)\right]. \tag{7.85}$$

By differentiating Eq. (7.80) in time and substituting the above evolution equation one obtains

$$\dot{\sigma}_{ij}(t) = C^0_{ijkl}\dot{\varepsilon}_{kl}(t) - \frac{1}{t_r}\left[\tilde{C}_{ijkl} D^0_{klmn}\left[\sigma_{mn}(t) + \alpha_{mn}(t)\right] - \alpha_{ij}(t)\right]. \tag{7.86}$$

This is the evolution equation for stress σ_{ij}. Defining a tensor function $\boldsymbol{H}(\boldsymbol{\sigma}, \boldsymbol{\alpha})$ as

$$H_{ij}(\sigma_{mn}, \alpha_{mn}) = -\frac{1}{t_r}\left[\tilde{C}_{ijkl} D^0_{klmn}(\sigma_{mn} + \alpha_{mn}) - \alpha_{ij}\right], \tag{7.87}$$

one can rewrite Eqs. (7.85) and (7.86) in the following concise form:

$$\dot{\boldsymbol{\sigma}}(t) = \boldsymbol{C}^0 \cdot \dot{\boldsymbol{\varepsilon}}(t) + \boldsymbol{H}\left(\boldsymbol{\sigma}(t), \boldsymbol{\alpha}(t)\right), \tag{7.88a}$$

$$\dot{\boldsymbol{\alpha}}(t) = -\boldsymbol{H}\left(\boldsymbol{\sigma}(t), \boldsymbol{\alpha}(t)\right). \tag{7.88b}$$

In summary, to compute the stress rate $\dot{\boldsymbol{\sigma}}$ at a given (current) time instant, it is necessary to know the current strain rate $\dot{\boldsymbol{\varepsilon}}$ and the current values of the fields $\boldsymbol{\sigma}$ and $\boldsymbol{\alpha}$ (as well as, obviously, the constant tensor moduli $\boldsymbol{C}^0$ and $\tilde{\boldsymbol{C}}$). In analogy, the rate $\dot{\boldsymbol{\alpha}}$ is determined from the current values of $\boldsymbol{\sigma}$ and $\boldsymbol{\alpha}$. These two tensor quantities play the role of *state parameters* in this formulation—they describe the current constitutive properties of the material. The constitutive formulation consists thus of the evolution equations for the two state parameters (7.88). These equations, after appropriate time integration performed in parallel, allow to determine the stress history for any given time history of the strain.

Discretization of the time axis leads to the incremental formulation of the constitutive equations in which at a typical time interval (step) $[t, t + \Delta t]$ one obtains the equations

$$\Delta\boldsymbol{\sigma} = \boldsymbol{C}^0 \cdot \Delta\boldsymbol{\varepsilon} + \boldsymbol{H}(\boldsymbol{\sigma}(t), \boldsymbol{\alpha}(t))\,\Delta t, \tag{7.89a}$$

$$\Delta\boldsymbol{\alpha} = -\boldsymbol{H}(\boldsymbol{\sigma}(t), \boldsymbol{\alpha}(t))\,\Delta t, \tag{7.89b}$$

i.e. the system of equations that allows to determine the state parameter increments given their initial values (i.e. referred to the beginning of the step) and the strain increment. The new state parameter values updated by the computed increments may then be used upon a simple recursive rule to compute the state parameter increments at the next time step, using the same Eq. (7.89).

The rate-type constitutive equations (7.88) may be presented in yet another form. Introducing the notations

$$\dot{\varepsilon}^{(c)} = -\boldsymbol{D}^0 \cdot \boldsymbol{H}\,, \qquad \dot{\varepsilon}^{(e)} = \dot{\varepsilon} - \dot{\varepsilon}^{(c)}, \tag{7.90}$$

one may rewrite Eq. (7.88a) as

$$\dot{\boldsymbol{\sigma}} = \boldsymbol{C}^0 \cdot (\dot{\varepsilon} - \dot{\varepsilon}^{(c)}) = \boldsymbol{C}^0 \cdot \dot{\varepsilon}^{(e)}. \tag{7.91}$$

In other words, the total strain rate $\dot{\varepsilon}$ can be expressed as a sum of two terms, $\dot{\varepsilon}^{(e)}$ and $\dot{\varepsilon}^{(c)}$, describing strains related to elastic and inelastic (creep) material behaviour, respectively. This distinction has important consequences for energy-related aspects of the viscoelastic constitutive equations that will be discussed in Sect. 7.3.3. In Eq. (7.91), the inelastic strain $\varepsilon^{(c)}$ may be considered as the state parameter instead of $\boldsymbol{\alpha}$—note that $\dot{\boldsymbol{\alpha}} = \boldsymbol{C}^0 \cdot \dot{\varepsilon}^{(c)}$ (cf. Eq. (7.90) and (7.88b)) and $\boldsymbol{C}^0$ is a known elastic stiffness tensor.

An example of a frequently utilized viscoelastic constitutive formulation is the so-called stationary creep model. In this formulation it is assumed that

$$\dot{\varepsilon}^{(c)}_{ij} = \frac{3}{2}\dot{\bar{\varepsilon}}^{(c)} \frac{\sigma^{\mathrm{D}}_{ij}}{\bar{\sigma}} \tag{7.92}$$

where σ^{D}_{ij} denotes the stress deviator and

$$\dot{\bar{\varepsilon}}^{(c)} = A\bar{\sigma}^n, \qquad \bar{\sigma} = \sqrt{\frac{3}{2}\sigma^{\mathrm{D}}_{ij}\sigma^{\mathrm{D}}_{ij}}. \tag{7.93}$$

A, n are material constants. Substituting Eq. (7.93) into Eq. (7.92) one can easily see that this model is a special case of the model defined by Eqs. (7.90)–(7.91).

7.3.3 Energy Dissipation in Viscoelastic Materials

Consider once again the problem of a stretched bar like that in Fig. 2.1, subject to cyclic load with the given stretching force history $P(\tau)$ depicted in Fig. 7.7a.

Let us first assume that the bar's material is linearly elastic with the Young modulus E (the bar stiffness coefficient is thus $k = EA/l$). The equations of linear elasticity imply that the bar's elongation (displacement of its end) is expressed by the same function of time, with the maximum value in each cycle $u_{\max} = P_{\max}/k$.

Let us look at the relation between force and elongation in the bar during its deformation process (Fig. 7.7b). Its graph is a straight segment whose points correspond to states of the bar's material particles cyclically repeated in the process. During each loading phase we move from left to right along this segment while during unloading we go back from right to left.

Recalling Eq. (5.41) defining the work performed by the force P on the displacement u, we observe that the work ΔW performed during the loading phase, say, between the displacement states $u = u_1$ and $u = u_2$, equals the area of the figure between the line $P(u)$ and the axis u within the interval $[u_1, u_2]$ (Fig. 7.7b):

$$\Delta W = \int_{u_1}^{u_2} P(u)\,\mathrm{d}u.$$

During the unloading phase, the work is the same but negative (the integration limits are then switched in the above formula). Since the material is elastic, the entire energy supplied to the bar by performing this work is accumulated in the material as the elastic strain energy. In the loading phase this energy increases while in the unloading phase it decreases (the material then returns, or releases, the accumulated energy by performing negative work). Note that the amount of energy returned is exactly equal to the energy accumulated during loading—the total work performed in each cycle is thus zero, as the total potential energy at the cycle's end. The total potential energy W at a given elongation u of the linear elastic bar is $W = \frac{1}{2}Pu = \frac{1}{2}ku^2$.

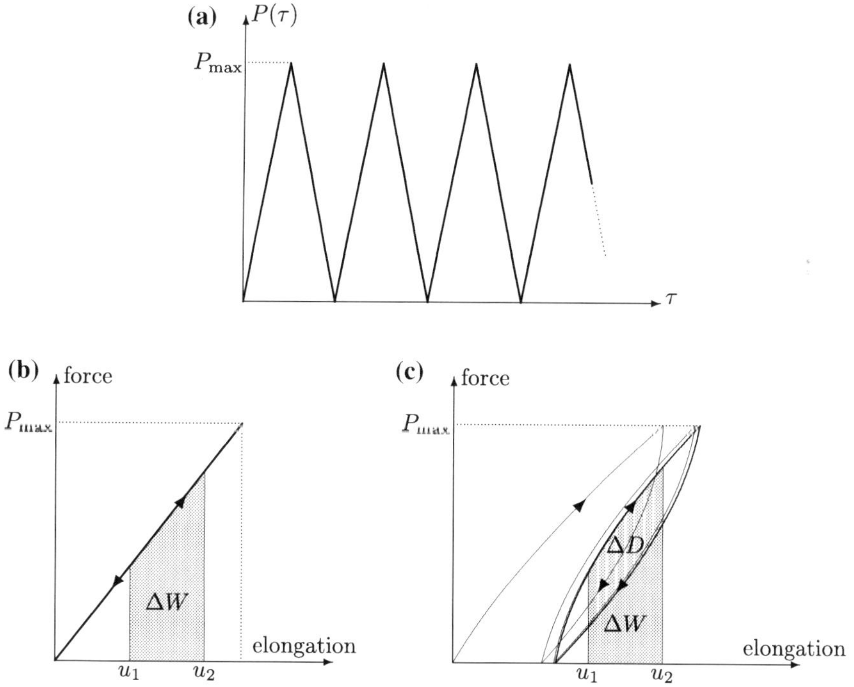

Fig. 7.7 Cyclic load of a linear elastic and a linear viscoelastic bar—the load history and the work performed

If the bar is made of a viscoelastic material, the relation between the load force and the bar's elongation is more complex, see Fig. 7.7c. It has a nonlinear form, dependent on the load rate. Besides, its time history is different for the loading and the unloading phases and after a number of initial cycles it forms a stabilized loop, called the hysteresis loop, along which the solution $P(u)$ moves in the subsequent time instants. The work ΔW performed by the force P on the displacement u in the given segment $[u_1, u_2]$ is, similarly as in the elastic case, equal to the area of the figure between the line $P(u)$ and the axis u in the interval $[u_1, u_2]$. The essential difference consists, however, in the fact that this area is different at loading and unloading, which means that during the unloading phase the material returns less energy than that supplied by the work performed at loading. The difference between the two amounts of energy, ΔD, equal to the area between the loading and unloading paths of the hysteresis loop within the segment $[u_1, u_2]$ (shaded in Fig. 7.7c), is the so-called dissipated energy, forfeited in each cycle due to inelastic phenomena occurring in the material. In our simplified models shown in Fig. 7.6 dissipation occurs in particular in hydraulic dampers.

Summarizing, in each load cycle the total work performed by the force P during loading is a sum of the elastic energy W accumulated in the material and the dissipated energy D, equal to the area of the hysteresis loop in Fig. 7.7c. In the unloading phase the negative work of the force P (the returned energy) equals only $-W$. Cyclic deformation of a viscoelastic bar requires thus supply of mechanical energy in each load cycle.

In the 3D case, the work of stresses on strains generates according to Eq. (5.51) increase of the energy density U in material volume unit. Generalization of the above observations leads to the conclusion that the work $\dot{U} = \sigma_{ij}\dot{\varepsilon}_{ij}$ in a viscoelastic material is also only partially accumulated as an elastic potential energy while the other part is dissipated due to inelastic phenomena. In practice, most of this energy is converted to heat and irreversibly released to the surroundings due to the temperature difference. Models and equations describing such phenomena will be discussed in more detail in Chap. 9.

7.4 Elastoplastic Materials

Several real materials—including metals and rocks—exhibit within the reasonable range of strains properties that do not fit the above discussed elastic or viscoelastic idealization. More strictly, they behave in an elastic way within only a limited range of strain values. Beyond this range, however, permanent and irreversible strains are induced in the material. The strains, called *plastic strains*, are not time-dependent (hence they do not have viscous character). In metals, an approximate strain limit (in uniaxial stretch test) beyond which plastic strains occur is about $\varepsilon = 0.002$.

7.4.1 One-Dimensional Models

Basic aspects of the stress–strain relations in the case of such inelastic and non-viscous processes can be best explained on an example of a one-dimensional bar stretched with a given time-dependent force $P(\tau)$ (see Fig. 2.1). Assuming small deformations, let us define the linearized nominal strain as

$$\varepsilon = \frac{\Delta l}{l} \tag{7.94}$$

and the nominal stress as

$$\sigma = \frac{P}{A}, \tag{7.95}$$

where again l and A denote the initial length and cross section area of the bar, respectively, while Δl—its elongation.

The bar's material behaviour may be schematically explained taking reference to the graphs in Fig. 7.8. On the left, characteristic time histories of the load are shown, expressed as the stress history $\sigma(\tau)$. On the right, the corresponding stress–strain (σ–ε) relations are depicted. It is clear from Fig. 7.8a that the relation between stress and strain is linear (linearly elastic) up to a certain limit stress value σ_y called the *yield stress*. This limit value initially equals σ_{y0} but—as we shall discuss soon—it may be subject to changes as the inelastic deformations develop in the material. At the stage corresponding to the OA segment in the σ–ε graph the material behaviour obeys the Hooke law (2.4). During further deformation (the AB segment), plastic flow is observed, characterized by a fast increase of strain at even a relatively low increase of stress above the initial yield stress value.

If at a certain moment (point B in Fig. 7.8a) the material gets unloaded by decreasing the stretching force value P, the material behaviour at this stage of the deformation process becomes again linearly elastic—note that the BC segment of the σ–ε graph is parallel to the OA segment corresponding to the initial elastic loading. If the load force vanishes, a permanent strain remains in the material. It is called plastic strain and denoted here by $\varepsilon^{(p)}$. It can thus be concluded that the axial strain in the material can be at each moment expressed as a sum of the elastic and the plastic strain,

$$\varepsilon = \varepsilon^{(e)} + \varepsilon^{(p)}, \tag{7.96}$$

of which the first is determined from the elasticity equation (2.4),

$$\sigma = E\varepsilon^{(e)}, \tag{7.97}$$

while the other is subject to a certain evolution law discussed further in this section. At the current stage of our considerations, the following rate form of the constitutive relation can be written:

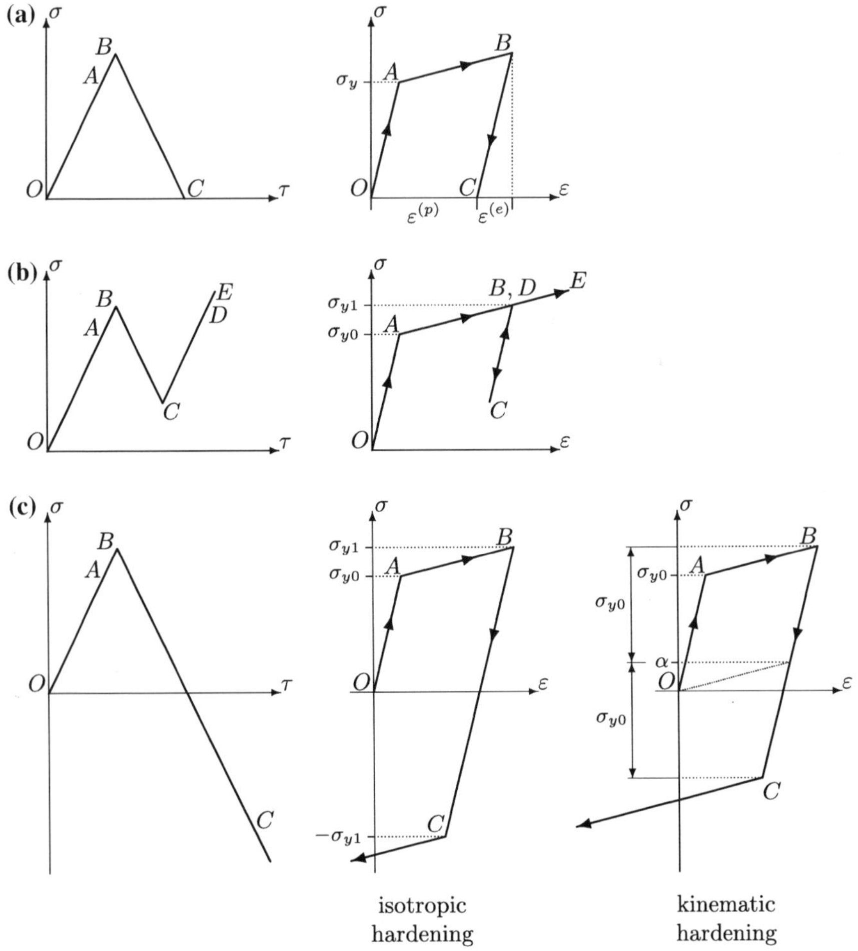

Fig. 7.8 Schematic histories of the σ–ε relations for various load paths for an elastoplastic material

$$\dot{\sigma} = \begin{cases} E\dot{\varepsilon} & \text{in the elastic process,} \\ E_T\dot{\varepsilon} & \text{in the inelastic process,} \end{cases} \tag{7.98}$$

where E_T denotes the tangent elastoplastic stiffness modulus.

Replacing in the σ–ε graph the stress by the force P and the strain by the elongation Δl, i.e. multiplying the variables on the graph axes by A and l, respectively, one obtains a graph of an identical shape in which (quite similarly to the viscoelastic case) the area of the tetragon $OABC$ equals the energy D dissipated during the inelastic deformation of the bar. The elastic energy U, accumulated in a volume unit of the material during loading and released at unloading, is equal to the area of the right-angled triangle between the BC segment and the graph abscissa.

If at the stage of elastic unloading the material gets again loaded (Fig. 7.8b), its behaviour is similar as it was during the initial loading—first elastic (the CD segment) and then elastoplastic (the DE segment). However, the yield stress σ_y is now higher than before and its current value σ_{y1} equals the stress assumed at the end of the previous elastoplastic loading stage. The σ–ε graph implies that the current value of the yield stress grows along with development of the plastic strain. This material feature is called plastic hardening. In the simplest, linear case hardening may be described by the following equation:

$$\sigma_y = \sigma_y(\varepsilon^{(p)}) = \sigma_{y0} + \zeta\varepsilon^{(p)}, \tag{7.99}$$

where the coefficient ζ (expressed in the units of stress) is called the *hardening modulus* and is related to the tangent modulus E_T as

$$\zeta = \frac{EE_T}{E + E_T}, \qquad E_T = \frac{E\zeta}{E - \zeta}. \tag{7.100}$$

It is noteworthy that stress in an elastoplastic material never exceeds the current yield limit σ_y. If it is lower than the limit ($\sigma < \sigma_y$), then the material is in the elastic regime of its behaviour. This is also the case when $\sigma = \sigma_y$ and $\dot{\sigma} \leq 0$. Otherwise, when $\sigma = \sigma_y$ and the load is increasing, the plastic strain is developing and the yield stress value is also increasing so that at each moment the equality $\sigma = \sigma_y$ holds true. One can conclude that at the plastic stage of deformation the stress rate can be expressed as, cf. Eqs. (7.98) and (7.99):

$$\dot{\sigma} = \dot{\sigma}_y = E_T\dot{\varepsilon} = \zeta\dot{\varepsilon}^{(p)}. \tag{7.101}$$

It turns out, however, that Eq. (7.99) describing the behaviour observed in Fig. 7.8a, b is not correct in the general case. To see this let us consider further unloading of the bar so that stress assumes negative values (compression), see Fig. 7.8c. One can expect that at this stage there should also be a limit compression stress value beyond which the inelastic phenomena occur. This is indeed true, although in such a case different materials may exhibit different detailed properties. Two most characteristic types of material behaviour at cyclic stretch–compression plastic deformation processes are the so-called isotropic and kinematic hardening models (Fig. 7.8c).

In the isotropic hardening model the current yield stress value is the same at stretch and compression, i.e. the plastic flows occurs if

$$|\sigma| = \sigma_y(\varepsilon^{(p)}). \tag{7.102}$$

Moreover, during flow at compression the yield stress is still increasing even if the plastic strain decreases. It becomes thus clear that, to be valid for both stretch and compression, Eq. (7.99) requires modification—at an arbitrary time instant $\tau = t$ it should rather read

$$\sigma_y(t) = \sigma_{y0} + \zeta \bar{\varepsilon}^{(p)}(t), \qquad \bar{\varepsilon}^{(p)}(t) = \int_0^t \left| \dot{\varepsilon}^{(p)} \right| \mathrm{d}\tau, \tag{7.103}$$

where $\bar{\varepsilon}^{(p)}$ denotes the effective (equivalent) plastic strain. The latter quantity is monotonically increasing in time during the plastic process, no matter whether the current value of plastic strain $\varepsilon^{(p)}$ is increasing or decreasing at the moment. One can thus conclude that the value of σ_y depends on the plastic strain history rather than on its current value.

In the kinematic hardening model, the current yield stress value is defined as

$$\sigma_y(\varepsilon^{(p)}) = \begin{cases} \sigma_{y0} + \alpha(\varepsilon^{(p)}) & \text{in stretching } (\dot{\sigma} > 0), \\ \sigma_{y0} - \alpha(\varepsilon^{(p)}) & \text{in compression } (\dot{\sigma} < 0), \end{cases} \tag{7.104}$$
$$\alpha(\varepsilon^{(p)}) = \zeta \varepsilon^{(p)},$$

where α is a quantity (in the units of stress) called the *back stress*, defined as the midway point between the current yield stress values at stretch and compression (the dotted line in Fig. 7.8c). The plastic flow condition can be more conveniently expressed in this model as

$$\left| \sigma - \alpha(\varepsilon^{(p)}) \right| = \sigma_y, \qquad \sigma_y \equiv \sigma_{y0}, \qquad \alpha(\varepsilon^{(p)}) = \zeta \varepsilon^{(p)}, \tag{7.105}$$

which means that the yield stress remains constant in the entire process, however, the transition from an elastic to elastoplastic deformation phase is not driven by the current stress level $|\sigma|$ but by the current difference between the stress and the back stress $|\sigma - \alpha|$.

The qualitative difference between the isotropic and kinematic hardening models can be explained on an example of cyclic stretch/compression of a material particle between the limit values $+\sigma_{\max}$ and $-\sigma_{\max}$ ($\sigma_{\max} > \sigma_{y0}$). In the case of isotropic hardening, according to the rules shown in Fig. 7.8, plastic flow only occurs in the first cycle's loading phase while later (during unloading and in the subsequent loading/unloading cycles) the σ–ε relation has purely elastic character, linearly oscillating around the point corresponding to $\sigma = 0$ and $\varepsilon = \varepsilon^{(p)}$ in the graph. In the case of kinematic hardening plastic flows occurs in each cycle, both at loading and unloading, and a stationary hysteresis loop appears in the σ–ε graph, whose area—similarly as in viscoelastic materials—corresponds to the density of energy dissipated in the material particle due to the plastic deformation. This property of cyclically loaded materials exhibiting the kinematic hardening property is called Bauschinger effect (Fig. 7.9).

It is noteworthy that hysteresis and cyclic energy dissipation may also appear in material with the isotropic hardening model. This happens if the value of $\sigma_{\max}$ is increasing in subsequent load cycles.

Note that in both the hardening models the plastic flow is activated if

$$\bar{\sigma} = \sigma_y \quad \text{and} \quad \dot{\bar{\sigma}} > 0, \tag{7.106}$$

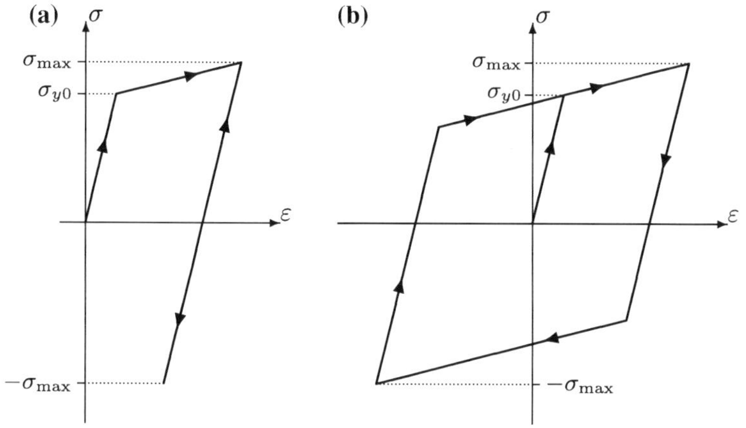

Fig. 7.9 σ–ε curves for cyclic symmetric load of an elastoplastic material with **a** isotropic and **b** kinematic hardening

where $\bar{\sigma}$ denotes the so-called effective (equivalent) stress defined as $\bar{\sigma} = |\sigma|$ for isotropic hardening and $\bar{\sigma} = |\sigma - \alpha|$ for kinematic hardening. In the former case, the yield stress is a given function of $\bar{\varepsilon}^{(p)}$ while in the latter it is a constant but the back stress α is a given function of $\bar{\varepsilon}^{(p)}$.

The model of material discussed above behaviour is obviously fairly simplified; the stress–strain graphs for real materials are usually different than those shown in Fig. 7.8. In particular, the hardening functions $\sigma_y(\bar{\varepsilon}^{(p)})$ or $\alpha(\varepsilon^{(p)})$ are frequently nonlinear (although they can be usually considered linear in the small strain range). In many materials the yield stress cannot be determined in a strict manner and the corner point denoted as A on the σ–ε curves in Fig. 7.8 is rather a smooth arch-type transition phase between elastic and inelastic ranges of material behaviour. The notions of isotropic and kinematic hardening should also be treated as certain extreme cases—in real materials both the hardening types may coexist in various proportions. The above discussion underlines, though, the most characteristic features of plastic deformation processes typical of many materials (including metals) and explains fundamental concepts of the elastoplasticity theory.

7.4.2 Three-Dimensional Formulation in Plastic Flow Theory

Let us now pass to the mathematical description of elastoplasticity in 3D material continuum. Let us start with the so-called plastic flow theory for small (linearized) strains. This description requires consistent definition of the following three notions: the initial yield stress, the hardening law and the plastic flow rule. These are necessary to further evaluate the plastic strain rate tensor $\dot{\varepsilon}_{ij}^{(p)}$ appearing in the postulated additive relation

$$\dot{\varepsilon}_{ij} = \dot{\varepsilon}^{(e)}_{ij} + \dot{\varepsilon}^{(p)}_{ij} , \tag{7.107}$$

where $\dot{\varepsilon}^{(e)}_{ij}$ is the elastic strain rate tensor related to stress σ_{ij} by the generalized Hooke law (7.3):

$$\sigma_{ij} = C_{ijkl}\varepsilon^{(e)}_{kl} , \qquad \dot{\sigma}_{ij} = C_{ijkl}\dot{\varepsilon}^{(e)}_{kl} . \tag{7.108}$$

Let us first limit ourselves to the simplest model of isotropic plastic hardening. The yield condition limiting the purely elastic range of material behaviour assumes then the form of a scalar inequality

$$f(\sigma_{ij}, \bar{\varepsilon}^{(p)}) \le 0, \tag{7.109}$$

which should be interpreted as follows: if $f < 0$ then only elastic deformation occurs, i.e. $\dot{\varepsilon}^{(p)} = 0$ (which does not mean that $\varepsilon^{(p)} = 0$; this state corresponds to e.g. the segments OA and BC on the one-dimensional graphs in Fig. 7.8), while $f = 0$ means that plastic flow may be activated and $\dot{\varepsilon}^{(p)} \neq 0$. By "may be" we mean the obvious (in view of our 1D considerations) constatation that plastic flow occurs at loading only—this notion is going to be formally defined in a while. Equation (7.109) may thus be understood as definition of a multidimensional hypersolid in the space of σ_{ij} components, whose boundary (the hypersurface $f = 0$, called the yield surface) consists of all the stress states at which the material is subject to plastic deformations while the hypersolid's interior contains the elastic stress states.

The yield surface changes during the deformation process, as it is parameterized by the scalar quantity $\bar{\varepsilon}^{(p)}$, called effective plastic strain, defined as

$$\bar{\varepsilon}^{(p)}(t) = \int_0^t \dot{\bar{\varepsilon}}^{(p)} \mathrm{d}\tau, \qquad \dot{\bar{\varepsilon}}^{(p)} = \sqrt{\frac{2}{3}\dot{\varepsilon}^{(p)}_{ij}\dot{\varepsilon}^{(p)}_{ij}} . \tag{7.110}$$

and is monotonically increasing as the plastic deformation proceeds.

The function f in Eq. (7.109) will be assumed in a general form

$$f(\sigma_{ij}, \bar{\varepsilon}^{(p)}) = F(\sigma_{ij}) - \sigma_y(\bar{\varepsilon}^{(p)}), \tag{7.111}$$

where $\sigma_y(\cdot)$ is the hardening function describing the current yield stress value. Assuming isotropic hardening, σ_y is only a function of $\bar{\varepsilon}^{(p)}$, frequently assumed in the linear form, $\sigma_y = \sigma_{y0} + \zeta\bar{\varepsilon}^{(p)}$, although this is not necessary for subsequent derivations. Let us define the current hardening modulus ζ as the derivative

$$\zeta = \frac{\mathrm{d}\sigma_y}{\mathrm{d}\bar{\varepsilon}^{(p)}} . \tag{7.112}$$

The function F directly depends on the stress deviator, i.e. the tensor

$$\sigma^{\mathrm{D}}_{ij} = \sigma_{ij} - \frac{1}{3}\delta_{ij}\,\sigma_{kk} , \tag{7.113}$$

cf. Eq. (3.68). For metals it is most frequently defined in the form

$$F(\sigma_{ij}) = \bar{\sigma} = \sqrt{\frac{3}{2}\sigma^{\mathrm{D}}_{ij}\sigma^{\mathrm{D}}_{ij}}\,, \tag{7.114}$$

where $\bar{\sigma}$ denotes the so-called effective (or equivalent) stress, often referred to as the Huber–Mises stress.

To complete the system of equations describing stress and strain rates one needs to describe the evolution of plastic strain. For most metals it appears justified to assume the so-called associated (with the yield surface (7.109)) flow rule, i.e.

$$\dot{\varepsilon}^{(p)}_{ij} = \lambda n_{ij}\,, \tag{7.115}$$

where n_{ij} represents the normal versor to the yield surface $f = 0$ in the stress component space,

$$n_{ij} = \frac{\dfrac{\partial f}{\partial \sigma_{ij}}}{\sqrt{\dfrac{\partial f}{\partial \sigma_{kl}}\dfrac{\partial f}{\partial \sigma_{kl}}}} \tag{7.116}$$

and λ is a positive scalar coefficient whose value will be derived below. Substituting Eqs. (7.109), (7.113) and (7.114) into Eq. (7.116) we obtain after transformations

$$n_{ij} = \frac{\dfrac{\partial \bar{\sigma}}{\partial \sigma_{ij}}}{\sqrt{\dfrac{\partial \bar{\sigma}}{\partial \sigma_{kl}}\dfrac{\partial \bar{\sigma}}{\partial \sigma_{kl}}}} = \frac{\sigma^{\mathrm{D}}_{ij}}{\sqrt{\sigma^{\mathrm{D}}_{kl}\sigma^{\mathrm{D}}_{kl}}} = \sqrt{\frac{3}{2}}\frac{\sigma^{\mathrm{D}}_{ij}}{\bar{\sigma}}\,. \tag{7.117}$$

The formulae (7.115) and (7.117) imply an important feature of plastic strain: it is deviatoric, i.e. $\dot{\varepsilon}^{(p)}_{ii} = 0$ (which after time integration is equivalent to $\varepsilon^{(p)}_{ii} = 0$). Thus, in view of Eq. (4.75), it does not change the material particle's volume. In other words, plastic deformation is incompressible. This obviously does not mean that the material is incompressible. However, the volume changes are related to elastic strains $\varepsilon^{(e)}_{ij}$ only.

During plastic deformation the stress σ_{ij} must remain on the yield surface. This means that any stress increment directed outwards of the surface must be associated with an appropriate motion of this surface so that the consistency condition (of the stress and the yield surface)

$$\dot{f} = 0 \tag{7.118}$$

is fulfilled. Upon assumption (7.111) this condition can be rewritten as

$$\frac{\partial F}{\partial \sigma_{ij}}\dot{\sigma}_{ij} - \frac{\partial \sigma_y}{\partial \bar{\varepsilon}^{(p)}}\dot{\bar{\varepsilon}}^{(p)} = 0, \tag{7.119}$$

i.e.

$$\sqrt{\frac{3}{2}}\,\dot{\sigma}_{ij}\,n_{ij} - \zeta\dot{\bar{\varepsilon}}^{(p)} = 0. \tag{7.120}$$

Since, in view of Eqs. (7.110), (7.115) and the obvious property $n_{ij}n_{ij} \equiv 1$ we have

$$\dot{\bar{\varepsilon}}^{(p)} = \sqrt{\frac{2}{3}\,\lambda^2 n_{ij}n_{ij}} = \sqrt{\frac{2}{3}}\,\lambda, \tag{7.121}$$

the following consistent value of λ can be derived from Eq. (7.120):

$$\lambda = \frac{3}{2\zeta}\,\dot{\sigma}_{ij}\,n_{ij}\,. \tag{7.122}$$

Denoting

$$h = \frac{2}{3}\,\zeta, \tag{7.123}$$

one may finally write down the plastic flow rule (7.115) as

$$\dot{\varepsilon}_{ij}^{(p)} = \frac{1}{h}(\dot{\sigma}_{kl}\,n_{kl})\,n_{ij}\,. \tag{7.124}$$

Remark. The space of symmetric tensor components σ_{ij} is 6-dimensional. It is thus hardly imaginable to geometrically interpret the above derived formulae in a graphical manner. However, knowing the principal directions of the stress tensor (i.e. the directions in which the tangent components vanish, $\sigma_{12} = \sigma_{23} = \sigma_{13} = 0$), we may project the three-dimensional objects considered above onto the space of principal stresses σ_{I}, σ_{II}, σ_{III}. Given the form of the function $F(\sigma_{ij})$ in which the deviatoric stress components are expressed as

$$\sigma_{\mathrm{I}}^{\mathrm{D}} = \frac{2\sigma_{\mathrm{I}} - \sigma_{\mathrm{II}} - \sigma_{\mathrm{III}}}{3}, \qquad \sigma_{\mathrm{II}}^{\mathrm{D}} = \frac{2\sigma_{\mathrm{II}} - \sigma_{\mathrm{III}} - \sigma_{\mathrm{I}}}{3}, \qquad \sigma_{\mathrm{III}}^{\mathrm{D}} = \frac{2\sigma_{\mathrm{III}} - \sigma_{\mathrm{I}} - \sigma_{\mathrm{II}}}{3},$$

we may conclude that the yield surface $f = 0$ has the form of an infinite cylinder of revolution (see Fig. 7.10), whose radius is σ_y and the symmetry axis is defined as $\sigma_{\mathrm{I}} = \sigma_{\mathrm{II}} = \sigma_{\mathrm{III}}$ (i.e. its points correspond to spherical stress states in material). As long as the stress state remains inside of the cylinder, the material only deforms in an elastic way. Approaching the cylinder's surface from inside activates plastic flow and a non-zero plastic strain rate appears. During this plastic deformation stage the stress state is moving over the yield surface which is gradually "swelling" (i.e. its radius σ_y is increasing along with the increase of $\bar{\varepsilon}^{(p)}$). At each time instant of the plastic deformation the plastic strain rate tensor is orthogonal to the cylinder's surface at the point corresponding to the current stress state.

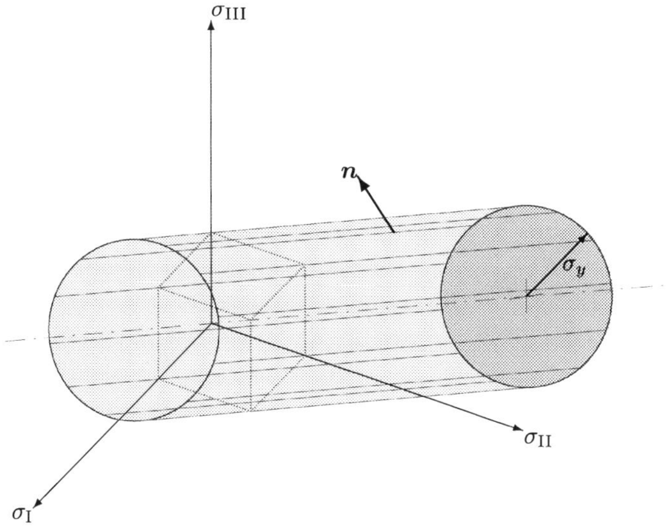

Fig. 7.10 The Huber–Mises yield surface in the space of principal stresses

Remark. As it can be seen from the equations derived above, a crucial from the point of view of plastic phenomena measure of stress is the equivalent Huber–Mises stress $\bar{\sigma}$. It is this quantity that—after exceeding of a certain limit value—drives the plastic strains in material. An interesting question is what does the equivalent stress actually measure. It is convenient to use for this purpose the deviatoric stress components expressed in principal stresses, shown in the previous remark box. Substituting them into Eq. $(7.114)_2$ we can obtain after transformations

$$\bar{\sigma} = \sqrt{\frac{1}{2}\left[(\sigma_I - \sigma_{II})^2 + (\sigma_{II} - \sigma_{III})^2 + (\sigma_{III} - \sigma_I)^2\right]}.$$

Comparing this result with Eq. (5.13), we can alternatively write

$$\bar{\sigma} = \sqrt{2\left[(\tau^{(1)})^2 + (\tau^{(2)})^2 + (\tau^{(3)})^2\right]},$$

where $\tau^{(1)}$, $\tau^{(2)}$ and $\tau^{(3)}$ denote extreme values of tangential stress in the material, appearing—as has been shown in Sect. 5.1.2—in the planes forming the angles of $\pm 45°$ with pairs of principal stress directions.

This result allows to conclude that the Huber–Mises stress $\bar{\sigma}$ is a measure of tangential stress level in a material particle. It is this class of stresses that drives those processes in the material microstructure that are manifested on the macro scale as the plastic flow.

We have so far determined the relation (7.124) between the stress and plastic strain rates. Combining it with Eq. (7.108) that determines the relation between the rates of stress and elastic strain and using the postulated additivity of the two strain types (7.107), we can come after several transformations at the following equation:

$$\dot{\sigma}_{ij} = \left(C_{ijkl} - \frac{C_{ijmn} n_{mn} n_{pq} C_{pqkl}}{h + n_{rs} C_{rstu} n_{tu}} \right) \dot{\varepsilon}_{kl} = C^{(ep)}_{ijkl} \dot{\varepsilon}_{kl} \,, \tag{7.125}$$

in which the tensorial modulus

$$C^{(ep)}_{ijkl} = C_{ijkl} - \frac{C_{ijmn} n_{mn} n_{pq} C_{pqkl}}{h + n_{rs} C_{rstu} n_{tu}} \tag{7.126}$$

is called the elastoplastic constitutive stiffness tensor. Note the analogy between Eq. (7.125) and the generalized Hooke law (7.3) expressed in terms of rates:

$$\dot{\sigma}_{ij} = C_{ijkl} \dot{\varepsilon}^{(e)}_{kl} \,. \tag{7.127}$$

Equation (7.125) is thus the rate-type constitutive equation of an elastoplastic material. Since

$$n_{ij} = n_{ij}(\sigma_{kl}), \qquad h = h(\bar{\varepsilon}^{(p)}), \tag{7.128}$$

the tangent constitutive tensor depends on the current stress state and the history of plastic strains,

$$C^{(ep)}_{ijkl} = C^{(ep)}_{ijkl} \left(\sigma_{mn}, \bar{\varepsilon}^{(p)} \right) . \tag{7.129}$$

The quantities σ_{ij} and $\bar{\varepsilon}^{(p)}$ become thus the *state parameters* of the elastoplastic constitutive equations. Evolution of the parameter $\bar{\varepsilon}^{(p)}$ may be also expressed as a linear function of the strain rate $\dot{\varepsilon}_{kl}$ as

$$\dot{\bar{\varepsilon}}^{(p)} = \sqrt{\frac{2}{3}} \lambda = \sqrt{\frac{2}{3}} \frac{1}{h} \dot{\sigma}_{ij} n_{ij} = \sqrt{\frac{2}{3}} \frac{1}{h} C^{(ep)}_{ijkl} n_{ij} \dot{\varepsilon}_{kl} = A_{kl} \left(\sigma_{mn}, \bar{\varepsilon}^{(p)} \right) \dot{\varepsilon}_{kl} \,. \tag{7.130}$$

Since the coefficient λ (cf. Eq. (7.122)) must be positive, plastic flow occurs if $\dot{\sigma}_{ij} n_{ij} > 0$, i.e. if the stress rate is directed outwards the yield surface. Only in such cases Eq. (7.125) is valid; otherwise the material obeys the elastic equation (7.127). The deformation process is called elastoplastic if

$$f(\sigma_{ij}, \bar{\varepsilon}^{(p)}) = 0 \quad \text{and} \quad n_{ij} \dot{\sigma}_{ij} > 0. \tag{7.131a}$$

Otherwise, i.e. if

$$f(\sigma_{ij}, \bar{\varepsilon}^{(p)}) < 0 \quad \text{or} \quad f(\sigma_{ij}, \bar{\sigma}^{(p)}) = 0 \;\; \text{and} \;\; n_{ij} \dot{\sigma}_{ij} \leq 0, \tag{7.131b}$$

the process is elastic (the limit case $n_{ij}\dot{\sigma}_{ij} = 0$ is sometimes referred to as neutral process), i.e. no new plastic strains are generated. The relations (7.131) are called the loading/unloading conditions. Together with Eqs. (7.127) and (7.125), they form the complete set of relations describing constitutive properties of the elastoplastic material.

We observe that a problem may appear in practical applications of the above constitutive formulation. Assume that the strain rate $\dot{\varepsilon}_{ij}$ is known and the corresponding stress rate $\dot{\sigma}_{ij}$ is sought. To find it, one of Eqs. (7.127) or (7.125) must be used, depending on the deformation process type. However, as long as $\dot{\sigma}_{ij}$ is not known, one cannot tell whether the process is elastic or elastoplastic as the conditions (7.131) require this rate as the testing quantity. For practical reasons it is more convenient to transform Eq. (7.131) to another form. To this aim, let us define the so-called trial elastic stress rate

$$\dot{\sigma}_{ij}^{(tr)} = C_{ijkl}\dot{\varepsilon}_{kl} , \tag{7.132}$$

which can be evaluated without knowing $\dot{\sigma}_{ij}$. No physical sense is assigned to this quantity. Also note that

$$C_{ijkl}\dot{\varepsilon}_{kl} = C_{ijkl}\dot{\varepsilon}_{kl}^{(e)} + C_{ijkl}\dot{\varepsilon}_{kl}^{(p)} = \dot{\sigma}_{ij} + \frac{1}{h} C_{ijkl}n_{kl}n_{mn}\dot{\sigma}_{mn} . \tag{7.133}$$

Multiplying side-wise by n_{ij} we obtain

$$n_{ij}\dot{\sigma}_{ij}^{(tr)} = n_{ij}\dot{\sigma}_{ij} + \frac{1}{h} n_{ij}C_{ijkl}n_{kl}n_{mn}\dot{\sigma}_{mn} , \tag{7.134}$$

which can be easily transformed to the form

$$n_{ij}\dot{\sigma}_{ij} = \frac{h}{h + n_{kl}C_{klmn}n_{mn}} n_{ij}\dot{\sigma}_{ij}^{(tr)} . \tag{7.135}$$

Equation (7.135) implies that the quantity $n_{ij}\dot{\sigma}_{ij}^{(tr)}$ is proportional to $n_{ij}\dot{\sigma}_{ij}$ with a positive coefficient. This allows to write the loading/unloading conditions as

$$f < 0 \quad \text{or} \quad f = 0 \ \text{and} \ n_{ij}\dot{\sigma}_{ij}^{(tr)} \le 0 , \quad \text{elastic process,} \tag{7.136a}$$

$$f = 0 \ \text{and} \ n_{ij}\dot{\sigma}_{ij}^{(tr)} > 0 , \quad \text{inelastic process.} \tag{7.136b}$$

The flow rule may be now rewritten in the form

$$\dot{\varepsilon}_{ij}^{(p)} = \frac{n_{pq}\dot{\sigma}_{pq}^{(tr)}n_{ij}}{h + n_{kl}C_{klmn}n_{mn}} . \tag{7.137}$$

The above relations assume a much simpler form in the case of isotropic materials, i.e. if the elastic stiffness tensor is given by Eq. (7.12). Keeping in mind symmetry and deviatoric character of the tensor n_{ij} (i.e. $n_{ij} = n_{ji}$, $n_{kk} = 0$) we can write

$$C_{ijkl}n_{kl} = 2\mu n_{ij}, \qquad C_{ijkl}n_{ij}n_{kl} = 2\mu, \qquad C^{(ep)}_{ijkl} = C_{ijkl} - \frac{4\mu^2}{2\mu + h} n_{ij}n_{kl}$$

and

$$\dot{\varepsilon}^{(p)}_{ij} = \frac{h}{2\mu + h}\left(\dot{\sigma}^{(tr)}_{kl} n_{kl}\right) n_{ij} = \frac{2\mu h}{2\mu + h}\left(n_{kl}\dot{\varepsilon}_{kl}\right) n_{ij} .$$

Let us now look at the energy-related aspects of elastoplastic deformations in view of the constitutive equations derived above. The additive relation (7.107) implies that the energy of internal forces (related to the volume unit) $\rho\dot{e}$ can also be expressed as as sum of two terms,

$$\rho\dot{e} = \sigma_{ij}\dot{\varepsilon}_{ij} = \sigma_{ij}\dot{\varepsilon}^{(e)}_{ij} + \sigma_{ij}\dot{\varepsilon}^{(p)}_{ij} . \tag{7.138}$$

It has been mentioned while discussing one-dimensional examples that elastoplastic deformations are accompanied by energy dissipation that manifests itself in the hysteresis loops on the cyclic σ–ε graphs. It is noteworthy that this fact is exactly reflected by the decomposition of the work rate of internal forces into two terms in Eq. (7.138). The first term, $\sigma_{ij}\dot{\varepsilon}^{(e)}_{ij}$, describes the elastic part of strain energy rate, i.e. the part that is accumulated in material as the potential strain energy. At unloading, when the strain decreases, this quantity is negative which means that the elastic energy is released. The second term, $\sigma_{ij}\dot{\varepsilon}^{(p)}_{ij}$, is always positive, since the plastic strain rate, being proportional to n_{ij}, ensures the positive sign of its full product with the stress tensor. This work is not accumulated—it is subject to dissipation (usually it is nearly fully converted to heat).

Exercise 7.3
Comparing the elastoplastic constitutive relations derived above for 3D continuum with the relations given in Sect. 7.4.1 for a 1D stretched bar, it is easy to notice analogies between Eqs. (7.110) and $(7.103)_2$ or between Eqs. (7.109), (7.111), (7.114) and (7.102). Show that the analogies are true, i.e. that Eqs. $(7.103)_2$ and (7.102) are special cases of their 3D counterparts.

Solution. Let us start by showing that the 1D yield condition $|\sigma| \leq \sigma_y$ is consistent with the 3D condition given by Eqs. (7.109), (7.111), (7.114) and concisely expressed as $\bar{\sigma} \leq \sigma_y$. Compute $\bar{\sigma}$ from Eq. (7.114) for the 1D stress state in which all stress components are zero except for $\sigma_{11} = \sigma$. The trace of such a stress tensor is $\sigma_{kk} = \sigma$ and it is easy to show that its deviator $\sigma^{\mathrm{D}}_{ij} = \sigma_{ij} - \frac{1}{3}\sigma_{kk}\delta_{ij}$ can be expressed by the following matrix of its components:

$$[\sigma^{\mathrm{D}}_{ij}] = \begin{bmatrix} \frac{2}{3}\sigma & 0 & 0 \\ 0 & -\frac{1}{3}\sigma & 0 \\ 0 & 0 & -\frac{1}{3}\sigma \end{bmatrix} .$$

Hence,

$$\bar{\sigma} = \sqrt{\frac{3}{2}\sigma^{\mathrm{D}}_{ij}\sigma^{\mathrm{D}}_{ij}} = \sqrt{\frac{3}{2}\left[\left(\frac{2}{3}\sigma\right)^2 + 2\left(-\frac{1}{3}\sigma\right)^2\right]} = |\sigma|.$$

In order to show consistency between the 1D definition of equivalent plastic strain $(7.103)_2$ with the definition (7.110), let us note that the plastic strain is deviatoric, i.e. in the case of 1D stretch/compression the longitudinal component rate $\dot{\varepsilon}^{(p)}_{11} \equiv \dot{\varepsilon}^{(p)}$ implies existence of transversal component rates $\dot{\varepsilon}^{(p)}_{22} = \dot{\varepsilon}^{(p)}_{33} = -\frac{1}{2}\dot{\varepsilon}^{(p)}$ (describing changes of transversal dimensions of the bar). Substituting this into Eq. (7.110), one obtains

$$\dot{\bar{\varepsilon}}^{(p)} = \sqrt{\frac{2}{3}\dot{\varepsilon}^{(p)}_{ij}\dot{\varepsilon}^{(p)}_{ij}} = \sqrt{\frac{2}{3}\left[\left(\dot{\varepsilon}^{(p)}\right)^2 + 2\left(-\frac{1}{2}\dot{\varepsilon}^{(p)}\right)^2\right]} = |\dot{\varepsilon}^{(p)}|.$$

It has been thus shown that the way we defined the quantities $\bar{\sigma}$ and $\bar{\varepsilon}^{(p)}$ in a 1D case of stretched/compressed bar is consistent with the general definitions for 3D continuum.

Exercise 7.4

Compute the equivalent Huber–Mises stress $\bar{\sigma}$ for a plane stress state with the components:

$$[\sigma_{ij}] = \begin{bmatrix} 1.2 & 0.6 & 0 \\ 0.6 & 0.3 & 0 \\ 0 & 0 & 0 \end{bmatrix} \text{MPa}.$$

Solution. The trace of this stress tensor is $\sigma_{kk} = 1.5$ MPa. Thus, its deviator $\sigma^{\mathrm{D}}_{ij} = \sigma_{ij} - \frac{1}{3}\sigma_{kk}\delta_{ij}$ is represented by the component matrix

$$[\sigma^{\mathrm{D}}_{ij}] = \begin{bmatrix} 0.7 & 0.6 & 0 \\ 0.6 & -0.2 & 0 \\ 0 & 0 & -0.5 \end{bmatrix} \text{MPa}$$

and—as can be seen—does not conform to the plane stress condition ($\sigma^{\mathrm{D}}_{33} \neq 0$). We can thus compute

$$\sigma^{\mathrm{D}}_{ij}\sigma^{\mathrm{D}}_{ij} = (0.7)^2 + (-0.2)^2 + (-0.5)^2 + 2(0.6)^2 = 1.5\,\text{MPa}^2$$

and

$$\bar{\sigma} = \sqrt{\frac{3}{2}\sigma^{\mathrm{D}}_{ij}\sigma^{\mathrm{D}}_{ij}} = 1.5\,\mathrm{MPa}.$$

The mathematical model of elastoplastic constitutive relations presented above is the simplest of those being employed in engineering practice. Other formulations, which are numerous indeed, are not going to be further discussed, but they seem at least worth a brief mentioning. The models differ by, among others, hardening models and yield surface forms. For instance, in the kinematic hardening model, as in the 1D example, the increasing stress $\bar{\sigma}$, while approaching the yield surface $f = 0$, will not make it swell (as $\sigma_y = \text{const}$ in this case) but rather move it in the space of stress components. Its transient location is defined by the deviatoric back stress tensor α_{ij} that indicates the position of the yield surface centre. In this formulation, the back stress tensor is an additional state parameter, i.e. Eqs. (7.109) and (7.114) are replaced by

$$f(\sigma_{ij}, \alpha_{ij}, \bar{\varepsilon}^{(p)}) \le 0, \tag{7.139}$$

$$F(\sigma_{ij}, \alpha_{ij}) = \bar{\sigma} = \sqrt{\tfrac{3}{2}\, s_{ij} s_{ij}}, \qquad s_{ij} = \sigma^{\mathrm{D}}_{ij} - \alpha_{ij}. \tag{7.140}$$

The evolution equation of α_{ij}, typically assumed in the linear form $\dot{\alpha}_{ij} = \zeta \dot{\varepsilon}^{(p)}_{ij}$, is a part of the constitutive equation system.

The cylindrical shape of the yield surface in the principal stress space (Fig. 7.10) is not the only one postulated in realistic constitutive models, either. In constitutive formulations typical of rocks and soils, for instance, the Drucker–Prager model is commonly used in which the yield surface resembles a cone that gets narrower as the positive (stretching) principal stress values increase. Besides, the yield surface does not need to be smooth—in the models of Tresca or Mohr–Coulomb it assumes prismatic or pyramidal shapes. In rocks and soils, the associated flow rule is not always justified and Eq. (7.115) must be then replaced by another relation.

Finally, it must be noted that the equations discussed above are limited to only small (linearized) strains. Beyond this range (which is much more realistic in terms of practical applications) the fundamental Eq. (7.107) is no longer valid. Instead of the strain additivity rule, the multiplicative decomposition of the deformation gradient is then postulated,

$$\boldsymbol{F} = \boldsymbol{F}^{(e)} \boldsymbol{F}^{(p)}, \qquad F_{ij} = F^{(e)}_{ik} F^{(p)}_{kj}, \tag{7.141}$$

in which elastic deformations described by the gradient $\boldsymbol{F}^{(e)}$ are defined on the configuration including large permanent plastic deformations described by $\boldsymbol{F}^{(p)}$. Upon such an assumption, all the equations derived in the small strain model have to be reformulated in quite a different way, with the use of a more advanced mathematical formalism.

All the constitutive models of elastoplasticity have a number of common features, though. These are: (i) rate type of constitutive relations, (ii) existence of a yield condition, i.e. a surface in the stress component space that limits the range of elastic material behaviour, (iii) existence of an evolution rule for plastic strain (or deformation gradient) on this surface and an evolution rule of the surface itself (both the rules must be related to each other by a certain consistency condition) and (iv) existence of state parameters—different in different formulations—that affect the shape of the yield surface and whose evolution is described by additional equations.

Finally, let us mention two types of plastic material behaviour that have not been included in the above discussion. These are the so-called elastoviscoplastic materials, in which development of plastic strains is additionally time-dependent, and the elastoplastic materials with microvoids, in which development of plastic strains causes nucleation and growth of microvoids. Interested readers are encouraged to study the broad literature on the subject.

Reference

1. Fung Y.C., 1981. *Biomechanics. Mechanical Properties of Living Tissues*. Springer-Verlag.

Chapter 8
Fundamental System of Solid Mechanics Equations

8.1 Field Equations and Initial-Boundary Conditions

Having presented in previous chapters various equations governing mechanical behaviour of material continuum, we are now in position to collect them together and write down the full system of equations describing the phenomena on hand. Along with appropriate initial-boundary conditions, the system will constitute the *initial-boundary formulation of solid mechanics.*

Consider the deformable body $\mathcal{B}$ in motion between an initial time instant $\tau = 0$ (at which the material points of $\mathcal{B}$ assume a known configuration C^0) and the final time instant $\tau = \bar{t}$ (at which the geometric configuration $C^{\bar{t}}$ of the body is unknown), cf. Fig. 4.1. Let us select an arbitrary known configuration C^r as the reference configuration in which each material point $P \in \mathcal{B}$ is assigned coordinates $\boldsymbol{x} \in \Omega^r$ in a selected Cartesian coordinate system (for the sake of clarity, the index r will be omitted at the coordinate notation). As before, Ω^r is assumed to be an open set in E^3 and its boundary $\partial\Omega^r$ is a piece-wise smooth surface. The body is subject to mechanical load in the form of mass forces $f_i(\boldsymbol{x}, \tau)$, $\boldsymbol{x} \in \Omega^r$, while on its surface $\partial\Omega^r$, either surface forces ${}_r\hat{t}_i(\boldsymbol{x}, \tau)$ or displacements $\hat{u}(\boldsymbol{x}, \tau)$ are imposed.

The nonlinear system of solid mechanics equations for an arbitrary point $\boldsymbol{x} \in \Omega^r$ at an arbitrary time instant $\tau \in [0, \bar{t}]$ has the form (8.1) below, which includes Eqs. (6.10), (6.24), (4.10), (4.15b), (5.24b) as well as a constitutive equation. The latter is written here in a general form in which p_α, $\alpha = 1, \ldots, n_p$, denote state parameters that may appear in some constitutive formulations (like equivalent plastic strain $\bar{\varepsilon}^{(p)}$ in elastoplasticity). All spatial gradients (unless explicitly stated) are understood as gradients with respect to x_i^r. We can thus write:

- continuity equation (1)

$$\dot{\rho}^\tau + \rho^\tau \frac{\partial \dot{u}_i^\tau}{\partial x_i^\tau} = \dot{\rho}^\tau + \rho^\tau \dot{u}_{i,j}^\tau ({}^rF_{ji}^\tau)^{-1} = 0, \tag{8.1a}$$

M. Kleiber and P. Kowalczyk, *Introduction to Nonlinear Thermomechanics of Solids*, Lecture Notes on Numerical Methods in Engineering and Sciences,
DOI 10.1007/978-3-319-33455-4_8

- equations of motion (3)

$$ {}_r\sigma^{\tau}_{ij,j} + \rho^r f^{\tau}_i - \rho^r \ddot{u}^{\tau}_i = 0, \tag{8.1b}$$

- geometric equations (9+6+9)

$$ {}^rF^{\tau}_{ij} = \delta_{ij} + {}^ru^{\tau}_{i,j}\,, \tag{8.1c}$$

$$ {}^0_r\varepsilon^{\tau}_{ij} = \frac{1}{2}\left({}^0u^{\tau}_{i,j} + {}^0u^{\tau}_{j,i} + {}^ru^{\tau}_{k,i}\,{}^ru^{\tau}_{k,j} - {}^ru^{0}_{k,i}\,{}^ru^{0}_{k,j}\right), \tag{8.1d}$$

$$ {}_r\sigma^{\tau}_{ij} = {}^rF^{\tau}_{ik}\,{}_r\tilde{\sigma}^{\tau}_{kj}\,, \tag{8.1e}$$

- constitutive equations (6+n_p)

$$ f_k(\rho^{\tau},\ {}_r\tilde{\sigma}^{\tau}_{ij},\ {}^0_r\varepsilon^{\tau}_{ij},\ p^{\tau}_{\alpha},\ {}_r\dot{\tilde{\sigma}}^{\tau}_{ij},\ {}_r\dot{\varepsilon}^{\tau}_{ij},\ \dot{p}^{\tau}_{\alpha}) = 0. \tag{8.1f}$$

The numbers of component equations have been given in parentheses. The number 6 at the geometric and constitutive equations indicates that symmetry of the strain tensor ${}^0_r\varepsilon^{\tau}_{ij}$ and the second Piola–Kirchhoff stress tensor ${}_r\tilde{\sigma}^{\tau}_{ij}$ has been taken into account and the repeating 3 out of 9 component equations have been therefore skipped.

The following unknown fields defined on the domain $\boldsymbol{x} \in \Omega^r$, $\tau \in [0, \bar{t}\,]$ appear in the above equations:

density	(1)	ρ^{τ}	(8.2a)
displacement	(3)	${}^0u^{\tau}_i$	(8.2b)
deformation gradient	(9)	${}^rF^{\tau}_{ij}$	(8.2c)
strain	(6)	${}^0_r\varepsilon^{\tau}_{ij}$	(8.2d)
stress (the first P–K tensor)	(9)	${}_r\sigma^{\tau}_{ij}$	(8.2e)
stress (the second P–K tensor)	(6)	${}_r\tilde{\sigma}^{\tau}_{ij}$	(8.2f)
constitutive state parameters	(n_p)	p^{τ}_k	(8.2g)

The numbers of unknown components have been again specified in parentheses. The displacement ${}^ru^{\tau}_i = {}^ru^0_i + {}^0u^{\tau}_i$ is not considered unknown as it linearly depends on ${}^0u^{\tau}_i$.

As can be seen, the total number of unknowns $34 + n_p$ equals the total number of equations. The system is thus complete and it is in theory possible to solve it. Note that the above problem solution meets all the conservation laws presented in Chap. 6:

- the mass conservation law is expressed by the continuity equation (8.1a),
- the momentum conservation law appears explicitly in the system as the vector equation of motion (8.1b),
- the angular momentum conservation law is fulfilled by imposing symmetry of the stress tensor ${}_r\tilde{\sigma}^{\tau}_{ij}$, and

- the mechanical energy conservation law is fulfilled as a consequence of the momentum conservation law.

The necessary condition for existence of the solution is a consistent set of initial and boundary conditions for the unknown fields. The initial conditions have the following form[1]:

$$\rho(\boldsymbol{x}, 0) = \rho^0(\boldsymbol{x}), \tag{8.3a}$$

$$u_i(\boldsymbol{x}, 0) = {}^r u_i^0(\boldsymbol{x}), \tag{8.3b}$$

$$\dot{u}_i(\boldsymbol{x}, 0) = v_i^0(\boldsymbol{x}), \tag{8.3c}$$

$${}_r\tilde{\sigma}_{ij}(\boldsymbol{x}, 0) = {}_r\tilde{\sigma}_{ij}^0(\boldsymbol{x}), \tag{8.3d}$$

$$p_\alpha(\boldsymbol{x}, 0) = p_\alpha^0(\boldsymbol{x}). \tag{8.3e}$$

All other fields at $\tau = 0$ can be determined from Eqs. (8.1) and (8.3).

The boundary conditions imposed on the surface[2] $\partial\Omega^r$ may assume various forms. Two basic forms of the conditions are the kinematic (Dirichlet) and dynamic (Neumann) ones. The kinematic condition has the following form

$${}^r u_i^\tau = \hat{u}_i(\boldsymbol{x}, \tau), \tag{8.4a}$$

while the dynamic condition

$${}_r\sigma_{ij}^\tau n_j^r = {}_r\hat{t}_i(\boldsymbol{x}, \tau), \tag{8.4b}$$

where $\hat{u}_i$ and $\hat{t}_i$ are known vector fields of displacement and stress, respectively, defined at $\boldsymbol{x} \in \partial\Omega^r$ i.e. on the boundary of $\mathcal{B}$. Conditions of the two types are contradictory—demanding a certain value of displacement at a selected material point, one automatically enforces a certain reaction stress value at this point and vice versa. Hence, only one of the conditions (8.4) may be imposed at each point (more strictly—exactly one of them must be imposed at each point as the system has no unique solution if there is at least one boundary point with no boundary condition defined). This means that the boundary surface $\partial\Omega^r$ may be divided into two (generally time-dependent) subdomains $\partial\Omega_u^r(\tau)$ and $\partial\Omega_\sigma^r(\tau)$ being open sets in $\partial\Omega^r$, such that at each time instant $\partial\Omega_u^r \cup \partial\Omega_\sigma^r = \partial\Omega^r$ and $\partial\Omega_u^r \cap \partial\Omega_\sigma^r = \emptyset$, and the conditions (8.4a) and (8.4b) are imposed on $\partial\Omega_u^r$ and $\partial\Omega_\sigma^r$, respectively (Fig. 8.1).

[1] Actual form of the initial conditions depends to some extent on the form of the differential equations in the system. In the case of elasticity initial stresses are uniquely defined by initial displacements and there is no need to explicitly define them. Besides, if dynamical effects are neglected ($\ddot{u}_i \equiv 0$) then the initial velocity condition is not necessary.

[2] If the boundary conditions in a real problem are defined in a configuration different from C^r, they must be transformed to the reference configuration with the use of appropriate geometrical formulae.

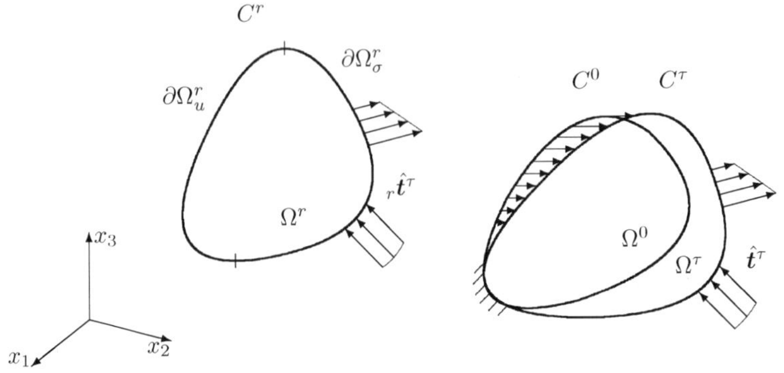

Fig. 8.1 Boundary conditions

The boundary conditions defined at $\tau = 0$ must be obviously additionally consistent with the initial conditions (8.3).

Remark. The form of the dynamic condition (8.4b) suggests that loads distributed on a body's surface are the only ones allowed. We know form engineering practice that bodies may be also loaded by concentrated forces acting at single points. The formalism adopted does not exclude such loads. Let us recall the Dirac delta function, introduced earlier in our one-dimensional discussion by Eq. (7.70). Assuming that a concentrated force $\boldsymbol{P}(\tau)$ is applied at a material point $\boldsymbol{x} = \boldsymbol{X} \in \partial\Omega^r_\sigma$, let us define

$$ {}_r\hat{\boldsymbol{t}}(\boldsymbol{x}, \tau) = \boldsymbol{P}(\tau)\,\delta(\boldsymbol{x} - \boldsymbol{X}), $$

where

$$ \delta(\boldsymbol{x}) = \begin{cases} 0 & \boldsymbol{x} \neq \boldsymbol{0}, \\ \infty & \boldsymbol{x} = \boldsymbol{0}, \end{cases} \qquad \forall f(\boldsymbol{x}) \;\; \forall \Gamma \ni \boldsymbol{x} \quad \int_\Gamma f(\boldsymbol{x})\,\delta(\boldsymbol{x})\,\mathrm{d}A = f(\boldsymbol{0}). \tag{8.5} $$

This formalism is also valid if the force is applied at a point that moves over the boundary surface of the body during its motion—in such a case one should simply assume $\boldsymbol{X} = \boldsymbol{X}(\tau)$.

Remark. The above discussion on separation of the boundary surface into the parts $\partial\Omega^r_u$ and $\partial\Omega^r_\sigma$ is unfortunately quite simplified. This is because it has been assumed for clarity that—at a given boundary point—either all displacement components or all stress components are imposed. However, consistent boundary conditions may be also defined in a mixed way, for instance by imposing at the same point the kinematic condition $u_1 = \hat{u}_1$ (obvious indices are skipped) and the dynamic condition $\sigma_{2i}n_i = \hat{t}_2$. Other, more complex formulations are possible as well—zero displacement value may be imposed in a selected spatial direction (i.e. the point may move in the plane orthogonal to this direction) and loads acting in this plane simultaneously applied at this point. In all the cases it is crucial that the imposed forces must not perform work on the imposed displacements. This condition is difficult to express in a concise way, though.

The two boundary condition types discussed above, although quite common in standard computational problems of solid mechanics, do not exhaust the variety of boundary interactions in real engineering problems. Let us briefly mention here one more and quite a typical case—the so-called contact boundary condition. This occurs when two bodies touch (or may potentially touch) each other with parts of their boundary surfaces, or if two parts of the deformed boundary surface of one body touch each other. This condition has a mixed character because it enforces certain bounds on relative displacements or velocities of boundary points in contact and simultaneously it implies certain unknown reaction forces appearing at these points. The contact condition can be written in the general form as

$$f(u_i) \leq 0, \qquad g_k(u_i, \dot{u}_i, \sigma_{ij}) \leq 0,$$

where the first inequality is the geometric condition of no interpenetration of the two potentially contacting surfaces (the sharp inequality $f < 0$ denotes no contact) while the second inequality is the "constitutive law of contact" describing bounds imposed on friction forces for instance. In the case of no contact, the second inequality obviously comes down to the trivial form of the dynamic boundary condition (8.4a) with zero stress $\hat{t}_i$.

8.2 Incremental Form of Equations

Equations of the system (8.1) must be fulfilled at each point $\boldsymbol{x} \in \Omega^r$ and at each time instant $\tau \in [0, \bar{t}]$. Approximate methods of their solution assume usually independent discretization in space and time domains. The discretization in space is quite a complex issue and will be discussed in more detail towards the end of this book. The discretization in time, to the contrary, consists usually in a simple division of the domain $[0, \bar{t}]$ into a series of relatively short intervals separated by points

$t_0 = 0, t_1, \dots, t_{n-1}, t_n = \bar{t}$ selected along the time axis. The solution of the system consists then in repeatedly solving it in the spatial domain only at subsequent time instants $\tau = t_{k+1}$ upon the assumptions that the solution at $\tau = t_k$ is known and that the solution within the interval $[t_k, t_{k+1}]$ is a linear function of time.

In such a case, the initial-boundary value problem (8.1) comes down to the series of boundary value problems formulated at a typical time interval (step) $[t, t+\Delta t]$, sequentially solved for subsequent intervals of the time discretization. Assuming the notation convention in which an increment of any quantity in such a typical interval is denoted as

$$\Delta(\cdot)^t = (\cdot)^{t+\Delta t} - (\cdot)^t,$$

one can derive the incremental form of the system (8.1) for the considered time interval by subtracting from each other appropriate equations written for the instants t and $t+\Delta t$. The following system of equations results then at each material point $\boldsymbol{x} \in \Omega^r$:

- continuity equation

$$(\rho^t + \Delta\rho^t)\, \Delta u^t_{i,j} ({}^rF^t_{ji})^{-1} + \Delta\rho^t = 0, \tag{8.6a}$$

- equations of motion

$$\Delta_r\sigma^t_{ij,j} + \rho^r \Delta f^t_i - \rho^r \Delta\ddot{u}^t_i = 0, \tag{8.6b}$$

- geometric equations

$$\Delta^r F^t_{ij} = \Delta u^t_{i,j}\,, \tag{8.6c}$$

$$\Delta_r\varepsilon^t_{ij} = \frac{1}{2}\left(\Delta u^t_{i,j} + \Delta u^t_{j,i} + \Delta u^t_{k,i}\, {}^r u^t_{k,j} + {}^r u^t_{k,i}\, \Delta u^t_{k,j} + \Delta u^t_{k,i}\, \Delta u^t_{k,j}\right), \tag{8.6d}$$

$$\Delta_r\sigma^t_{ij} = \Delta_r\tilde{\sigma}^t_{ij} + u_{i,k}\, \Delta_r\tilde{\sigma}^t_{kj} + \Delta u_{i,k}\, ({}_r\tilde{\sigma}^t_{kj} + \Delta_r\tilde{\sigma}^t_{kj}), \tag{8.6e}$$

- constitutive equations (in a general incremental form)

$$\Delta_r\tilde{\sigma}^t_{ij} = C^t_{ijkl}({}_r\tilde{\sigma}^t_{mn}, p^t_\alpha)\, \Delta_r\varepsilon^t_{kl} + H^t_{ij}({}_r\tilde{\sigma}^t_{mn}, p^t_\alpha)\Delta t, \tag{8.6f}$$

$$\Delta p^t_\beta = Z^t_{\beta kl}({}_r\tilde{\sigma}^t_{mn}, p^t_\alpha)\, \Delta_r\varepsilon^t_{kl} + Y^t_\beta({}_r\tilde{\sigma}^t_{mn}, p^t_\alpha)\Delta t \tag{8.6g}$$

which have to be considered together with the boundary conditions

$$\Delta_r\sigma^t_{ij} n_j = \Delta\hat{t}^t_i\,, \qquad x_i \in \partial\Omega^r_\sigma\,, \tag{8.6h}$$

$$\Delta u^t_i = \Delta\hat{u}^t_i\,, \qquad x_i \in \partial\Omega^r_u\,. \tag{8.6i}$$

The choice of the reference configuration C^r is arbitrary but may strongly influence effectiveness of the solution process. In the case of the so-called *updated Lagrangian description* the reference configuration is always the known configuration at the beginning of the considered interval, i.e. C^t. It is easy to see that the form of Eq. (8.6) gets much more simple upon this assumption. For instance, the terms ${}^{r}u_{i}^{t}$ and their gradients vanish and ${}^{r}F_{ij}^{t} \equiv \delta_{ij}$. On the other hand, a difficulty is that at the end of the interval all the tensorial quantities computed (stresses, strains, etc.) have to be transformed to the new updated reference configuration before the computations for the next time interval can start.

Nevertheless, the updated Lagrangian description is a commonly accepted way to use the incremental formulation of nonlinear mechanics equations.

8.3 Some Special Cases

The system of equations (8.1) along with the initial and boundary conditions (8.3), (8.4) has obviously several possible simplified forms, selection of which depends on the character of a particular problem. If dynamical effects (inertial forces) are negligibly small, then the accelerations $\ddot{u}_i$ may be neglected in the equations of motion, for instance. In this case, the quasi-static problem equations are obtained, in which the time variable is nothing else but a parameter that orders the sequence of events, called therefore a pseudo-time.

The form of the equations is fairly simplified in the case of elastic constitutive equations. Equation (8.1f) can be then expressed in the form ${}_{r}\tilde{\sigma}_{ij}^{\mathcal{T}} = f_{ij}({}_{r}^{0}\varepsilon_{kl}^{\mathcal{T}})$ and no state parameters p_α appear, which greatly reduces the number of equations and unknowns. Besides, as the given displacement field determines the stress field in elastic materials, the initial conditions (8.3b) and (8.3d) are mutually dependent and the second of them may be skipped. The elastic problem formulation has also a significant advantage: it is quite easy to eliminate some of the system equations (8.1) by substituting them into others to finally obtain a system of three equations of motion expressed in gradients of three unknown displacement components. Having solved the system, one is able to express all the remaining fields by the remaining equations (geometric, constitutive and mass conservation).

If the problem is simultaneously elastic and static, then the time variable may be entirely removed from the system which then becomes a boundary value problem that may be solved independently at each time instant τ (in particular $\tau = 0$) with the given boundary conditions. Initial conditions are needless in this case, except for the initial density distribution (8.3a).

Let us formulate our system of equations for a case even more simple yet. Assume the problem is static, linearly elastic and that deformations are small (which allows to neglect configuration changes and describe strains by the linearized Cauchy tensor). Upon these assumptions, the equations at an arbitrary $\boldsymbol{x} \in \Omega$ can be written as (again, the numbers of component equations are given in parentheses),

equations of motion	(3)	$\sigma_{ij,j} + \rho f_i = 0\,,$	(8.7a)
geometric equations	(6)	$\varepsilon_{ij} = \frac{1}{2}(u_{i,j} + u_{j,i}),$	(8.7b)
constitutive equations	(6)	$\sigma_{ij} = C_{ijkl}\varepsilon_{kl}\,,$	(8.7c)

while the boundary conditions have the form

$$u_i = \hat{u}_i\,, \qquad \boldsymbol{x} \in \partial\Omega_u\,, \tag{8.8a}$$

$$\sigma_{ij} n_j = \hat{t}_i\,, \qquad \boldsymbol{x} \in \partial\Omega_\sigma\,. \tag{8.8b}$$

Obviously, $\partial\Omega_u \cup \partial\Omega_\sigma = \partial\Omega$ and $\partial\Omega_u \cap \partial\Omega_\sigma = \emptyset$. Unknown in this formulation are: u_i (3), ε_{ij} (6) and σ_{ij} (6); all the other quantities are assumed known. Thus, there are 15 equations and 15 unknowns. This system can again be easily reduced to three equations with three unknowns u_i by subsequent substitution of Eq. (8.7b) into Eq. (8.7c) and then the latter into Eq. (8.7a).

Even for such a simplified system of differential equations, its analytical solution turns out to be impossible in most of the engineering applications. There are only few examples of such closed solutions available for very simple geometries and boundary conditions. One of them will be shown below in which material isotropy will be additionally assumed, i.e.

$$\sigma_{ij} = \lambda\,\delta_{ij}\varepsilon_{kk} + 2\mu\,\varepsilon_{ij} = \frac{E}{1+\nu}\left(\frac{\nu}{1-2\nu}\,\delta_{ij}\varepsilon_{kk} + \varepsilon_{ij}\right),$$

and mass forces will be neglected, i.e. $\rho f_i = 0$.

8.4 Example of Analytical Solution

Consider an elastic cuboid with the dimensions $l \times 2b \times 2h$ shown in Fig. 8.2. The cuboid is fixed at the wall $x_1 = 0$ and loaded by a bending moment M at the opposite wall $x_1 = l$. More strictly, the boundary conditions are expressed as follows:

- Kinematic conditions:

$$u_1 = 0 \quad \text{for } x_1 = 0,$$
$$u_2 = 0 \quad \text{for } x_1 = x_2 = 0,$$
$$u_3 = 0 \quad \text{for } x_1 = x_2 = x_3 = 0.$$

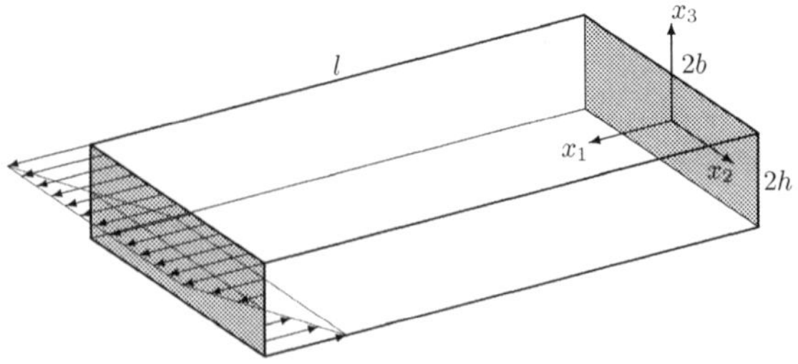

Fig. 8.2 Example: bending of beam with rectangular cross-section

The conditions mean that the material points on the wall $x_1 = 0$ may only move in its plane, except for the points located on the axis x_3 that may only move along the axis. The wall's central point is the only one fully fixed, with no displacement allowed.

- Dynamic conditions:

$$
\begin{aligned}
&t_1 = t_2 = t_3 = 0 && \text{for } x_3 = \pm h,\\
&t_1 = t_2 = t_3 = 0 && \text{for } x_2 = \pm b,\\
&t_1 = \hat{t}_{\max}\frac{x_3}{h} \quad t_2 = t_3 = 0, && \text{for } x_1 = l,\\
&t_2 = 0 && \text{for } x_1 = 0,\ x_2 \neq 0,\\
&t_3 = 0 && \text{for } x_1 = 0,\ x_2^2 + x_3^2 \neq 0.
\end{aligned}
$$

The conditions mean that at all the boundary points where no kinematic conditions are imposed stress is zero, except for the wall $x_1 = l$ where a specified distribution of normal stress is given. If $\hat{t}_{\max} = \frac{3}{4}\frac{M}{bh^2}$ then it is easy to show that the resultant moment of load forces acting on the wall is M. Since $t_i = \sigma_{ij}n_j$ and the normal versor components n_i at all the walls have trivial zero/one values, the above dynamic conditions can be rewritten in the form:

$$
\begin{aligned}
&\sigma_{13} = \sigma_{23} = \sigma_{33} = 0 && \text{for } x_3 = \pm h,\\
&\sigma_{12} = \sigma_{22} = \sigma_{23} = 0 && \text{for } x_2 = \pm b,\\
&\sigma_{11} = \hat{t}_{\max}\frac{x_3}{h} \quad \sigma_{12} = \sigma_{13} = 0, && \text{for } x_1 = l,\\
&\sigma_{12} = 0 && \text{for } x_1 = 0,\ x_2 \neq 0,\\
&\sigma_{13} = 0 && \text{for } x_1 = 0,\ x_2^2 + x_3^2 \neq 0.
\end{aligned}
$$

The example on hand is known as the so-called pure bending of a beam with a rectangular cross-section. The displacement field being its analytical solution turns out to have the form of a second order polynomial with respect to x_i. Skipping the details of the solution procedure, we may write the final result in the form

$$
u_1 = Cx_1x_3, \qquad u_2 = -C\nu x_2x_3, \qquad u_3 = -\tfrac{1}{2}C(x_1^2 - \nu x_2^2 + \nu x_3^2),
$$

where $C = \hat{t}_{\max}/(Eh) = \frac{3}{4}\frac{M}{Ebh^3}$.

It is easy to show that such a displacement field fulfills the kinematic boundary conditions. Let us now verify the remaining equations, starting from computing strains:

$$
\begin{aligned}
&\varepsilon_{11} = u_{1,1} = Cx_3, && \varepsilon_{12} = \tfrac{1}{2}(u_{1,2} + u_{2,1}) = 0,\\
&\varepsilon_{22} = u_{2,2} = -C\nu x_3, && \varepsilon_{13} = \tfrac{1}{2}(u_{1,3} + u_{3,1}) = 0,\\
&\varepsilon_{33} = u_{3,3} = -C\nu x_3, && \varepsilon_{23} = \tfrac{1}{2}(u_{2,3} + u_{3,2}) = 0
\end{aligned}
$$

and then stresses:

$$\sigma_{11} = \frac{E[(1-\nu)\varepsilon_{11} + \nu(\varepsilon_{22}+\varepsilon_{33})]}{(1+\nu)(1-2\nu)} = ECx_3 = \hat{t}_{\max}\frac{x_3}{h},$$
$$\sigma_{22} = \frac{E[(1-\nu)\varepsilon_{22} + \nu(\varepsilon_{33}+\varepsilon_{11})]}{(1+\nu)(1-2\nu)} = 0,$$
$$\sigma_{33} = \frac{E[(1-\nu)\varepsilon_{33} + \nu(\varepsilon_{11}+\varepsilon_{22})]}{(1+\nu)(1-2\nu)} = 0,$$
$$\sigma_{12} = \frac{E\varepsilon_{12}}{1+\nu} = 0,$$
$$\sigma_{13} = \frac{E\varepsilon_{13}}{1+\nu} = 0,$$
$$\sigma_{23} = \frac{E\varepsilon_{23}}{1+\nu} = 0.$$

Substituting the stresses into the dynamic boundary conditions, we can easily make sure that they are fulfilled as well. We can also note that the distribution of σ_{11} imposed on the wall $x_1 = l$ as an external load remains the same in all other cross-sections $x_1 = \text{const}$ along the beam. Equations of motion are fulfilled in a trivial way as all the stress gradient components that appear in the equations are zero.

Deformed shape of the considered cuboid (beam) is shown in Fig. 8.3. Displacements have been obviously scaled so that they are clearly visible (the assumption of linearized strains does not allow for analysis of deformations that are as large as those shown in the figure). Since material particles located above the plane $x_3 = 0$ are

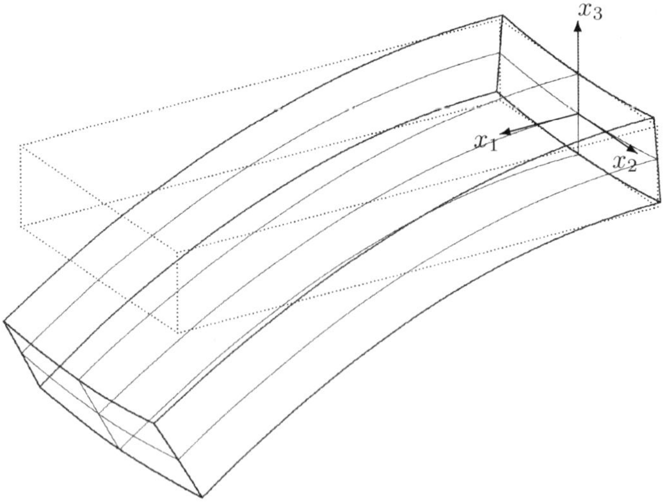

Fig. 8.3 Deformation of beam with rectangular cross-section at pure bending

stretched in the direction x_1, the upper part of the beam is elongated in this direction. For an analogous reason, its lower part is compressed, which results in bending of the entire cuboid in the plane orthogonal to the x_2 axis, with convexity on the upper side. Besides, if the Poisson ratio $\nu > 0$, the uniaxial stretch of particles in the upper part of the beam results in decrease of their transversal dimensions (in contrast to particles in the lower part). This effect is responsible for the slight bending of the beam cross-sections $x_1 = \text{const}$ in the opposite direction to the axial bending, visible in the figure.

As already mentioned, the above solution belongs to very few analytical solutions of real solid mechanics problems. Other examples of such solutions are pure stretching/compression of a bar with constant cross-section, torsion of an axisymmetric beam, or some problems defined on infinite domains Ω (like space, half-space, plane, etc.). In all the cases, assumption of small strains and isotropic elasticity is also required. In most other cases of practical importance, the effective solution of the problem requires application of approximate numerical methods. These methods will be briefly described in subsequent chapters of this book.

Chapter 9
Fundamentals of Thermomechanics and Heat Conduction Problem

Heat is another—apart from mechanical—form of energy accumulated and transferred in material. The necessary condition for heat transfer between two interacting systems (bodies or subdomains in a body) is a temperature difference between the systems. In this chapter, various forms of heat energy transfer, such as conduction, convection or radiation will be discussed. In mechanical systems, these processes are usually accompanied by a heat production due to inelastic deformation. Besides, in a body subject to temperature changes, thermal stresses and strains appear. These phenomena are the subject of research in thermomechanics of deformable bodies. Let us thus start from short introduction of two fundamental rules governing this branch of physics, called laws of thermodynamics.

9.1 Laws of Thermodynamics

9.1.1 The First Law of Thermodynamics

The first law of thermodynamics is a natural generalization of the mechanical energy conservation law discussed in Sect. 6.4. Let us recall its form (6.42) as

$$\dot{L} = \dot{K} + \dot{W}, \tag{9.1}$$

where the left-hand side $\dot{L}$ expresses the power supplied to the system (i.e. the body $\mathcal{B}$) through the work of external forces while the right-hand side stands for the power accumulated in the system in the form of kinetic K and internal W energy rates. (All equations in this chapter refer to the time instant $\tau = t$, with respect to the current configuration C^t, thus the indices t and r will be consequently skipped.)

M. Kleiber and P. Kowalczyk, *Introduction to Nonlinear Thermomechanics of Solids*, Lecture Notes on Numerical Methods in Engineering and Sciences,
DOI 10.1007/978-3-319-33455-4_9

Let us now extend this energy balance to include effects related to heat. Additional terms will then appear on both sides of the equation. The left-hand side must be completed by the heat power supplied to the system, denoted by $\dot{Q}$. There are two ways the heat may be supplied: (i) flow through the external surface $\partial\Omega$ bounding the body's domain Ω and (ii) generation by internal heat sources. This can be expressed analytically in the form

$$\dot{Q} = -\int_{\partial\Omega} q_i n_i \,\mathrm{d}A + \int_{\Omega} \rho r \,\mathrm{d}V, \tag{9.2}$$

where $\boldsymbol{q}$ is a vector field of *heat flux* related to the unit time and the unit area, $\boldsymbol{n}$ is the versor normal to the boundary and r is the internal heat generation rate per unit mass. The minus sign in Eq. (9.2) is due to the fact that the versor $\boldsymbol{n}$ is directed outwards the body while $\dot{Q}$ is supposed positive when heat flows inwards.

An additional term should also appear on the right-hand side to describe the rate of thermal energy accumulated in the body. It has the form

$$\dot{E}_T = \int_{\Omega} \rho \dot{e}_T \,\mathrm{d}V,$$

where $\dot{e}_T$ denotes the rate of thermal energy density per unit mass ($\rho\dot{e}_T$ is thus its density per unit volume of material).

Denoting by $\dot{E} = \dot{W} + \dot{E}_T$ the total internal energy rate one can express the first law of thermodynamics as

$$\dot{L} + \dot{Q} = \dot{K} + \dot{E}, \tag{9.3}$$

where

$$\dot{L} = \int_{\partial\Omega} t_i v_i \,\mathrm{d}A + \int_{\Omega} \rho f_i v_i \,\mathrm{d}V, \tag{9.4a}$$

$$\dot{Q} = -\int_{\partial\Omega} q_i n_i \,\mathrm{d}A + \int_{\Omega} \rho r \,\mathrm{d}V, \tag{9.4b}$$

$$\dot{K} = \int_{\Omega} \rho \dot{v}_i v_i \,\mathrm{d}V, \tag{9.4c}$$

$$\dot{E} = \int_{\Omega} \rho \dot{e} \,\mathrm{d}V, \qquad \rho\dot{e} = \sigma_{ij}\dot{\varepsilon}_{ij} + \rho\dot{e}_T\,. \tag{9.4d}$$

Substituting Eqs. (9.4) into Eq. (9.3) and applying the Gauss–Ostrogradski theorem to all the boundary integrals (keeping in mind that $t_i = \sigma_{ij} n_j$ and $\dot{\varepsilon}_{ij} = \mathrm{sym}(v_{i,j})$), one obtains:

$$\int_{\Omega} \left[\left(\sigma_{ij,j} + \rho f_i - \rho \dot{v}_i \right) v_i - \rho \dot{e}_T - q_{i,i} + \rho r \right] \mathrm{d}V = 0. \tag{9.5}$$

Since the expression in parentheses vanishes in view of the momentum conservation law (6.24), this equation can be reduced to the form

$$\int_{\Omega} \left(-\rho \dot{e}_T - q_{i,i} + \rho r\right) \mathrm{d}V = 0. \tag{9.6}$$

Besides, since the integration domain is arbitrary (the energy balance is true for any subdomain of the body $\mathcal{B}$, too), the integrand must vanish at each material point, i.e.

$$\rho \dot{e}_T = -q_{i,i} + \rho r, \qquad \boldsymbol{x} \in \Omega. \tag{9.7}$$

This is the local form of the first law of thermodynamics. It implies that the internal energy rate related to unit volume of a material particle, cf. Eq. (9.4d), is expressed as

$$\rho \dot{e} = \sigma_{ij}\dot{\varepsilon}_{ij} - q_{i,i} + \rho r = \boldsymbol{\sigma} \cdot \dot{\boldsymbol{\varepsilon}} - \operatorname{div} \boldsymbol{q} + \rho r, \tag{9.8}$$

i.e. it is the sum of three components: (i) work of stress on strain rates, (ii) rate of heat supplied from the particle's surrounding (equal to minus heat flux divergence) and (iii) rate of heat supplied by internal heat source in the particle.

Let us now recall our discussion on energy accumulation in the material presented in Sect. 6.4. As we have mentioned there, if material is inelastic then only a part of work performed by internal forces, described by the expression $\sigma_{ij}\dot{\varepsilon}_{ij}$, is accumulated in the form of potential energy. The remaining part is subject to dissipation. For elastoplastic materials, cf. Eq. (7.138), we have for instance

$$\sigma_{ij}\dot{\varepsilon}_{ij} = \sigma_{ij}\dot{\varepsilon}^{(e)}_{ij} + \sigma_{ij}\dot{\varepsilon}^{(p)}_{ij}, \tag{9.9}$$

where the terms on the right-hand side describe the potential energy rate and the dissipated energy rate. We recall that the latter term cannot be considered as a potential energy rate; in fact it is the work of internal forces (per unit time) converted to heat, i.e. the energy form that did not appear in the balance equation (6.42).

Since Eq. (9.8) contains thermal energy terms, this problem can now be presented in a strict mathematical form. In an inelastic material the two components of the term $\sigma_{ij}\dot{\varepsilon}_{ij}$, see Eq. (9.9), correspond to two different types of internal energy. The definition of the quantity $\rho\dot{e}_T$ in Eq. (9.4d) should be thus modified by adding the mechanical dissipation term. Equation (9.8) is thus rewritten in the form:

$$\rho \dot{e} = \underbrace{\sigma_{ij}\dot{\varepsilon}^{(e)}_{ij}}_{\rho \dot{e}_M} + \underbrace{\sigma_{ij}\dot{\varepsilon}^{(p)}_{ij} - q_{i,i} + \rho r}_{\rho \dot{e}_T}. \tag{9.10}$$

We observe that only the first of the two terms describes the rate of the potential energy (i.e. energy that may be released at unloading, when the term $\sigma_{ij}\dot{\varepsilon}^{(e)}_{ij}$ is negative). Thermal energy cannot be converted back to mechanical work, which results from the second law of thermodynamics as discussed in the next section.

Equation (9.10) is still not fully adequate to precisely describe the actual energy transformations in the material element. This is because the work $\sigma_{ij}\dot{\varepsilon}_{ij}^{(p)}$ is not fully converted to heat so that one should rather write

$$\rho \dot{e}_T = \chi \sigma_{ij} \dot{\varepsilon}_{ij}^{(p)} - q_{i,i} + \rho r, \tag{9.11}$$

where the coefficient $\chi < 1$ assumes for metals values ranging from 0.85 to 0.95. The remaining part of the work of internal forces is dissipated in other ways.

Note also that the terms $\chi\sigma_{ij}\dot{\varepsilon}_{ij}^{(p)}$ and ρr in Eq. (9.11) describe in fact similar phenomena—the first describes the density of heat sources related to inelastic deformations while the second other heat sources of an undefined nature. Without affecting generality of the discussion, one may thus collect the two terms into one and turn back to the initial form of Eq. (9.7) in which the term ρr would now include, among others, heat sources of the mechanical type like $\chi\sigma_{ij}\dot{\varepsilon}_{ij}^{(p)}$, for instance. In many problems of thermomechanics of solids these are in fact the only material heat sources needed to be considered in analysis.

Concluding, we emphasize again that to correctly describe energy balance in material in which inelastic deformations occur, thermal energy terms should be taken into account.

9.1.2 *The Second Law of Thermodynamics*

All thermomechanical processes must obey the first law of thermodynamics. Experience shows, however, that several processes that fulfill this law are not allowed in reality, and this fact results from the second law of thermodynamics. This law expresses our common beliefs supported by observations that e.g. the heat flow in a system with no heat sources may only occur from warmer to colder parts and never oppositely, or that mechanical energy in friction processes can be converted to heat but never oppositely.

We are not going to utilize the second law of thermodynamics in our further discussed equations. Thus, we will only present it shortly, just for the sake of the lecture's completeness. In more advanced studies of thermomechanics of solids, this law is e.g. a basis to formulate limitations on possible forms of constitutive equations.

A key role in the formulation of the second law is played by the function called *entropy* (which is a free Greek translation of "towards"). Entropy may be defined as a quantitative measure of microscopic randomness and disorder in the system; its physical interpretation can be derived on the basis of statistical mechanics.

Let us denote by $\eta = \eta(\boldsymbol{x}^t, t) = \eta^t(\boldsymbol{x}^t)$ entropy related to unit mass in the current configuration at $\tau = t$. Entropy of the body (or its part) occupying in this configuration the domain Ω is thus

$$S = \int_{\Omega} \rho \eta \, \mathrm{d}V, \tag{9.12}$$

where indices t indicating the current time instant and the corresponding reference configuration have been skipped at the symbols for the sake of transparency. The entropy supply into the domain Ω is a sum of supply through the external surface $\partial\Omega$ and internal entropy sources distributed in Ω, i.e., cf. Eq. (9.2),

$$R = -\int_{\partial\Omega^t} h_i n_i \,\mathrm{d}A + \int_{\Omega} \rho s \,\mathrm{d}V \geq 0, \tag{9.13}$$

where s denotes the entropy sources density while $\boldsymbol{h}$ is the entropy flux vector related to unit area in the current configuration. The minus sign in the first term on the right-hand side is due to the fact that the versor $\boldsymbol{n}$ is directed outwards of the body while R is assumed positive when entropy flows inwards.

The difference between the entropy rate $\dot{S}$ and its supply rate R is denoted by Γ and called the total production of entropy in the unit time. It is postulated that this quantity is never negative, i.e.

$$\Gamma = \dot{S} - R \geq 0. \tag{9.14}$$

Making use of the definitions (9.12) and (9.13) and Eq. (6.13) resulting from the mass conservation law, we can rewrite Eq. (9.14) in the form

$$\Gamma = \int_{\Omega} \rho\dot{\eta} \,\mathrm{d}V + \int_{\partial\Omega} h_i n_i \,\mathrm{d}A - \int_{\Omega} \rho s \,\mathrm{d}V \geq 0. \tag{9.15}$$

This inequality is called the second law of thermodynamics or the law of entropy increase. It establishes the direction of possible energy transformations between its various forms and postulates irreversibility of certain thermodynamical processes. It states that entropy (contrary to mass or energy) does not obey the conservation law—instead, its rate must fulfill the inequality $\Gamma \geq 0$.

A thermodynamical process is called reversible if no entropy is produced in it, i.e. $\Gamma = 0$. Such processes should only be treated as idealizations of real physical processes which are irreversible by their nature.

In classical thermodynamics, it is assumed that the entropy flux $\boldsymbol{h}$ and the density of entropy sources s are related to the heat flux $\boldsymbol{q}$ and density of heat sources r, respectively, by the following formulae,

$$\boldsymbol{h} = \frac{\boldsymbol{q}}{\theta}, \qquad s = \frac{r}{\theta}, \tag{9.16}$$

where θ is the (always positive) *absolute temperature*. The above relations imply the inequality

$$\Gamma = \int_{\Omega} \rho\dot{\eta} \,\mathrm{d}V + \int_{\partial\Omega} \frac{q_i}{\theta} n_i \,\mathrm{d}A - \int_{\Omega} \rho \frac{r}{\theta} \,\mathrm{d}V \geq 0, \tag{9.17}$$

called the Clausius–Duhem inequality in its global form. In view of the Gauss–Ostrogradski theorem, the inequality (9.17) can be transformed to the form

$$\int_{\Omega}\left[\rho\dot{\eta}+\left(\frac{q_i}{\theta}\right)_{,i}-\rho\frac{r}{\theta}\right]\mathrm{d}V\geq 0. \tag{9.18}$$

Since the integration domain is arbitrary, this inequality implies the local form of the Clausius–Duhem inequality,

$$\rho\dot{\eta}+\left(\frac{q_i}{\theta}\right)_{,i}-\rho\frac{r}{\theta}\geq 0, \tag{9.19}$$

or, written equivalently,

$$\theta\rho\dot{\eta}+q_{i,i}-\frac{1}{\theta}q_i\theta_{,i}-\rho r\geq 0. \tag{9.20}$$

9.2 Heat Conduction Problem

When discussing the second law of thermodynamics we have mentioned the absolute temperature θ, i.e. a measurable scalar quantity that describes the thermal state of a material particle. It is proportional to the amount of thermal energy e_T accumulated in unit mass of material; $\theta = 0$ means the absolute zero temperature which corresponds to zero thermal energy. The temperature unit in SI is kelvin [K].

In practical engineering applications a relative temperature T is frequently used. It may be defined with respect to the absolute zero temperature (i.e. $T = \theta$) or to another arbitrary reference temperature—this choice has no importance in view of this section's considerations.

Wherever temperature differences between material particles exist (i.e. temperature gradients are not equal to zero), heat flow occurs. Heat can be exchanged through conduction, convection or radiation. Heat flow is called nonstationary if the temperature field is a function of both spatial location and time, $T = T(\boldsymbol{x}, \tau)$. If $T = T(\boldsymbol{x})$, i.e. $\dot{T} \equiv \frac{\partial T}{\partial \tau} = 0$, then the heat flow is stationary.

Heat conduction in a solid body is the internal heat exchange between material particles of different temperatures. In a thermally isotropic material this phenomenon can be described by the macrosopic Fourier law,

$$\boldsymbol{q} = -\lambda\nabla T, \qquad q_i = -\lambda T_{,i}\,. \tag{9.21}$$

It states that heat flux related to unit area, $\boldsymbol{q}$, is proportional to the temperature gradient ∇T. The minus sign underlines the consequence of the second law of thermodynamics, implying that heat may only flow from places with higher temperature to those with lower temperature. The parameter $\lambda \geq 0$ is the thermal conductivity coefficient, specific to the particular material and assuming values from a very wide spectrum for various materials.

It is easy to generalize the Fourier law to thermally anisotropic materials, i.e. such that the thermal conductivity is different in different spatial directions. In such a case we have

$$\boldsymbol{q} = -\boldsymbol{\lambda}\nabla T, \qquad q_i = -\lambda_{ij} T_{,j}\,, \tag{9.22}$$

where $\boldsymbol{\lambda}$ is the second order tensor of thermal conductivity whose component λ_{ij} denotes the conductivity coefficient in the direction x_i through the surface orthogonal to x_j. According to the second law of thermodynamics, the tensor must be positive definite (for each vector ∇T there must be $\boldsymbol{q} \cdot \nabla T \leq 0$). In the case of orthotropy, there are three orthogonal principal directions in which this tensor is expressed by a diagonal matrix of its components,

$$[\lambda_{ij}] = \begin{bmatrix} \lambda_1 & 0 & 0 \\ 0 & \lambda_2 & 0 \\ 0 & 0 & \lambda_3 \end{bmatrix}, \qquad \lambda_k > 0, \tag{9.23}$$

i.e. we have $q_1 = -\lambda_1 T_{,1}$, $q_2 = -\lambda_2 T_{,2}$, $q_3 = -\lambda_3 T_{,3}$. Isotropy is a special case of orthotropy in which $\boldsymbol{\lambda} = \lambda \boldsymbol{I}$, i.e. $\lambda_{ij} = \lambda \delta_{ij}$.

The equation governing the heat flow in the solid body $\mathcal{B}$ can be constructed by combining the equations of the first law of thermodynamics and the Fourier law of conductivity. Let us first restrict ourselves to the case of a fixed rigid body (or at least a body whose deformations and geometry changes are negligibly small).

As we indicated, thermal energy accumulated in a material particle is proportional to its temperature. This relation is expressed by the following rate formula

$$\dot{e}_T = c\dot{T}, \tag{9.24}$$

where $c = \mathrm{d}e_T/\mathrm{d}T$ denotes the specific heat coefficient, being a material constant. Substitution of this formula into Eq. (9.7) describing the rate of thermal energy, leads to the relation

$$\rho c\dot{T} = -q_{i,i} + \rho r, \tag{9.25}$$

which, after using the Fourier law (9.21), can be transformed to the form

$$\rho c\dot{T} = \mathrm{div}(\lambda\, \mathrm{grad} T) + \rho r = (\lambda T_{,i})_{,i} + \rho r. \tag{9.26}$$

The resulting relation, called the Fourier–Kirchhoff equation, describes non-stationary temperature field in a fixed, thermally isotropic solid. It can be generalized to the case of thermal anisotropy as follows,

$$\rho c\dot{T} = \mathrm{div}(\boldsymbol{\lambda}\, \mathrm{grad} T) + \rho r = (\lambda_{ij} T_{,j})_{,i} + \rho r. \tag{9.27}$$

There are several special cases of the general form (9.26) of the Fourier–Kirchhoff equation. If the conductivity λ and the specific heat ρc are homogeneous (the same in all material particles) and the volumetric heat source output ρr is temperature-independent, the equation assumes the simple linear form

$$\frac{\rho c}{\lambda}\dot{T} = \nabla^2 T + \frac{\rho r}{\lambda}, \tag{9.28}$$

where ∇^2 is the Laplacian (divergence of gradient) operator

$$\nabla^2 = \frac{\partial^2}{\partial x_1^2} + \frac{\partial^2}{\partial x_2^2} + \frac{\partial^2}{\partial x_3^2}. \tag{9.29}$$

In absence of heat sources Eq. (9.28) is reduced to the so-called standard diffusion equation (Poisson equation)

$$\frac{\rho c}{\lambda}\dot{T} = \nabla^2 T, \tag{9.30}$$

which, in the case stationary heat exchange ($\dot{T} = 0$), becomes the Laplace equation

$$\nabla^2 T = 0. \tag{9.31}$$

To determine the temperature field form the Fourier–Kirchhoff equation, it is necessary to account for appropriate initial and boundary conditions. The initial condition defines the temperature distribution in the domain Ω at the time instant $\tau = 0$:

$$T(\boldsymbol{x}, 0) = \hat{T}_0(\boldsymbol{x}), \qquad \boldsymbol{x} \in V. \tag{9.32}$$

The boundary conditions may appear in a number of forms:

1. condition imposed on the temperature distribution on a part of the boundary $\partial\Omega_T$,

$$T = \hat{T}(\boldsymbol{x}, \tau), \qquad \boldsymbol{x} \in \partial\Omega_T, \quad \tau \in [0, t]. \tag{9.33}$$

2. condition imposed on the heat flux distribution on the remaining part of the boundary $\partial\Omega_q = \partial\Omega - \partial\Omega_T$:

$$-q_i n_i = \lambda_{ij} T_{,j} n_i = \hat{q}(\boldsymbol{x}, \tau) + q^{(c)}(\boldsymbol{x}, \tau, T) + q^{(r)}(\boldsymbol{x}, \tau, T),$$
$$\boldsymbol{x} \in \partial\Omega_q, \quad \tau \in [0, t], \tag{9.34a}$$

where $\hat{q}$ is a prescribed inward heat flux through the boundary surface, $q^{(c)}$ a heat flux corresponding to convective heat exchange with an external fluid and $q^{(r)}$ a heat flux corresponding to radiation. Equation (9.34a) is a general form in which some of the fluxes are usually zero. It is stressed that—except for $\hat{q}$—the remaining fluxes are functions of local temperature at the boundary (thus making the boundary condition a mixed one). Namely,

$$q^{(c)} = \alpha[T_\infty(\tau) - T(\boldsymbol{x}, \tau)], \tag{9.34b}$$

where T_∞ is a given temperature of the external fluid and α is the convection coefficient (dependent on the fluid properties and, frequently, also on temperature) and

$$q^{(r)} = \tilde{h}(T_r^4 - T^4(\boldsymbol{x}, \tau)), \tag{9.34c}$$

where T_r is a known temperature of the external radiation source and $\tilde{h}$ is a coefficient dependent on the Stefan–Boltzmann constant and the surface emissivity.

Other, more complex boundary condition formulations may also appear in practical problems, but we limit our discussion to only those listed above.

The Fourier–Kirchhoff equation derived above, along with its initial and boundary conditions, describe the heat flow in a body whose particles are assumed fixed. In thermomechanical problems considered in this study, displacements of material particles are frequently so small that their influence on the form of heat conduction equations is negligible and thus accuracy of such an analysis may be considered satisfactory. However, this is not always the case—several problems including large and fast deformations require formulations that include geometry changes. In such cases all the above derived equations are still valid, but one must be aware that the geometric configuration in which the equations are formulated is the current one, i.e. related to the time instant τ. Hence, all the gradients appearing in the equations should be understood as differentiation with respect to this configuration coordinates, i.e. $(\cdot)_{,i} = \partial(\cdot)/\partial x_i^\tau$, while the derivative $\dot{T}$ in Eq. (9.26) should be understood as the material derivative, cf. Eq. (4.54),

$$\frac{\mathrm{d}T(x(\tau), \tau)}{\mathrm{d}\tau} = \frac{\partial T}{\partial \tau} + \frac{\partial T}{\partial x_i}\frac{\mathrm{d}x_i}{\mathrm{d}\tau} = \frac{\partial T}{\partial \tau} + T_{,i}v_i\,,$$

describing temperature changes in a material point in motion, i.e. a point occupying different locations $\boldsymbol{x}^\tau$ at different time instants τ.

Exercise 9.1 A flat infinite plate with the thickness $2h$, the thermal conductivity coefficient λ and the constant internal heat source density ρr has been placed between two domains of different temperatures T_1 and $T_2 > T_1$ and infinite ability to absorb/release heat (which means in practice that the temperatures on the plate boundaries are given and equal T_1 and T_2, respectively). Compute the temperature distribution along the plate thickness $T(x)$ and values of the heat flux on the boundaries.

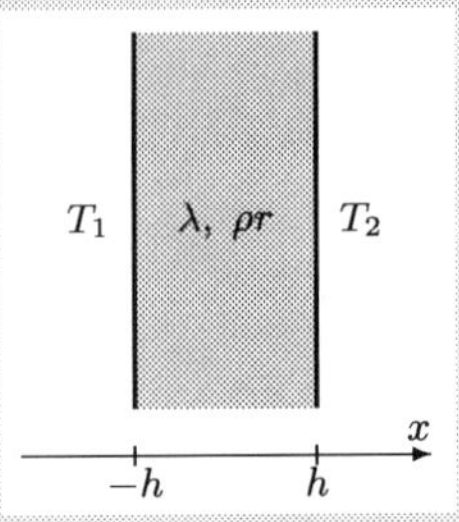

Solution. The Fourier–Kirchhoff equation (9.26) has the following form for the problem on hand:

$$\lambda \frac{\mathrm{d}^2 T}{\mathrm{d}x^2} + \rho r = 0.$$

Its general solution is a quadratic polynomial

$$T(x) = -\frac{\rho r}{2\lambda} x^2 + bx + c.$$

The coefficients b and c can be determined from the boundary conditions:

$$T(-h) = T_1\,, \qquad T(h) = T_2\,,$$

which allows to compute

$$b = \frac{T_2 - T_1}{2h}\,, \qquad c = \frac{T_2 + T_1}{2} + \frac{\rho r}{2\lambda} h^2.$$

It is thus clear that if no heat sources are present then the temperature distribution is linear, otherwise, the temperature within the plate grows higher and assumes a parabolic distribution. The heat flux as a function of x is expressed as

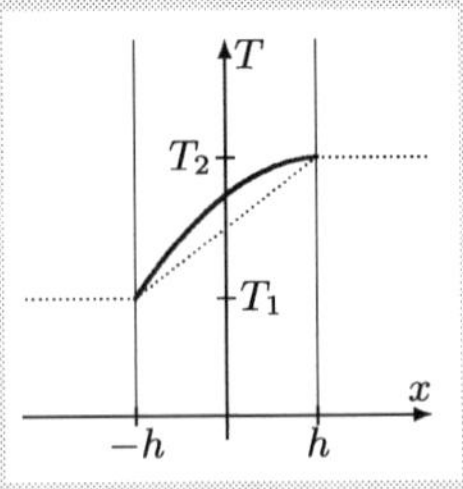

$$q(x) = -\lambda \frac{\mathrm{d}T}{\mathrm{d}x} = \rho r x - \frac{T_2 - T_1}{2h}\,,$$

and its values at the plate boundaries are

$$q_1 = q(-h) = -\frac{T_2 - T_1}{2h} - \rho r h, \qquad q_2 = q(h) = -\frac{T_2 - T_1}{2h} + \rho r h.$$

If no heat sources exist, the heat flux is constant in the entire plate, proportional to the temperature difference $T_2 - T_1$ and directed towards the lower temperature side T_1. Otherwise, additional emission of heat appears on both the boundaries, which increases the outward heat flux on the colder side and decreases the inward heat flux on the warmer one. If $\rho r > (T_2 - T_1)/(2h^2)$ then heat is emitted on the both plate sides, but the emission on the colder side is higher than on the other one.

Exercise 9.2 A flat infinite plate with the thickness $2h$, the thermal conductivity coefficient λ and no internal heat sources has been placed between two domains of different temperatures T_1 and $T_2 > T_1$ that exchange heat with the plate through convection with the coefficient α. Compute the temperature distribution along the plate thickness $T(x)$ and values of heat flux on the boundaries.

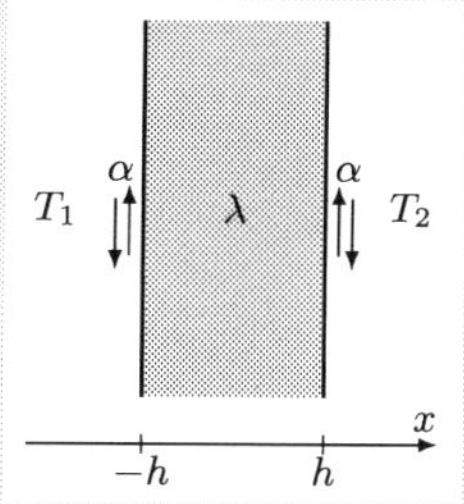

Solution. The Fourier–Kirchhoff equation (9.26) assumes now the following form:

$$\lambda \frac{\mathrm{d}^2 T}{\mathrm{d}x^2} = 0.$$

Its general solution is a linear function $T(x) = bx + c$ which means that the heat flux in the plate is constant and equals $q = -\lambda(\mathrm{d}T/\mathrm{d}x) = -\lambda b$. The coefficients b and c can be determined from the boundary conditions in which the normal versor to the boundary should be assumed 1 at $x = h$ and -1 at $x = -h$,

$$q(-h) = \alpha(T_1 - \tilde{T}_1), \qquad -q(h) = \alpha(T_2 - \tilde{T}_2),$$

Here, $\tilde{T}_1 = T(-h)$ and $\tilde{T}_2 = T(h)$ denote unknown actual temperatures on the plate boundaries. Substituting the general solution form we come at

$$b = \alpha(T_1 + bh - c), \qquad -b = \alpha(T_2 - bh - c),$$

that allow to compute

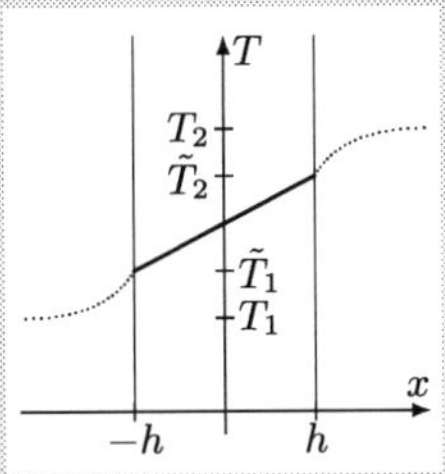

$$b = \frac{T_2 - T_1}{2(h + \frac{\lambda}{\alpha})}, \qquad c = \frac{T_2 + T_1}{2}.$$

In can be seen that the boundary temperatures

$$\tilde{T}_1 = \frac{T_2 + T_1}{2} - \Delta T, \qquad \tilde{T}_2 = \frac{T_2 + T_1}{2} + \Delta T, \qquad \Delta T = \frac{T_2 - T_1}{2(1 + \frac{\lambda}{h\alpha})}$$

differ from the temperatures of external fluids and that the difference is the higher, the lower the convection coefficient α is. The heat flux in the plate equals

$$q = -\lambda b = -\frac{T_2 - T_1}{2(\frac{h}{\lambda} + \frac{1}{\alpha})}.$$

9.3 Fundamental System of Solid Thermomechanics Equations. Thermomechanical Couplings

Let us recall the fundamental system of equations of solid mechanics derived in Chap. 8. It consists of differential equations (8.1) with initial and boundary conditions (8.3) and (8.4) in which the fields listed in (8.2) are unknown.

The initial-boundary formulation of solid thermomechanics, i.e. the fundamental system of differential equations with initial and boundary conditions for this problem, consists of the above-mentioned equations accompanied by the heat conduction equations discussed in this chapter, namely Eqs. (9.21) and (9.25) together with the conditions (9.33) and (9.34). Four equations with four unknowns can be obviously reduced by a simple substitution to one equation—the Fourier–Kirchhoff Eq. (9.26) with the unknown T. These equations, derived for a fixed body, are also valid in a moving deformable body provided they are written in the current configuration and upon additional limitations indicated in the previous section.

Let us now write down the full system of equations that govern thermomechanical phenomena at each material point $\boldsymbol{x} \in \Omega^r$ and at each time instant $\tau \in [0, \bar{t}\,]$:

$$\dot{\rho}^\tau + \rho^\tau \dot{u}^\tau_{i,i} = 0, \tag{9.35a}$$

$$_r\sigma^\tau_{ij,j} + \rho^r f^\tau_i - \rho^r \ddot{u}^\tau_i = 0\,, \tag{9.35b}$$

$$^rF^\tau_{ij} = \delta_{ij} + {}^ru^\tau_{i,j}\,, \tag{9.35c}$$

$$^0_r\varepsilon^\tau_{ij} = \tfrac{1}{2}\left({}^rF^\tau_{ki}\,{}^rF^\tau_{kj} - {}^rF^0_{ki}\,{}^rF^0_{kj}\right), \tag{9.35d}$$

$$_r\sigma^\tau_{ij} = {}^rF^\tau_{ik}\,{}_r\tilde{\sigma}^\tau_{kj}\,, \tag{9.35e}$$

$$f_K(\rho^\tau,\ {}_r\tilde{\sigma}^\tau_{ij},\ {}^0_r\varepsilon^\tau_{ij},\ p^\tau_\alpha,\ {}_r\dot{\tilde{\sigma}}^\tau_{ij},\ {}_r\dot{\varepsilon}^\tau_{ij},\ \dot{p}^\tau_\alpha) = 0\,, \tag{9.35f}$$

$$q^\tau_i = -\lambda T^\tau_{,i}\,, \tag{9.36a}$$

$$\rho^\tau c\dot{T}^\tau = -q^\tau_{i,i} + \rho^\tau r^\tau\,. \tag{9.36b}$$

Gradients in all the above equations are understood as partial derivatives with respect to coordinates in the current configuration C^τ and the time derivatives as material derivatives,

$$(\cdot)_{,i} = \frac{\partial(\cdot)}{\partial x^\tau_i} = \frac{\partial(\cdot)}{\partial x^r_k}\,({}^rF^\tau_{ki})^{-1}\,, \qquad \dot{(\cdot)} = \frac{\partial(\cdot)(x^r,\tau)}{\partial\tau}\,. \tag{9.37}$$

The system (9.35) and (9.36) splits naturally into two parts. Equations (9.35) describe mechanical phenomena while equations (9.36)—thermal phenomena in the behaviour of the body $\mathcal{B}$. In the simplest case of an elastic body subject to very small deformations and small temperature changes, the two parts of the system are independent—they can be solved separately and the solution of one of them does not affect the solution of the other. It is said that in such a case the mechanical and thermal parts of the system are decoupled. This idealized situation occurs seldom; it most cases at least some quantities in the "mechanical" equations depend on temperature and heat flow or some quantities in the "thermal" equations depend on the body's deformation.

The most common sources of couplings that may appear between the two groups of equations are briefly discussed below.

Influence of Geometry Changes on the Temperature Field

In the presence of large deformations, changes in geometric configurations affect gradients and rates of the unknown fields, cf. Eqs. (9.37). Applying the formulae in Eqs. (9.36) one can see that gradients and rates of velocity, i.e. the unknowns in

Eqs. (9.35), appear there. Equations (9.36) cannot therefore be solved without first solving Eqs. (9.35). This kind of coupling may also appear in boundary conditions—geometry changes imply in particular changes of the external surface on which heat exchange occurs which also affects the values of temperature, heat flux or convection conditions defined on this surface by Eqs. (9.33) and (9.34).

Heat Generation Due to Inelastic Deformations

This phenomenon has already been mentioned a few times before. If inelastic deformations occur in a material then the work of stresses on these deformations increases the thermal energy in the material. According to Eq. (9.11), the right-hand side of Eq. (9.36b) should be thus increased by the term $\chi \sigma_{ij} \dot{\varepsilon}_{ij}^{(p)}$ (its precise form depends on the choice of the reference configuration and the conjugate stress and strain measures). Alternatively, this term may be treated as a part of the term ρr—as an additional heat source per unit volume. Similarly as in the previous case, solution of Eqs. (9.36) is only possible if the system of Eqs. (9.35) is solved first.

In Chap. 8, contact boundary conditions were mentioned. They were not discussed in more detail but it is worth remembering that if friction sliding occurs on the contact surfaces then an additional heat source related to this phenomenon appears in the system. This source must also be included in the boundary conditions accompanying Eqs. (9.36).

Temperature Dependence of Material Constants

Material parameter values depend frequently on temperature. If these are coefficients of the thermal conductivity λ or convection α then nonlinearity in Eqs. (9.36) becomes more significant and their solution may become more difficult, but no thermomechanical couplings are generated. The latter is the case, however, when material parameters in the constitutive equations (9.35f) (the Young modulus E, yield stress σ_y, etc.) are temperature-dependent. Then, it is necessary to solve first the system (9.36) in order to find the solution of the system (9.35).

Thermal Expansion of Materials

The phenomenon of thermal expansion is commonly known—increase of the material's temperature implies increase of its volume. If the temperature changes are sufficiently high, the expansion strains are comparable to those resulting from the constitutive equations and thus cannot be neglected. In such cases, constitutive equations have to be modified to include this effect while temperature T must appear as an additional argument of the function f_k in Eq. (9.35f). This issue will be discussed in more detail in Sect. 9.4.

As can be seen from the above summary, thermomechanical couplings related to the physical phenomena discussed have often "unilateral" character—either the heat flux and temperature distribution affect the form of Eqs. (9.35) or the deformations determined from these equations affect the form of Eqs. (9.36). It frequently happens in particular engineering applications that only one of the two unilateral coupling directions is dominant while the other may be neglected. The system of equations (9.35) and (9.36) may be then solved in a staggered way—first its independent part is solved and then the other, dependent on the solution of the first one follows. For example, if the dominant coupling phenomena are the thermal expansion and temperature-dependence of material parameters, then the solution procedure of the system has to start from solving the heat conduction equations (9.36). If the problem is time-dependent and it is necessary to solve the system several times for subsequent time steps, this staggered sequence should be repeated at each of the steps.

In problems in which couplings in both directions occur simultaneously, it is necessary to solve the full system of equations (9.35) and (9.36) with appropriate boundary conditions at each time step.

9.4 Thermal Expansion in Constitutive Equations of Linear Elasticity

In the previous section a series of physical phenomena have been listed that induce couplings between the formulations of mechanics and heat flow in solids. Many of them cannot be straightforwardly represented in a concise mathematical form and this is why we have limited ourselves to just their verbal description. It is relatively easy, however, to mathematically express the thermal expansion phenomena. Such a formulation will now be discussed in more detail for small deformations of linear elastic materials.

Suppose elastic material is simultaneously subject to internal forces and thermal expansion deformations. The local deformation gradient at a particular time instant with respect to an undeformed configuration may then be expressed (obvious indices skipped) as

$$\boldsymbol{F} = \boldsymbol{F}^{(T)}\boldsymbol{F}^{(\sigma)}. \tag{9.38}$$

The gradients $\boldsymbol{F}^{(T)}$ and $\boldsymbol{F}^{(\sigma)}$ describe here the deformations related to thermal and mechanical effects, respectively. It is thus assumed that the two deformation types are mutually independent and—if they appear simultaneously—the total deformation is the combination of the two contributions of different physical nature. Recalling Eq. (4.10), one may also write:

$$u_{i,j} = u_{i,j}^{(T)} + u_{i,j}^{(\sigma)} + u_{i,k}^{(T)}\,u_{k,j}^{(\sigma)} \approx u_{i,j}^{(T)} + u_{i,j}^{(\sigma)}\,. \tag{9.39}$$

Under the assumption of small deformations the second order term in the above equation may be neglected and the Cauchy strain components (4.19) written as

$$\varepsilon_{ij} = \varepsilon_{ij}^{(T)} + \varepsilon_{ij}^{(\sigma)}\,, \tag{9.40}$$

i.e. as the sum of thermal and mechanical strains.

Thermal strains depend on the material temperature according to the formula

$$\varepsilon_{ij}^{(T)} = a_{ij} T, \qquad T = \theta - \theta_0 , \tag{9.41}$$

where T denotes the relative temperature (measured with respect to a reference temperature θ_0, i.e. a temperature of a "natural" state at which there are no thermal strains) and a_{ij} are components of the symmetric thermal expansion tensor. If the material is isotropic then its expansion is the same in all directions and described by a spherical tensor $a_{ij} = a\delta_{ij}$, i.e.

$$\varepsilon_{ij}^{(T)} = a\delta_{ij} T, \tag{9.42}$$

where a denotes the scalar thermal expansion coefficient.

Mechanical strains in a linear elastic material are described by the constitutive equation (7.10) known from Chap. 7 as

$$\varepsilon_{ij}^{(\sigma)} = D_{ijkl}\, \sigma_{kl} \, . \tag{9.43}$$

Substituting the above relations into Eq. (9.40) one obtains

$$\varepsilon_{ij} = D_{ijkl}\, \sigma_{kl} + aT\delta_{ij} \, , \tag{9.44}$$

which, after transformations using the alternative form of the constitutive equation (7.3), can be expressed as

$$\sigma_{ij} = C_{ijkl}(\varepsilon_{kl} - aT\delta_{kl}) = C_{ijkl}\varepsilon_{kl} - C_{ijkk} aT. \tag{9.45}$$

If the material is isotropic then, according to Eqs. (7.12) and (7.13),

$$\begin{aligned} \sigma_{ij} &= \lambda\, \delta_{ij}(\varepsilon_{kk} - 3aT) + 2\mu\,(\varepsilon_{ij} - aT\delta_{ij}) \\ &= \lambda\, \delta_{ij}\varepsilon_{kk} + 2\mu\, \varepsilon_{ij} - (3\lambda + 2\mu) aT\, \delta_{ij} \, . \end{aligned} \tag{9.46}$$

The latter term on the right-hand side of Eq. (9.46), i.e. $-(3\lambda + 2\mu)aT\,\delta_{ij}$, describes the thermal stress, i.e. stress induced in material constrained so as to enforce zero strain and heated up to the temperature T. The minus sign at this term indicates that heating up leads to compression stress which is negative according to our convention. The quantity $(3\lambda + 2\mu)$ is, cf. Eqs. (7.20), (7.21), the bulk modulus (stiffness to volumetric strains), also denoted as $3K$.

Let us present the above equations in the matrix form, analogous to that used in Sect. 7.2.1. Replacing the Lamé constants by the engineering constants, cf. Eq. (7.20), we obtain

$$\begin{Bmatrix} \sigma_{11} \\ \sigma_{22} \\ \sigma_{33} \\ \sigma_{12} \\ \sigma_{23} \\ \sigma_{31} \end{Bmatrix} = \frac{E}{(1+\nu)(1-2\nu)} \begin{bmatrix} 1-\nu & \nu & \nu & & & \\ \nu & 1-\nu & \nu & & & \\ \nu & \nu & 1-\nu & & & \\ & & & \frac{1-2\nu}{2} & & \\ & & & & \frac{1-2\nu}{2} & \\ & & & & & \frac{1-2\nu}{2} \end{bmatrix} \begin{Bmatrix} \varepsilon_{11} \\ \varepsilon_{22} \\ \varepsilon_{33} \\ 2\varepsilon_{12} \\ 2\varepsilon_{23} \\ 2\varepsilon_{31} \end{Bmatrix} - \frac{EaT}{1-2\nu} \begin{Bmatrix} 1 \\ 1 \\ 1 \\ 0 \\ 0 \\ 0 \end{Bmatrix}, \tag{9.47}$$

$$\begin{Bmatrix} \varepsilon_{11} \\ \varepsilon_{22} \\ \varepsilon_{33} \\ 2\varepsilon_{12} \\ 2\varepsilon_{23} \\ 2\varepsilon_{31} \end{Bmatrix} = \begin{bmatrix} 1/E & -\nu/E & -\nu/E & & & \\ -\nu/E & 1/E & -\nu/E & & & \\ -\nu/E & -\nu/E & 1/E & & & \\ & & & 1/G & & \\ & & & & 1/G & \\ & & & & & 1/G \end{bmatrix} \begin{Bmatrix} \sigma_{11} \\ \sigma_{22} \\ \sigma_{33} \\ \sigma_{12} \\ \sigma_{23} \\ \sigma_{31} \end{Bmatrix} + aT \begin{Bmatrix} 1 \\ 1 \\ 1 \\ 0 \\ 0 \\ 0 \end{Bmatrix}. \tag{9.48}$$

In each of the equations the first terms on the right-hand side describe the mechanical effects and are identical to the corresponding terms derived in Sect. 7.2.1, while the second terms describe the thermal expansion effects. Note that both the thermal strains and stresses are spherical.

In the plane stress formulation, i.e. the formulation in which $\sigma_{33} = \sigma_{23} = \sigma_{13} = 0$ (a planar plate loaded in its plane and subject to temperature changes), we have to remove from Eq. (9.48) three equations, leaving only those corresponding to the in-plane components. We then arrive at

$$\begin{Bmatrix} \varepsilon_{11} \\ \varepsilon_{22} \\ 2\varepsilon_{12} \end{Bmatrix} = \begin{bmatrix} 1/E & -\nu/E & \\ -\nu/E & 1/E & \\ & & 1/G \end{bmatrix} \begin{Bmatrix} \sigma_{11} \\ \sigma_{22} \\ \sigma_{12} \end{Bmatrix} + aT \begin{Bmatrix} 1 \\ 1 \\ 0 \end{Bmatrix}. \tag{9.49}$$

Inversion of this relation leads to the formula

$$\begin{Bmatrix} \sigma_{11} \\ \sigma_{22} \\ \sigma_{12} \end{Bmatrix} = \frac{E}{1-\nu^2} \begin{bmatrix} 1 & \nu & 0 \\ \nu & 1 & 0 \\ 0 & 0 & \frac{1-\nu}{2} \end{bmatrix} \begin{Bmatrix} \varepsilon_{11} \\ \varepsilon_{22} \\ 2\varepsilon_{12} \end{Bmatrix} - \frac{EaT}{1-\nu} \begin{Bmatrix} 1 \\ 1 \\ 0 \end{Bmatrix}. \tag{9.50}$$

We can thus see that thermal stresses appearing in a heated plate constrained so as to block in-plane strains equal $\sigma_{11} = \sigma_{22} = -\frac{EaT}{1-\nu}$. They are lower than in an analogous 3D case (where they equal $-\frac{EaT}{1-2\nu}$) because here the plate can freely expand in the transversal direction.

The transversal strain can be determined from Eqs. (9.48) and (9.50) as

$$\varepsilon_{33} = -\frac{\nu}{E}(\sigma_{11} + \sigma_{22}) + aT = -\frac{\nu}{1-\nu}(\varepsilon_{11} + \varepsilon_{22}) + \frac{1+\nu}{1-\nu}aT. \tag{9.51}$$

In particular, if the heated plate cannot deform in its plane then the transversal strain is $\varepsilon_{33} = \frac{1+\nu}{1-\nu} aT$, i.e. higher than at free 3D expansion when the strains are $\varepsilon_{11} = \varepsilon_{22} = \varepsilon_{33} = aT$.

In the plane strain formulation, in which $\varepsilon_{33} = \varepsilon_{23} = \varepsilon_{13} = 0$, we proceed in a similar way, this time starting from the reduction of Eq. (9.47) to only three in-plane component equations,

$$\begin{Bmatrix} \sigma_{11} \\ \sigma_{22} \\ \sigma_{12} \end{Bmatrix} = \frac{E}{(1+\nu)(1-2\nu)} \begin{bmatrix} 1-\nu & \nu & 0 \\ \nu & 1-\nu & 0 \\ 0 & 0 & \frac{1-2\nu}{2} \end{bmatrix} \begin{Bmatrix} \varepsilon_{11} \\ \varepsilon_{22} \\ 2\varepsilon_{12} \end{Bmatrix} - \frac{EaT}{1-2\nu} \begin{Bmatrix} 1 \\ 1 \\ 0 \end{Bmatrix}. \tag{9.52}$$

Inversion of this relation leads to the formula

$$\begin{Bmatrix} \varepsilon_{11} \\ \varepsilon_{22} \\ 2\varepsilon_{12} \end{Bmatrix} = \frac{1+\nu}{E} \begin{bmatrix} 1-\nu & -\nu & 0 \\ -\nu & 1-\nu & 0 \\ 0 & 0 & 2 \end{bmatrix} \begin{Bmatrix} \sigma_{11} \\ \sigma_{22} \\ \sigma_{12} \end{Bmatrix} + (1+\nu)aT \begin{Bmatrix} 1 \\ 1 \\ 0 \end{Bmatrix}. \tag{9.53}$$

Thus, if a material in plane strain is heated up at no mechanical loads then its in-plane strains equal $\varepsilon_{11} = \varepsilon_{22} = (1+\nu)aT$, i.e. they are higher than in the case of free 3D expansion (which seems natural as the transversal component ε_{33} is blocked).

The transversal stress can be determined from Eqs. (9.47) and (9.53) as

$$\sigma_{33} = \frac{E\nu(\varepsilon_{11} + \varepsilon_{22})}{(1-2\nu)(1+\nu)} - \frac{EaT}{1-2\nu} = \nu(\sigma_{11} + \sigma_{22}) - EaT. \tag{9.54}$$

In particular, if no in-plane stress is present then the transversal thermal stress is $\sigma_{33} = -EaT$.

Exercise 9.3 Consider an isotropic elastic deformable solid with the material properties $E = 80\,000$ MPa, $\nu = \frac{1}{3}$ and $a = 10^{-5}$ [1/K]. Assume it has been heated up to $T = 100$ K. Compute thermal stresses for the three cases: (i) 3D deformation at fully constrained strains, (ii) plane stress at constrained in-plane strains and (iii) plane strain at zero in-plane stresses. In each case, compute how much should the temperature T raise to make the thermal stresses exceed the yield limit $\sigma_y = 160$ MPa.

Solution. In the 3D case Eq. (9.47) should be used. At zero strain we obtain $\sigma_{11} = \sigma_{22} = \sigma_{33} = -\frac{EaT}{1-2\nu} = -240$ MPa (the remaining components are zero).

In the plane stress case Eq. (9.50) should be applied. At zero in-plane strain we obtain $\sigma_{11} = \sigma_{22} = -\frac{EaT}{1-\nu} = -120$ MPa (the remaining components are zero).

In the plane strain case Eq. (9.54) should be used. At zero in-plane stress we obtain $\sigma_{33} = -EaT = -80\,\text{MPa}$. In all the three cases compression stress is obtained.

According to the yield condition (7.111), the yield limit value σ_y must be compared against the Huber–Mises equivalent stress $\bar{\sigma}$, Eq. (7.114), corresponding to the current stress state. Since stress is proportional to T, it is easy to show that $\bar{\sigma}$ is a linear function of T. The limit temperature increment leading to initiation of plastic deformation is thus expressed as $\gamma \cdot 100\,\text{K}$ where $\gamma = \sigma_y/\bar{\sigma}$.

In the 3D case the stress state is spherical, i.e. its deviator is zero. Thus, $\bar{\sigma} = 0$ and further heating of the material will not lead to plastic deformation, no matter how high T is.

In the plane stress case the deviatoric stress components are computed from Eq. (7.111). They are: $\sigma^{\text{D}}_{11} = \sigma^{\text{D}}_{22} = -40$ and $\sigma^{\text{D}}_{33} = 80\,\text{MPa}$ (the remaining components are zero). Note that even upon the plane stress assumption the transversal deviatoric stress component is non-zero. Thus,

$$\bar{\sigma} = \sqrt{\frac{3}{2}(40^2 + 40^2 + 80^2)} = 120\,\text{MPa}.$$

This is 75 % of the yield limit ($\gamma = \frac{4}{3}$). The limit could be exceeded if the temperature increment exceeded the value of $T = \frac{4}{3} \cdot 100 \approx 133\,\text{K}$.

In the plane strain case the deviatoric stress components are computed as $\sigma^{\text{D}}_{11} = -\frac{2}{3} \cdot 80$, $\sigma^{\text{D}}_{22} = \sigma^{\text{D}}_{33} = \frac{1}{3} \cdot 80\,\text{MPa}$. Simple calculations lead to the result $\bar{\sigma} = 80\,\text{MPa}$, which is the half of the yield limit ($\gamma = 2$), i.e. to induce plastic flow, the temperature should exceed the value $T = 200\,\text{K}$.

Chapter 10
Variational Formulations in Solid Thermomechanics

In the previous chapters, the so-called local formulations of typical problems related to solid thermomechanics have been discussed. Their characteristic common feature is that the derived differential equations must hold true at each material point and at each time instant and thus the unknown fields sought must fulfill them in the entire domain $\Omega \times [0, \bar{t}\,]$. We have also mentioned, without coming into detail, that solution for the presented equations is usually difficult—effective analytical methods do not exist while approximate methods require advanced numerical tools and procedures.

This chapter's objective is thus to acquaint the reader with fundamentals of modern approximate solution methods of boundary and initial-boundary value problems of nonlinear solid thermomechanics.

10.1 Variational Principles—Introduction

As an alternative to the already presented local (so-called strong) formulation we will now briefly introduce the so-called variational formulation (also refereed to as weak).[1] Its essence is to construct a functional whose minimum (or sometimes only stationarity) is achieved for the function being a solution of the corresponding boundary value problem. Although formally equivalent, the weak formulation has a number of essential advantages compared to the local approach. First, the concept of a variational functional makes it possible to collect in one expression many different equations describing the problem considered. Second, the functional has frequently a certain physical sense and its value is independent of the selected coordinate system (to obtain local equations of the problem in another coordinate system it is sufficient to express the functional in this system and apply the variational procedure). Third, variational principles are frequently a basis to investigating stability and uniqueness

[1] Variational formulations are sometimes defined as only special cases of a more general class of weak formulations. This distinction of notions is not essential in view of our further discussion.

M. Kleiber and P. Kowalczyk, *Introduction to Nonlinear Thermomechanics of Solids*, Lecture Notes on Numerical Methods in Engineering and Sciences,
DOI 10.1007/978-3-319-33455-4_10

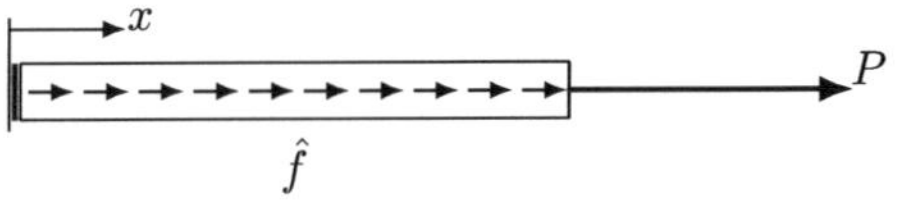

Fig. 10.1 One-dimensional example, geometry and load of a bar

of solutions. Fourth, the variational formulations form a natural basis to construct methods of approximate problem solution and analyse their accuracy.

As it is frequently practised in this book, we start our presentations of variational principles from the simplest, one-dimensional example of linear statics of a stretched bar, shown in Fig. 10.1. The load consists of a force P applied at the right end and a mass load f uniformly distributed along its length (to eliminate the unknown ρ in the equations, the notation $\hat{f} = \rho f =$ const is introduced here for the volumetric forces). At the left bar end the displacement $\hat{u}$ is imposed (to be more general it is assumed non-zero). Under the assumption of small deformations, the problem is described by:

- the equilibrium equations

$$\sigma_{,x} + \hat{f} = 0, \qquad x \in (0, l), \tag{10.1a}$$

- the geometric equation

$$\varepsilon = u_{,x}\,, \qquad x \in (0, l), \tag{10.1b}$$

- the constitutive equation

$$\sigma = E\varepsilon, \qquad x \in (0, l), \tag{10.1c}$$

- the stress boundary condition

$$\sigma = \frac{P}{A} = \hat{t}, \qquad x = l, \tag{10.1d}$$

- the kinematic boundary condition

$$u = \hat{u}, \qquad x = 0. \tag{10.1e}$$

A simplified notation has been introduced in which tensorial quantities are replaced by scalars according to the rule:

$$\sigma_{ij} \longrightarrow \sigma_{xx} = \sigma, \quad \varepsilon_{ij} \longrightarrow \varepsilon_{xx} = \varepsilon, \quad u_i \longrightarrow u_x = u, \quad \rho f_i \longrightarrow \rho f_x = \hat{f}.$$

Eliminating the unknowns σ, ε from the equations and denoting additionally

$$\tilde{f} = \frac{\hat{f}}{E}, \qquad \tilde{t} = \frac{\hat{t}}{E}, \tag{10.2}$$

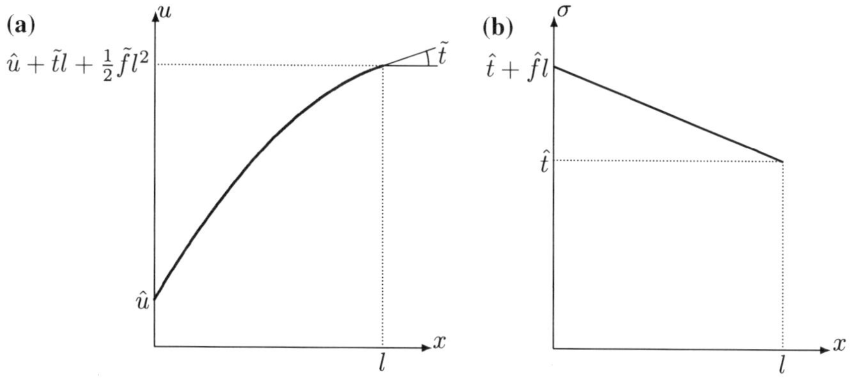

Fig. 10.2 One-dimensional example, analytical solution: **a** displacement, **b** stress

one may rewrite the system of equations in the form

$$u_{,xx} + \tilde{f} = 0, \qquad x \in (0, l), \tag{10.3a}$$

$$u(0) = \hat{u}, \tag{10.3b}$$

$$u_{,x}(l) = \tilde{t}, \tag{10.3c}$$

which contains only one unknown $u(x)$. It is easy to verify that the analytical solution of the system (10.3) has the following form,

$$u(x) = \hat{u} + (\tilde{t} + \tilde{f}l)x - \frac{1}{2}\tilde{f}x^2, \tag{10.4}$$

and the corresponding stress distribution is linear,

$$\sigma(x) = E(\tilde{t} + \tilde{f}(l - x)) = \hat{t} + \hat{f}(l - x). \tag{10.5}$$

The graphs of both the functions are depicted in Fig. 10.2.

It was obviously not our objective to just obtain this trivial result. Rather, we will now indicate ways to obtain an approximate solution of the problem (10.3) that could be followed in more complex situations. Even if some numerical methods (like the finite difference method) are typically formulated directly in terms of local equations, several others (like the finite element method) require quite a different formulation.

Let us define the set $\mathcal{P}$ of functions $u(x)$, $x \in [0, l]$, called trial functions, fulfilling the kinematic boundary condition, i.e.

$$\mathcal{P} = \{u : [0, l] \rightarrow R, \quad u(0) = \hat{u}\}. \tag{10.6}$$

The set $\mathcal{P}$ is called the set of *kinematically admissible displacement fields*, i.e. those which satisfy by the definition the kinematic boundary conditions (10.3b).

Let us also define the set $\mathcal{W}$ of *weight functions* $\delta u(x), x \in [0, l]$, called variations of the function $u(x)$, the only difference being that they must fulfill homogeneous conditions at the end $x = 0$, i.e.

$$\mathcal{W} = \{\delta u : [0, l] \longrightarrow R, \quad \delta u(0) = 0\}. \tag{10.7}$$

The way we defined the sets $\mathcal{P}$ and $\mathcal{W}$ does not impose any additional limitations on functions u and δu. Thus, in the broadest meaning, they may be understood as sets of all kinematically admissible displacement fields and their variations. In such a case it is clear that the solution of the problem (10.3) (if exists) belongs to $\mathcal{P}$ and

$$\forall_{u \in \mathcal{P}} \forall_{\delta u \in \mathcal{W}} \quad u + \delta u \in \mathcal{P}. \tag{10.8}$$

However, one may formulate additional requirements for the forms of $u(x)$ making $\mathcal{P}$ contain only a certain class of kinematically admissible fields, not necessarily including the local problem solution. It is then important that definitions of the classes $\mathcal{P}$ and $\mathcal{W}$ are compatible, i.e. the condition (10.8) is fulfilled. We shall discuss this issue later; at the moment we only assume that the trial functions contain the solution of Eqs. (10.3) and require that they are sufficiently smooth (we do not formulate the corresponding conditions explicitly and assume that the functions $u(x)$ fulfill the differentiability requirements necessary to perform all further transformations).

The system of equations (10.3) will now be transformed as follows. If a function $u \in \mathcal{P}$ fulfills Eqs. (10.3), it must also fulfill the following equation:

$$\int_0^l (u_{,xx} + \tilde{f})\delta u \, \mathrm{d}x - \left(u_{,x}(l) - \tilde{t}\right) \delta u(l) = 0, \tag{10.9}$$

in which δu is an arbitrary variation of the set $\mathcal{W}$. Using the equality

$$u_{,xx}\delta u = (u_{,x}\delta u)_{,x} - u_{,x}\delta u_{,x}$$

and denoting, in analogy to the definition (10.1b), $\delta\varepsilon = \delta u_{,x}$, one may transform the above integral equation to the form

$$[u_{,x}\delta u]_0^l - \int_0^l u_{,x}\delta\varepsilon \, \mathrm{d}x + \int_0^l \tilde{f}\delta u \, \mathrm{d}x - u_{,x}(l)\delta u(l) + \tilde{t}\delta u(l) = 0.$$

Since $\delta u(0) = 0$, the final result takes the form

$$\int_0^l u_{,x}\delta\varepsilon \, \mathrm{d}x = \int_0^l \tilde{f}\delta u \, \mathrm{d}x + \tilde{t}\delta u(l) \, . \tag{10.10}$$

The *weak formulation* of the problem may be now defined as follows:

> *for given values of $\tilde{f}$, $\hat{u}$ and $\tilde{t}$, a function $u \in \mathcal{P}$ is sought such that Eq. (10.10) is fulfilled for each variation $\delta u \in \mathcal{W}$.*

This function is called the weak (or generalized) solution of the problem (10.3) as the above variational equation is not the only possible for the problem considered (which will be discussed later in the context of 3D problems).

Equation (10.10) is called the *virtual work principle* and the variation δu appearing in it—the virtual displacement. The physical sense of Eq. (10.10) is that the work performed by external forces on virtual displacements must be equal to the work performed by internal forces on strains related to virtual displacements (in this particular equation, all the work terms have been scaled by $1/E$).

Considering the variational formulation as an alternative to the local formulation makes only sense if its solution $u(x)$ is also the local problem solution. We have just shown the inverse relation—if $u(x)$ fulfills Eq. (10.3) then it also fulfills Eq. (10.10). Let us show that the opposite implication is also true.

Assume that $u(x)$ is the weak solution, i.e. for an arbitrary $\delta u \in \mathcal{W}$ and $u(0) = \hat{u}$ Eq. (10.10) holds true. Integrating by parts and using the condition $\delta u(0) = 0$, we arrive at

$$-\int_0^l u_{,xx}\delta u \,\mathrm{d}x + u_{,x}(l)\delta u(l) = \int_0^l \hat{f}\delta u \,\mathrm{d}x + \hat{t}\delta u(l), \tag{10.11}$$

i.e. the equation

$$\int_0^l (u_{,xx} + \hat{f})\delta u \,\mathrm{d}x + [\hat{t} - u_{,x}(l)]\delta u(l) = 0, \tag{10.12}$$

that must be fulfilled for arbitrary $\delta u \in \mathcal{W}$. In particular, Eq. (10.12) must hold true for the following variation δu:

$$\delta u(x) = x(l - x)(u_{,xx} + \hat{f}), \tag{10.13}$$

fulfilling the required condition $\delta u(0) = 0$ and the condition $\delta u(l) = 0$. Substituting Eq. (10.13) into Eq. (10.12), we obtain

$$\int_0^l x(l - x)(u_{,xx} + \hat{f})^2 \,\mathrm{d}x = 0. \tag{10.14}$$

Noting that $(u_{,xx} + \hat{f})^2 \geq 0$ and that for $x \in (0, l)$ we have also $x(l - x) > 0$, we finally get

$$u_{,xx} + \hat{f} = 0, \qquad x \in (0, l). \tag{10.15}$$

In view of the above equality, Eq. (10.12) may be simplified to the form

$$\delta u(l)[\hat{t} - u_{,x}(l)] = 0, \tag{10.16}$$

which, upon arbitrary choice of $\delta u(l)$, implies

$$u_{,x} = \hat{t} \ \text{ for } \ x = l. \tag{10.17}$$

In other words, if u is the variational problem solution then it fulfills all the required local equations: the kinematic boundary condition $u(0) = 0$ (by definition) and, as has been shown, the equilibrium equation (10.15) and the stress boundary condition (10.17). The equivalence of the two formulations, local and variational, has thus been demonstrated.

It is noteworthy that the above conclusion is only true if the functions appearing in both the formulations are smooth enough to assure that the appropriate operations are possible. Each of the formulations has different requirements in this context. As will be shown soon, imposed limitations on the functions' smoothness may result in the situation in which the strong formulation has no solution while the weak formulation has one. This is obviously a convincing argument for using the latter formulation in computational practice.

It is also worth pointing out that the variational formulation does not explicitly include the requirement $u_{,x}(l) = \tilde{t}$. This condition results implicitly from the variational principle and is called the *natural condition*. The condition $u(0) = \hat{u}$ is treated in a different way—the trial functions u must satisfy it by definition. The conditions of this type are called the *essential conditions*. The fact that the variational problem solution fulfills natural boundary conditions is important for construction of variational principles for more complex problems, which will be discussed in more detail later.

Let us now go back to our example. The strong solution of the considered problem has the form (10.3). The procedure of obtaining the corresponding weak solution goes as follows. Assume first that the set of trial functions $\mathcal{P}$ contains functions of the general form

$$u(x) = a + bx + cx^2, \tag{10.18}$$

in which, due to the kinematic boundary condition $u(0) = \hat{u}$, one has to assume $a = \hat{u}$, i.e.

$$u(x) = \hat{u} + bx + cx^2. \tag{10.19}$$

The coefficients b and c are as of now unknown. The set of weight functions $\mathcal{W}$ is defined as

$$\delta u(x) = dx + ex^2, \tag{10.20}$$

which assures that the functions δu satisfy the homogeneous boundary condition at $x = 0$ as well as the condition $u + \delta u \in \mathcal{P}$. Substituting Eqs. (10.19) and (10.20) into Eq. (10.10) we obtain the following equation:

$$\int_0^l (b + 2cx)(d + 2xe)\mathrm{d}x = \int_0^l \tilde{f}(dx + ex^2)\mathrm{d}x + \tilde{t}(dl + el^2), \qquad (10.21)$$

which must hold true for each variation $\delta u \in \mathcal{W}$, i.e. for arbitrary d and e. Integration leads to the result

$$bdl + (be + cd)l^2 + \frac{4}{3}cel^3 = \tilde{f}\left(\frac{1}{2}dl^2 + \frac{1}{3}el^3\right) + \tilde{t}(dl + el^2),$$

which can be reordered to the form

$$d\left(bl + cl^2 - \frac{1}{2}\tilde{f}l^2 - \tilde{t}l\right) + e\left(bl^2 + \frac{4}{3}cl^3 - \frac{1}{3}\tilde{f}l^3 - \tilde{t}l^2\right) = 0. \qquad (10.22)$$

Since Eq. (10.22) must be true for any d and e, the two expression in parentheses must vanish. This leads to the system of two linear equations with two unknowns,

$$\begin{aligned} bl + cl^2 &= \frac{1}{2}\tilde{f}l^2 + \tilde{t}l\,, \\ bl^2 + \frac{4}{3}cl^3 &= \frac{1}{3}\tilde{f}l^3 + \tilde{t}l^2, \end{aligned}$$

whose solution is

$$c = -\frac{1}{2}\tilde{f}, \qquad b = \tilde{f}l + \tilde{t}\,. \qquad (10.23)$$

Substituting the values of b and c into Eq. (10.19), we arrive at the result in the form

$$u(x) = \hat{u} + (\tilde{t} + \tilde{f}l)x - \frac{1}{2}\tilde{f}x^2, \qquad (10.24)$$

which is identical to the strong solution (10.4) obtained previously.

This conclusion is not surprising—as we have just demostrated, the two formulations, strong and weak, are equivalent. However, no essential advantage of the weak formulation in the context of solution methods is visible yet. To see this, we now repeat the above procedure assuming different definitions of the sets $\mathcal{P}$ and $\mathcal{W}$. We namely consider the case in which the set $\mathcal{P}$ does not contain the strong problem solution (10.4).

Consider two definitions of such sets of trial functions and their variations:

$$u(x) = \hat{u} + ax, \qquad \delta u(x) = cx \qquad (10.25)$$

and

$$u(x)=\begin{cases}\hat{u}+ax, & x\in[0,\frac{l}{2}),\\ \hat{u}+a\frac{l}{2}+b(x-\frac{l}{2}), & x\in[\frac{l}{2},l],\end{cases}$$
$$\delta u(x)=\begin{cases}cx, & x\in[0,\frac{l}{2}),\\ c\frac{l}{2}+d(x-\frac{l}{2}), & x\in[\frac{l}{2},l],\end{cases} \tag{10.26}$$

i.e. the classes of linear functions and continuous piece-wise linear functions (on two halves of the segment $[0,l]$). In both the cases the functions u and δu fulfill the required kinematic conditions. It is also clear that none of them may fulfill Eq. (10.3a). Moreover, the functions (10.26) do not even meet the condition of second order differentiability required in this equation. Nevertheless, none of the two forms, (10.25) and (10.26), makes seeking the weak solution impossible—to compute the integrals in Eq. (10.10) the first order differentiability is only required, and not even at all the points of the function's domain.

Substitution of the function (10.25) into Eq. (10.10) leads immediately to the simple equation

$$cl\left(a-\frac{1}{2}\tilde{f}l-\tilde{t}\right)=0,$$

true for each c if only $a=\tilde{t}+\frac{1}{2}\tilde{f}l$. The weak solution has thus the form

$$u(x)=\hat{u}+\left(\tilde{t}+\frac{1}{2}\tilde{f}l\right)x, \tag{10.27}$$

and the corresponding stress distribution becomes

$$\sigma(x)=E\left(\tilde{t}+\frac{1}{2}\tilde{f}l\right)=\hat{t}+\frac{1}{2}\hat{f}l=\text{const.} \tag{10.28}$$

Substituting the functions (10.26) into the integrals in Eq. (10.10) one obtains

$$\int_0^l u_{,x}\delta u_{,x}\,\mathrm{d}x=\int_0^{\frac{l}{2}}ac\,\mathrm{d}x+\int_{\frac{l}{2}}^l bd\,\mathrm{d}x=\frac{1}{2}(ac+bd)l,$$
$$\int_0^l \tilde{f}\delta u\,\mathrm{d}x=\int_0^{\frac{l}{2}}\tilde{f}cx\,\mathrm{d}x+\int_{\frac{l}{2}}^l\tilde{f}\left[\frac{cl}{2}+d\left(x-\frac{l}{2}\right)\right]\mathrm{d}x=\frac{3}{8}\tilde{f}cl^2+\frac{1}{8}\tilde{f}dl^2.$$

After reordering of terms, Eq. (10.10) assumes the form

$$\frac{1}{2}cl\left(a-\frac{3}{4}\tilde{f}l-\tilde{t}\right)+\frac{1}{2}dl\left(b-\frac{1}{4}\tilde{f}l-\tilde{t}\right)=0.$$

Since the variation coefficients c and d are arbitrary, the two expressions in parentheses must vanish which leads to the result

$$a = \tilde{t} + \frac{3}{4}\tilde{f}l, \qquad b = \tilde{t} + \frac{1}{4}\tilde{f}l.$$

The displacement field has thus the form

$$u(x) = \begin{cases} \hat{u} + \left(\tilde{t} + \frac{3}{4}\tilde{f}l\right)x, & x \in [0, \frac{l}{2}), \\ \hat{u} + \frac{1}{4}\tilde{f}l^2 + \left(\tilde{t} + \frac{1}{4}\tilde{f}l\right)x, & x \in [\frac{l}{2}, l], \end{cases} \tag{10.29}$$

while the stress field the form

$$\sigma_x(x) = \begin{cases} E\left(\tilde{t} + \frac{3}{4}\tilde{f}l\right) = \hat{t} + \frac{3}{4}\hat{f}l, & x \in [0, \frac{l}{2}), \\ E\left(\tilde{t} + \frac{1}{4}\tilde{f}l\right) = \hat{t} + \frac{1}{4}\hat{f}l, & x \in [\frac{l}{2}, l]. \end{cases} \tag{10.30}$$

Graphs of the derived weak solutions are shown in Fig. 10.3. Both differ from the analytical solution. It is noteworthy, however, that in the predefined class of function each of the solutions is a very good approximate of the analytical solution. Our intuition leads to the conclusion (easy to verify numerically) that if the segment $[0, l]$ were divided into $n > 2$ equal sub-segments in the definition of the piece-wise linear function class (10.26), then the graph of the computed solution $u(x)$ in the form of a broken line would approximate the analytical solution's parabola the better, the higher were the number of segments n. Moreover, if the bar's load were more complex than in the example (one can think of $\hat{f}(x)$ as a trigonometric or exponential function, for instance) and the analytical solution wouldn't have that simple form of

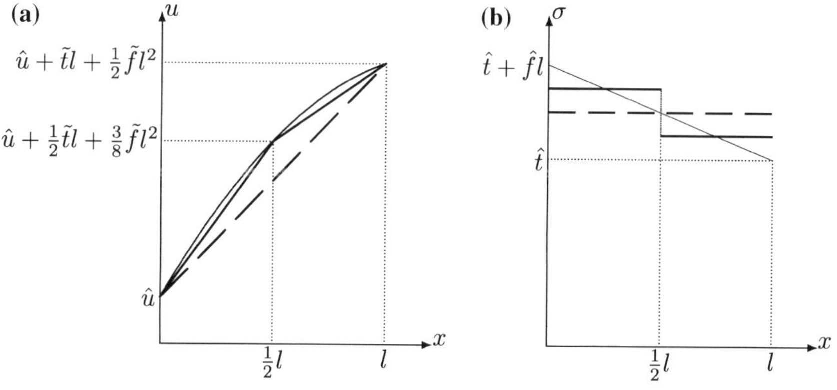

Fig. 10.3 One-dimensional example: approximate (weak) solutions, **a** displacements, **b** stresses; *dashed line* for the trial functions (10.25), *solid line* for the trial functions (10.26), *thin line*—the analytical solution

a low order polynomial, the problem—from the viewpoint of the variational problem solution methodology—would not be much more complicated. In other words, the formulation allows to obtain approximate solutions of mechanics problems that may be sought in classes of functions with low differentiability requirements indeed.

Approximate character of the computed solutions has an important consequence made visible in the stress graphs in Fig. 10.3b. Approximate stress values do not accurately fulfill the natural boundary conditions (10.1d). Stress at the bar's right end, that is required to equal $\hat{t}$, equals in the weak solutions $\hat{t} + \frac{1}{2}\hat{f}l$ for the linear function and $\hat{t} + \frac{1}{4}\hat{f}l$ for the piece-wise linear function, respectively. In the latter case discontinuity of the stress field at $x = l/2$ may also be observed implying that the equilibrium equation cannot be fulfilled at this point. This is not surprising—after all, we are still talking about approximate solutions. Numerical practice shows (which can also be formally proved) that the better the selected class of trial functions allows to approximate the unknown exact solution, the lower the errors in equilibrium equations and natural boundary conditions are. For example, increasing the number of sub-segments n in dividing the segment $[0, l]$ at the piece-wise linear function class definition leads to more dense and lower steps in the "stairs" in the $\sigma(x)$ graph making errors smaller.

To complete our discussion regarding the simple 1D example, let us define the following functional,

$$J[u] = \int_0^l \left(\frac{1}{2} E (u_{,x})^2 - \hat{f} u \right) \mathrm{d}x - \hat{t} u(l), \tag{10.31}$$

i.e. expression that assumes a scalar value for an arbitrary function $u(x) \in \mathcal{P}$. It is called the *potential energy functional* of the mechanical system on hand. It is interesting to analyse the form of the functional's argument $u(x)$ at which it assumes its extreme value.

The variational calculus will be helpful to this aim. Its fundamentals can be found in [1], for instance. Up to now, while defining the variations $\delta u \in \mathcal{W}$, we have not limited their values. Let us note that all conclusions drawn in our discussion remain valid in particular if one assumes $|\delta u| < \epsilon$ for each $x \in [0, l]$ where $\epsilon > 0$ is an arbitrary number whose value can be selected as very small. In view of this new requirement, the variations $\delta u(x)$ become infinitesimal, kinematically admissible perturbations of the functions $u(x)$ (in a similar sense as differentials $\mathrm{d}x$ are infinitesimal perturbations of the variable x). With this definition, stationarity of the functional J is ensured if, for each variation $\delta u(x)$ (infinitesimal perturbation of the function $u(x)$), the functional variation δJ (perturbation of its value) is zero.

The variation δJ can be determined from Eq. (10.31) following rules similar to those of function differentiation. Since the integration limits are not subject to variation, we can write

$$\delta J = \int_0^l (E u_{,x} \delta u_{,x} + \hat{f} \delta u) \,\mathrm{d}x - \hat{t} \delta(l).$$

Requiring this expression to vanish, we may divide the equation side-wise by E, make use of (10.2) and finally arrive at Eq. (10.10), i.e. the weak formulation of our problem. Since, as shown earlier, this formulation is equivalent to the strong formulation (10.3), Eqs. (10.3a) and (10.3c) are the so-called *Euler equations* for the functional (10.31)—their fulfillment is equivalent to the stationarity condition of this functional.

Let us compute the second variation of the potential energy functional. Remembering that $\delta(\delta u) \equiv 0$ and $E > 0$ we obtain the result

$$\delta^2 J = \int_0^l E(\delta u_{,x})^2 \, \mathrm{d}x \geq 0,$$

whose non-negative sign (positive for each nonzero variation δu) indicates that the extremum sought is in fact the minimum.

The above derivations allow to conclude that he weak problem solution minimizes the potential energy of the system defined by Eq. (10.31). If the function class $\mathcal{P}$ includes the analytical solution then the functional reaches its minimum among all kinematically admissible displacement fields. If this is not the case, as it was in the considered classes (10.25) and (10.26), we obtain the minimum functional value in the class of the given trial functions. Determination of the form of J and calculation of its values for the analytical solution (10.4) and the approximate solutions (10.27) and (10.29) is left to the reader as an exercise.

10.2 Variational Formulations for Linear Mechanics Problems

10.2.1 Virtual Work Principle and Potential Energy

Consider the problem of linear statics known from Chap. 8 in the form (8.7)–(8.8),

$$\left.\begin{aligned} \sigma_{ij,j} + \hat{f}_i &= 0 \\ \sigma_{ij} &= C_{ijkl}\varepsilon_{kl} \\ \varepsilon_{ij} &= \tfrac{1}{2}(u_{i,j} + u_{j,i}) \end{aligned}\right\} \quad \boldsymbol{x} \in \Omega, \tag{10.32a}$$

$$u_i = \hat{u}_i\,, \qquad \boldsymbol{x} \in \partial\Omega_u\,, \tag{10.32b}$$

$$\sigma_{ij} n_j = \hat{t}_i\,, \qquad \boldsymbol{x} \in \partial\Omega_\sigma\,, \tag{10.32c}$$

where, similarly as in the previous section, the notation $\hat{f}_i = \rho f_i$ has been adopted. Equations (10.32a) must be fulfilled at all the material points of the body occupying the domain Ω while the boundary conditions (10.32c)–(10.32b)—at all the material points of the two disjoint and complementary parts of its boundary $\partial\Omega = \partial\Omega_u \cup \partial\Omega_\sigma$, $\partial\Omega_u \cap \partial\Omega_\sigma = \emptyset$. In other words, Eqs. (10.32) define the local (strong) formulation for the problem considered.

The system of 15 equations (10.32a) with 15 unknowns u_i, ε_{ij}, σ_{ij} may be easily reduced by elimination of variables to a system of three equations with three unknowns u_i. To do this, let us only leave the first three of the 15 equations, $\sigma_{ij,j} + \hat{f}_i = 0$, substituting for stresses σ_{ij} the expression $\sigma_{ij} = C_{ijkl}\, u_{k,l}$ resulting from the remaining equations (symmetry of C_{ijkl} with respect to the last two indices has been utilized).

The set of kinematically admissible displacement fields (trial functions) is defined as

$$\mathcal{P} = \{\boldsymbol{u}(\boldsymbol{x}) : \Omega \to V^3, \quad u_i = \hat{u}_i \text{ for } \boldsymbol{x} \in \partial\Omega_u\} \tag{10.33}$$

while the set of kinematically admissible variations of the fields as

$$\mathcal{W} = \{\delta\boldsymbol{u}(\boldsymbol{x}) : \Omega \to V^3, \quad \delta u_i = 0 \text{ for } \boldsymbol{x} \in \partial\Omega_u\,, \quad |\delta\boldsymbol{u}| < \epsilon\}. \tag{10.34}$$

(The condition $|\delta\boldsymbol{u}| < \epsilon$, is not necessary for the derivations below, but it will play an important role in the discussion of stationarity conditions of the potential energy functional.) If the field $\boldsymbol{u}(\boldsymbol{x})$ is the solution of the problem (10.32) then (i) $\boldsymbol{u} \in \mathcal{P}$ and (ii) for each variation $\delta\boldsymbol{u} \in \mathcal{W}$ the following integral equation is fulfilled,

$$\int_\Omega (\sigma_{ij,j} + \hat{f}_i)\, \delta u_i \,\mathrm{d}V - \int_{\partial\Omega_\sigma} (\sigma_{ij} n_j - \hat{t}_i)\, \delta u_i \,\mathrm{d}A = 0. \tag{10.35}$$

Let us first transform the second integral in the above equation. This is a surface integral over a part of the boundary $\partial\Omega$. Note that the definition (10.34) implies the variation δu_i in the integrand to vanish on the remaining part of the boundary. This allows us to extend the integration domain to the entire domain $\partial\Omega$ and transform the integral according to the Gauss–Ostrogradski theorem to the form

$$\int_{\partial\Omega_\sigma} (\sigma_{ij} n_j - \hat{t}_i)\, \delta u_i \,\mathrm{d}A = \int_\Omega (\sigma_{ij,j}\delta u_i + \sigma_{ij}\delta u_{i,j}) \,\mathrm{d}V - \int_{\partial\Omega_\sigma} \hat{t}_i \delta u_i \,\mathrm{d}A.$$

Substituting the above result into Eq. (10.35) and introducing the notation $\delta\varepsilon_{ij} = \operatorname{sym} \delta u_{i,j}$, one obtains

$$\int_\Omega \sigma_{ij}\delta\varepsilon_{ij} \,\mathrm{d}V = \int_\Omega \hat{f}_i \delta u_i \,\mathrm{d}V + \int_{\partial\Omega_\sigma} \hat{t}_i \delta u_i \,\mathrm{d}A. \tag{10.36}$$

This is the virtual work principle for the problem of linear solid statics. It states that the sum of works performed by the external loads on the kinematically admissible displacement variations (called *virtual displacements*) equals the total work of the internal forces (stresses) on the corresponding strain variations.

This principle defines the weak (variational) formulation of the problem considered. In this formulation, a displacement field $\boldsymbol{u} \in \mathcal{P}$ is sought such that Eq. (10.36) is satisfied for each variation $\delta\boldsymbol{u} \in \mathcal{W}$. In other words, it is the *weak solution* of the problem. It can be shown (in a similar way to that demonstrated in the previous section for the 1D example) that if the strong problem solution (10.32) belongs to the set $\mathcal{P}$ then the weak solution coincides with it. This means that in a sufficiently wide class of functions (meeting appropriate smoothness requirements) the two formulations, strong (10.32) and weak (10.36), are equivalent.

Let us finally define the potential energy functional for the considered problem of linear solid statics as

$$J[\boldsymbol{u}] = \int_\Omega \left(\frac{1}{2} C_{ijkl} \varepsilon_{ij} \varepsilon_{kl} - \hat{f}_i u_i \right) \mathrm{d}V - \int_{\partial\Omega_\sigma} \hat{t}_i u_i \, \mathrm{d}A . \tag{10.37}$$

Using the known symmetry properties of the tensor C_{ijkl}, one may alternatively write

$$J[\boldsymbol{u}] = \int_\Omega \left(\frac{1}{2} C_{ijkl} u_{i,j} u_{k,l} - \hat{f}_i u_i \right) \mathrm{d}V - \int_{\partial\Omega_\sigma} \hat{t}_i u_i \, \mathrm{d}A . \tag{10.38}$$

The following *potential energy theorem* holds true:

> *Among all sufficiently smooth displacement fields $u_i(\boldsymbol{x})$ fulfilling the kinematical boundary conditions (10.32b), the actual displacement field (i.e. the one ensuring fulfillment of Eqs. (10.32)) corresponds to the minimum value of the total potential energy defined by Eq. (10.38).*

It is easy to prove the above statement by taking an infinitesimal variation of the displacement field and computing the corresponding variation of the functional J. The variation δJ turns out equal to the difference between the right- and left-hand sides of the virtual work equation (10.36) which means that it must be zero. Hence, the displacement field being the weak solution of the problem (and also its strong solution, in accordance with the previous discussion) ensures stationarity of the potential energy. In order to show that the stationarity condition is in this case also the minimum condition, we compute the second variation $\delta^2 J$ (remembering that $\delta(\delta u_i) = 0$) as

$$\delta^2 J = \int_\Omega C_{ijkl} \delta u_{i,j} \delta u_{k,l} \, \mathrm{d}V = \int_\Omega C_{ijkl} \delta\varepsilon_{ij} \delta\varepsilon_{kl} \, \mathrm{d}V . \tag{10.39}$$

Recalling properties of the elastic stiffness tensor discussed in Chap. 7, resulting from the physical interpretation of the elastic energy, we conclude that this expression is positive for each non-zero strain variation $\delta\boldsymbol{\varepsilon}$. This means that the displacement field being the solution of the problem (10.32) ensures the minimum of the potential energy.

Remark. Careful readers might observe a weak point in this reasoning. One can imagine a non-zero variation of the displacement field $\delta\boldsymbol{u} \in \mathcal{W}$ for which the strain variation $\delta\boldsymbol{\varepsilon}$ is zero. This is a variation that corresponds to (infinitesimal) rigid motion of the body. This variation, although non-zero, does not affect the potential energy value. Thus, the minimum value of the energy may correspond to several different displacement fields $\boldsymbol{u} \in \mathcal{P}$, which differ between each other by a rigid motion of the body only. Note, however, that if the kinematical boundary conditions in the problem of statics allowed rigid motion of the body, the local formulation would not have a unique solution, either—it would also be defined only up to a rigid motion accuracy. For this reason, it is required that the kinematical boundary conditions in problems of statics are formulated in a way excluding such ambiguities. One may then guarantee that exactly one displacement field $\boldsymbol{u} \in \mathcal{P}$ minimizes the potential energy of the system.

In the above the problem of linear statics, the right-hand side of the equations of motion $(10.32\text{a})_1$ has been assumed zero. Inclusion of dynamical phenomena, i.e. considering the equations of motion in the form

$$\sigma_{ij,j} + \hat{f}_i = \rho \ddot{u}_i \tag{10.40}$$

(with ρ assumed known), can also be similarly considered, although it makes the variational equations slightly more complex. Following the same transformations as in the case of statics, one arrives at the following form of the virtual work principle:

$$\int_\Omega \sigma_{ij} \delta\varepsilon_{ij} \, \mathrm{d}V + \int_\Omega \rho \ddot{u}_i \delta u_i \, \mathrm{d}V = \int_\Omega \hat{f}_i \delta u_i \, \mathrm{d}V + \int_{\partial\Omega_\sigma} \hat{t}_i \delta u_i \, \mathrm{d}A \tag{10.41}$$

or, introducing the constitutive equation and using the symmetry of the stiffness tensor, at

$$\int_\Omega C_{ijkl} u_{k,l} \delta u_{i,j} \, \mathrm{d}V + \int_\Omega \rho \ddot{u}_i \delta u_i \, \mathrm{d}V = \int_\Omega \hat{f}_i \delta u_i \, \mathrm{d}V + \int_{\partial\Omega_\sigma} \hat{t}_i \delta u_i \, \mathrm{d}A. \tag{10.42}$$

The functional J, whose stationary point is the solution in question, assumes the following form:

$$J[\boldsymbol{u}] = \int_\Omega \left(\frac{1}{2} C_{ijkl} u_{i,j} u_{k,l} + \rho \ddot{u}_i u_i - \hat{f}_i u_i \right) \mathrm{d}V - \int_{\partial\Omega_\sigma} \hat{t}_i u_i \, \mathrm{d}A. \tag{10.43}$$

We have shown above that as long as all fields appearing in the formulations fulfill appropriate smoothness requirements (in the strong formulation these requirements are stronger due to higher order derivatives appearing in its equations) the strong and weak solutions are the same. In any specific boundary value problem the smoothness requirements may be enforced by the physical sense of the quantities represented by the functions appearing in the problem equations as well as by the way the trial functions in an approximate solution method are defined. It may happen that, in a selected class of trial functions, the local formulation has no solution at all (because the functions are not sufficiently smooth, for instance) while the variational formulation can still be solved.

To highlight the issue in more detail, let us note that the solution $u_i(\boldsymbol{x})$ of the system (10.32) must be a twice differentiable function in the entire domain Ω. In the variational formulation based on the potential energy functional, it is enough if the solution $u_i(\boldsymbol{x})$ is (i) continuous in Ω and (ii) differentiable in Ω, possibly except for a set of points of the zero measure. It is thus clear that solution of the variational formulation (10.37) of a solid mechanics problem may exist in a class of trial functions but it may not be the solution of the local boundary value problem (10.32).

We are going to illustrate this issue on an example. Let us go back to the beginning of this book and consider once again the problem of statics discussed in Chap. 2 and presented in Fig. 2.7. This is the problem of a plane square plate (cantilever) loaded in its plane by a concentrated force applied at a square corner. This problem was previously discussed to highlight inconsistency of notions of 1D mechanics with the concept of material continuum mechanics.

Assume the plate's side length is l, its thickness equals h and its elastic properties are described by the Young modulus E and the Poisson ratio ν. Statics of the system is described by the system of equations (10.32) which in the case of plane stress is reduced to the form

$$\begin{aligned} &\sigma_{11,1}+\sigma_{12,2}=0, \quad &&\varepsilon_{11}=u_{1,1}, \quad &&\sigma_{11}=D(\varepsilon_{11}+\nu\varepsilon_{22}),\\ &\sigma_{12,1}+\sigma_{22,2}=0, \quad &&\varepsilon_{22}=u_{2,2}, \quad &&\sigma_{22}=D(\varepsilon_{22}+\nu\varepsilon_{11}),\\ & &&2\varepsilon_{12}=u_{1,2}+u_{2,1}, \quad &&\sigma_{12}=2G\varepsilon_{12} \end{aligned} \tag{10.44}$$

for $x_1 \in (0, l)$, $x_2 \in (0, l)$, where the notation

$$D=\frac{E}{1-\nu^2}, \qquad G=\frac{E}{2(1+\nu)}.$$

has been introduced. Boundary conditions on the square plate sides can be written as follows:

$$\begin{aligned} &\sigma_{12}=0, \quad \sigma_{22}=\tfrac{1}{h}P\,\delta(x_1-l,x_2) &&\text{for } x_1\in(0,l], && x_2=0,\\ &\sigma_{11}=0, \quad \sigma_{12}=0 &&\text{for } x_1=l, && x_2\in(0,l), \quad (10.45)\\ &\sigma_{12}=0, \quad \sigma_{22}=0 &&\text{for } x_1\in(0,l], && x_2=l,\\ &u_1=u_2=0 &&\text{for } x_1=0, && x_2\in[0,l], \quad (10.46) \end{aligned}$$

where $\delta(\cdot,\cdot)$ denotes the 2D Dirac pulse function.

Even if the form of the equations appears simple, their analytical solution would be difficult. Let us try, instead, to find a weak solution for a very simple class of trial functions, namely bilinear functions with respect to the coordinates x_1, x_2. Taking into account the boundary conditions (10.46), we define

$$\mathcal{P} = \left\{ u_i(x_1, x_2) = A_i \frac{x_1}{l} + B_i \frac{x_1 x_2}{l^2} \right\} \quad i = 1, 2, \tag{10.47a}$$

$$\mathcal{W} = \left\{ \delta u_i(x_1, x_2) = a_i \frac{x_1}{l} + b_i \frac{x_1 x_2}{l^2} \right\} \quad i = 1, 2, \tag{10.47b}$$

where A_1, A_2, B_1, B_2 are independent coefficients whose values are the subject of our search while a_1, a_2, b_1, b_2 denote arbitrary variation coefficients. The variational equation of virtual work (10.36) assumes in this case the form

$$\int_{\Omega} \sigma_{ij} \delta\varepsilon_{ij} \, \mathrm{d}V - (-P) \, \delta u_2(l, 0) = 0,$$

i.e., written for the particular problem geometry, the form

$$h \int_0^l \int_0^l (\sigma_{11}\delta\varepsilon_{11} + \sigma_{22}\delta\varepsilon_{22} + 2\sigma_{12}\delta\varepsilon_{12}) \, \mathrm{d}x_1 \mathrm{d}x_2 + P\delta u_2(l, 0) = 0. \tag{10.48}$$

According to the definition (10.47) one can write

$$\begin{aligned} \varepsilon_{11} &= \tfrac{1}{l}\left(A_1 + B_1 \tfrac{x_2}{l}\right), & \delta\varepsilon_{11} &= \tfrac{1}{l}\left(a_1 + b_1 \tfrac{x_2}{l}\right), \\ \varepsilon_{22} &= \tfrac{1}{l} B_2 \tfrac{x_1}{l}, & \delta\varepsilon_{22} &= \tfrac{1}{l} b_2 \tfrac{x_1}{l}, \\ 2\varepsilon_{12} &= \tfrac{1}{l}\left(A_2 + B_1 \tfrac{x_1}{l} + B_2 \tfrac{x_2}{l}\right), & 2\delta\varepsilon_{12} &= \tfrac{1}{l}\left(a_2 + b_1 \tfrac{x_1}{l} + b_2 \tfrac{x_2}{l}\right), \\ & & \delta u_2(l, 0) &= a_2, \end{aligned}$$

while the constitutive equations become

$$\begin{aligned} \sigma_{11} &= \frac{D}{l}\left(A_1 + \nu B_2 \frac{x_1}{l} + B_1 \frac{x_2}{l}\right), \\ \sigma_{22} &= \frac{D}{l}\left(\nu A_1 + B_2 \frac{x_1}{l} + \nu B_1 \frac{x_2}{l}\right), \\ \sigma_{12} &= \frac{G}{l}\left(A_2 + B_1 \frac{x_1}{l} + B_2 \frac{x_2}{l}\right). \end{aligned}$$

Thus,

$$\begin{aligned} &\sigma_{11}\delta\varepsilon_{11} + \sigma_{22}\delta\varepsilon_{22} + 2\sigma_{12}\delta\varepsilon_{12} = \\ &= \frac{1}{l^2}\left(K_0 + K_1 \frac{x_1}{l} + K_2 \frac{x_2}{l} + K_3 \frac{x_1^2}{l^2} + K_4 \frac{x_2^2}{l^2} + K_5 \frac{x_1 x_2}{l^2}\right), \end{aligned}$$

where

$$
\begin{aligned}
K_0 &= DA_1a_1 + GA_2a_2\,,\\
K_1 &= D\nu(B_2a_1 + A_1b_2) + G(B_1a_2 + A_2b_1),\\
K_2 &= D(A_1b_1 + B_1a_1) + G(A_2b_2 + B_2a_2),\\
K_3 &= DB_2b_2 + GB_1b_1\,,\\
K_4 &= DB_1b_1 + GB_2b_2\,,\\
K_5 &= (D\nu + G)(B_2b_1 + B_1b_2).
\end{aligned}
$$

Substituting this into the virtual work equation and performing the integration, we arrive at

$$
hK_0 + \frac{h}{2}(K_1 + K_2) + \frac{h}{3}(K_3 + K_4) + \frac{h}{4}K_5 + Pa_2 = 0.
$$

Reordering the terms with respect to the coefficients a_i, b_i, we obtain

$$
\begin{aligned}
D\left(A_1 + \tfrac{1}{2}B_1 + \tfrac{1}{2}\nu B_2\right)a_1 &+ \left[G\left(A_2 + \tfrac{1}{2}B_1 + \tfrac{1}{2}B_2\right) + \tfrac{1}{h}P\right]a_2\\
&+ \left[\tfrac{1}{2}DA_1 + \tfrac{1}{2}GA_2 + \tfrac{1}{3}(D{+}G)B_1 + \tfrac{1}{4}(D\nu{+}G)B_2\right]b_1\\
&+ \left[\tfrac{1}{2}D\nu A_1 + \tfrac{1}{2}GA_2 + \tfrac{1}{4}(D\nu{+}G)B_1 + \tfrac{1}{3}(D{+}G)B_1\right]b_2 = 0.
\end{aligned}
$$

Since the above equation must be satisfied for arbitrary values of the variation coefficients a_i, b_i, the following system of 4 equations with 4 unknowns is obtained:

$$
\begin{bmatrix}
1 & 0 & \frac{1}{2} & \frac{1}{2}\nu\\
0 & 1 & \frac{1}{2} & \frac{1}{2}\\
\frac{1}{2}D & \frac{1}{2}G & \frac{1}{3}(D{+}G) & \frac{1}{4}(D\nu{+}G)\\
\frac{1}{2}D\nu & \frac{1}{2}G & \frac{1}{4}(D\nu{+}G) & \frac{1}{3}(D{+}G)
\end{bmatrix}
\begin{Bmatrix} A_1\\ A_2\\ B_1\\ B_2 \end{Bmatrix}
=
\begin{Bmatrix} 0\\ -\frac{P}{hG}\\ 0\\ 0 \end{Bmatrix}.
$$

Skipping the simple but tedious solution procedure, the final result becomes

$$
\begin{aligned}
A_1 &= -\frac{P}{h}\left(\frac{3}{G{+}D} + \frac{3\nu}{G{+}D{+}3E}\right), & B_1 &= \frac{6P}{h(G{+}D)}\,,\\
A_2 &= -\frac{P}{h}\left(\frac{1}{G} + \frac{3}{G{+}D} + \frac{3}{G{+}D{+}3E}\right), & B_2 &= \frac{6P}{h(G{+}D{+}3E)}\,.
\end{aligned}
$$

The above coefficients are in fact the solution of the weak formulation. Substituting them into Eq. (10.47a) we get the form of the displacement field components $u_i(\boldsymbol{x})$ in the entire domain of the square plate. Let us once again underline that this is an approximate solution obtained in a function class that does not include the solution of the strong formulation (10.44). Let us note, however, that (i) although the assumed class of trial functions was extremely simple, we were still able to obtain the solution and (ii) whatever the quantitative inaccuracy of this solution may be, its form (see

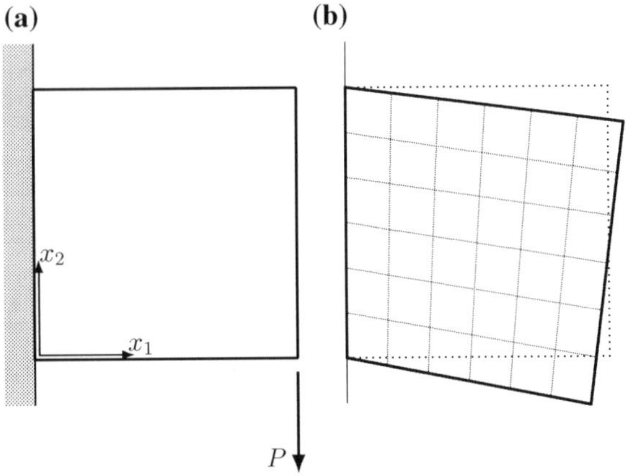

Fig. 10.4 Plane plate: **a** geometry and boundary conditions, **b** approximate displacement form in the bilinear function class

Fig. 10.4b) corresponds well to our intuition about deformation expected in this situation. Indeed, among all kinematically admissible displacement fields of the form (10.47a), the one we have obtained is the best approximate of the exact solution.

10.2.2 Extended Variational Formulations

In a process of searching for approximate solutions, it often appears convenient to further extend the class of admissible functions in the variational formulation by allowing discontinuities of some fields, e.g. displacements $u_i(\boldsymbol{x})$ in the domain Ω. The question how the variational functional of the type (10.37) should be defined to allow for some discontinuity surfaces in the considered fields will be briefly addressed below.

Let us denote all such surfaces by the symbol $\mathcal{N}$ (Fig. 10.5). Each of the surfaces has two sides, arbitrarily numbered by the counter $s = 1, 2$. For an arbitrary field $\boldsymbol{a}(\boldsymbol{x})$, $\boldsymbol{x} \in \Omega$, we denote by $\boldsymbol{a}^{(s)}(\boldsymbol{x})$, $\boldsymbol{x} \in \mathcal{N}$, the limit values of the field $\boldsymbol{a}$ for

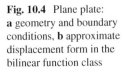

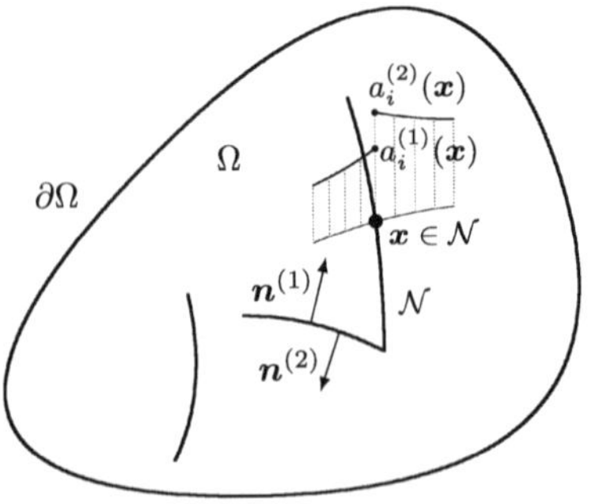

Fig. 10.5 Deformable solid with discontinuity surfaces

a series of material points convergent to $\boldsymbol{x}$ from inside the domain on the side s of the surface $\mathcal{N}$. Let us further denote by

$$\langle \boldsymbol{a}(\boldsymbol{x})\rangle = \boldsymbol{a}^{(1)}(\boldsymbol{x}) - \boldsymbol{a}^{(2)}(\boldsymbol{x}), \qquad \boldsymbol{x} \in \mathcal{N}, \tag{10.49}$$

the discontinuity of the field $\boldsymbol{a}(\boldsymbol{x})$ when crossing the surface $\mathcal{N}$. Let us also introduce the notation $\boldsymbol{n}^{(1)}(\boldsymbol{x})$ and $\boldsymbol{n}^{(2)}(\boldsymbol{x})$, $\boldsymbol{x} \in \mathcal{N}$, for unit normal vectors to the surface $\mathcal{N}$ on both its sides (directed outwards the corresponding sides). Obviously, at each $\boldsymbol{x} \in \mathcal{N}$ we have $\boldsymbol{n}^{(1)} = -\boldsymbol{n}^{(2)}$.

Consider the variational functional

$$\bar{J}[\boldsymbol{u}, \boldsymbol{\lambda}] = \int_\Omega \left(\frac{1}{2} C_{ijkl} u_{i,j} u_{k,l} - \hat{f}_i u_i\right) \mathrm{d}V - \int_{\partial\Omega_\sigma} \hat{t}_i u_i \,\mathrm{d}A - \int_{\mathcal{N}} \lambda_i \langle u_i \rangle \,\mathrm{d}A, \tag{10.50}$$

defined on two independent argument vector fields, $\boldsymbol{u}(\boldsymbol{x}), \boldsymbol{x} \in \Omega \cup \partial\Omega$ and $\boldsymbol{\lambda}(\boldsymbol{x}), \boldsymbol{x} \in \mathcal{N}$. The field $\boldsymbol{\lambda}$ is a kind of the functional Lagrange multiplier while the formulation (10.38) is a special case of the formulation (10.50) in which an additional condition is imposed, of the form $\langle u_i(\boldsymbol{x})\rangle = 0, \bar{x} \in \mathcal{N}$. In Eq. (10.50) it is assumed as before that the displacement field $u_i(\boldsymbol{x})$ fulfills by its definition the boundary condition (10.32b). Computing variation of the functional (10.50) as

$$\delta\bar{J} = \frac{\partial\bar{J}}{\partial u_i}\delta u_i + \frac{\partial\bar{J}}{\partial\lambda_i}\delta\lambda_i \tag{10.51}$$

and enforcing the stationarity condition $\delta\bar{J} = 0$, one arrives at

$$\begin{aligned} &\int_\Omega \sigma_{ij}\delta u_{i,j}\,\mathrm{d}V - \int_\Omega \hat{f}_i \delta u_i \,\mathrm{d}V - \int_{\partial\Omega_\sigma} \hat{t}_i \delta u_i \,\mathrm{d}A \\ &\qquad - \int_{\mathcal{N}} \delta\lambda_i \langle u_i\rangle \,\mathrm{d}A - \int_{\mathcal{N}} \lambda_i \left(\delta u_i^{(1)} - \delta u_i^{(2)}\right) \mathrm{d}A = 0. \end{aligned}$$

Using the property $\sigma_{ij}\delta u_{i,j} = (\sigma_{ij}\delta u_i)_{,j} - \sigma_{ij,j}\delta u_i$, one may apply the Gauss–Ostrogradski theorem to the volume integral of the expression $(\sigma_{ij}\delta u_i)_{,j}$ (remembering that here the boundary of the domain Ω must be extended to include the surfaces $\mathcal{N}$). After transformations, the following equation is obtained,

$$\begin{aligned} &\int_{\mathcal{N}} \left[\left(\sigma_{ij}^{(1)} n_j^{(1)} - \lambda_i\right)\delta u_i^{(1)} + \left(\sigma_{ij}^{(2)} n_j^{(2)} + \lambda_i\right)\delta u_i^{(2)}\right] \mathrm{d}A - \int_{\mathcal{N}} \delta\lambda_i \langle u_i\rangle \,\mathrm{d}A \\ &\qquad - \int_\Omega (\sigma_{ij,j} + \hat{f}_i)\delta u_i \,\mathrm{d}V + \int_{\partial\Omega_\sigma} (\sigma_{ij} n_j - \hat{t}_i)\delta u_i \,\mathrm{d}A = 0, \end{aligned} \tag{10.52}$$

This equation must be fulfilled for each displacement variation $\delta\boldsymbol{u}$ and multiplier $\delta\boldsymbol{\lambda}$. This implies the system of local equations consisting of Eqs. (10.32), i.e.

$$\sigma_{ij,j} + \hat{f}_i = 0, \quad \boldsymbol{x} \in \Omega, \qquad\qquad \sigma_{ij} n_j - \hat{t}_i = 0, \quad \boldsymbol{x} \in \partial\Omega_\sigma ,$$

and, additionally, of equations in the form

$$\left.\begin{array}{l} \langle u_i \rangle = 0 \\ \lambda_i = \sigma_{ij}^{(1)} n_j^{(1)} = -\sigma_{ij}^{(2)} n_j^{(2)} \end{array}\right\} \quad \boldsymbol{x} \in \mathcal{N}. \tag{10.53}$$

Since $n_j^{(1)} = -n_j^{(2)}$, the above conditions are equivalent to

$$\left.\begin{array}{l} u_i^{(1)} = u_i^{(2)} \\ \sigma_{ij}^{(1)} n_j^{(1)} = \sigma_{ij}^{(2)} n_j^{(1)} \\ \lambda_i = \sigma_{ij}^{(1)} n_j^{(1)} \end{array}\right\} \quad \boldsymbol{x} \in \mathcal{N}. \tag{10.54}$$

This means that the displacement field is continuous at points on the surfaces $\mathcal{N}$ and the stress vectors on both their sides are in equilibrium. Besides, we can see that the vector fields of Lagrange multipliers $\boldsymbol{\lambda}$ may be interpreted as values of stress vectors on one of the surface sides.

The latter conclusion allows to write the functional $\bar{J}$ in the following, alternative form,

$$\begin{aligned} \bar{J}[\boldsymbol{u}, \boldsymbol{\sigma}] = & \int_\Omega \left(\frac{1}{2} C_{ijkl} u_{i,j} u_{k,l} - \hat{f}_i u_i \right) \mathrm{d}V - \int_{\partial\Omega_\sigma} \hat{t}_i u_i \, \mathrm{d}A \\ & - \int_{\mathcal{N}} \sigma_{ij} n_j \langle u_i \rangle \, \mathrm{d}A, \end{aligned} \tag{10.55}$$

in which the indices $^{(1)}$ at the stress and normal vector symbols are skipped. Stationarity of this functional is ensured if both the equilibrium equations (10.32) and displacement continuity conditions at $\boldsymbol{x} \in \mathcal{N}$ are fulfilled.

In the variational formulation, the continuity conditions for $\boldsymbol{u}(\boldsymbol{x})$ on the surface $\mathcal{N}$ are included in the functional (10.50) in an implicit way—they are said to be fulfilled through variational solution of the problem. In other words, the conditions expressing the continuity of $\boldsymbol{u}(\boldsymbol{x})$ on $\mathcal{N}$ are treated in the same way as the remaining equations of the problem, say the equilibrium equations or stress boundary conditions. This fact has far reaching consequences for approximate solutions of the problem.

Let us also stress that the modified variational formulation with the functional J in the form (10.50) or (10.55) does not have the character of minimum (with $\delta \bar{J} = 0$ and $\delta^2 \bar{J} \geq 0$), since it only leads to the stationarity condition $\delta \bar{J} = 0$. In other words, the Euler equations for the functional $\bar{J}$ are the appropriate differential equations of the linear elasticity theory and the continuity conditions for $u_i(\boldsymbol{x})$. Nothing can be said about the sign of the second functional's variation.

Let us note that, in both the cases considered above, the exact satisfaction of the kinematical boundary conditions (10.32b) by the sought solution $\boldsymbol{u}(\boldsymbol{x})$ was required. These conditions are in fact nothing else but just continuity conditions of the solution

$\boldsymbol{u}(\boldsymbol{x})$ imposed on the boundary $\partial\Omega_u$. Thus, it should be possible to include these conditions in a variational manner, as it has just been done inside the domain Ω. To verify this, consider the following functional,

$$\bar{J}[\boldsymbol{u},\boldsymbol{\sigma}] = \int_{\Omega}\left(\frac{1}{2}C_{ijkl}u_{i,j}u_{k,l} - \hat{f}_i u_i\right)\mathrm{d}V - \int_{\partial\Omega_\sigma}\hat{t}_i u_i\,\mathrm{d}A$$
$$- \int_{\partial\Omega_u}\sigma_{ij}n_j(\hat{u}_i - u_i)\,\mathrm{d}A, \tag{10.56}$$

defined for arbitrary continuous fields $\boldsymbol{u}(\boldsymbol{x})$, $\boldsymbol{x} \in \Omega$. It is not required that the fields meet the boundary conditions at $\partial\Omega_u$. Compared to the functional (10.38), an additional term appears here being an integral over $\partial\Omega_u$. Skipping the transformations (similar as before), we can conclude that the stationarity condition of the functional (10.56) is equivalent to fulfillment of Eq. (10.32a) and the boundary conditions for stresses (10.32c) and displacements (10.32b). The latter condition

$$u_i = \hat{u}_i\,, \qquad \boldsymbol{x} \in \partial\Omega_u\,, \tag{10.57}$$

results from arbitrary choice of the variation components $\delta\sigma_{ij}$ in the expression

$$\int_{\partial\Omega_u}(\hat{u}_i - u_i)\delta\sigma_{ij}n_j\,\mathrm{d}(\partial\Omega) = 0. \tag{10.58}$$

Recalling the terminology introduced earlier, one can say that the boundary conditions (10.57) are for the functional (10.56) the natural boundary conditions (similarly as the conditions (10.32c)) while for the functional (10.38) they were the essential conditions. In the functional (10.56), no essential boundary conditions are required to be fulfilled by the sought functions $\boldsymbol{u}(\boldsymbol{x})$.

10.3 Variational Formulations for Nonlinear Mechanics Problems

The above presented variational formulations have been derived under very strong simplifying assumptions of linear elasticity. The question may thus arise whether such an attractive solution method may also be employed for solving nonlinear problems. The answer is positive, although weak formulations for this kind of problems are more complex and may not feature all the advantages specified in the previous section. Two examples of such formulations will now be presented: statics problem of nonlinear elasticity and a quasi-statical problem for inelastic material behaviour in an incremental formulation.

10.3.1 Elasticity at Large Deformations

The system of equations (8.1) will be considered as a basis for the strong formulation for this problem. It is additionally assumed that $\rho^r \ddot{u}_i \approx 0$ (problem of statics) and the variable ρ is eliminated by assuming the forces $\hat{f}_i = \rho f_i$ known. In such a formulation, initial conditions do not matter—the system may be solved independently at each time instant $\tau = t$ and the boundary conditions (8.4) for this time instant are the only ones necessary. In summary, the strong formulation at each $\boldsymbol{x} \in \Omega^r$ and $\tau = t \in [0, \bar{t}]$ for the problem on hand has the form

$$ {}_r\sigma^t_{ij,j} + \hat{f}^t_i = 0, \tag{10.59a}$$

$$ {}_r\sigma^t_{ij} = (\delta_{ik} + {}^r u^t_{i,k})\, {}_r\tilde{\sigma}^t_{kj}\,, \tag{10.59b}$$

$$ {}_r\tilde{\sigma}^t_{ij} = \frac{\partial\, {}_r U^t}{\partial\, {}^0_r\varepsilon^t_{ij}}\,, \tag{10.59c}$$

$$ {}^0_r\varepsilon^t_{ij} = \frac{1}{2}\left({}^0 u^t_{i,j} + {}^0 u^t_{j,i} + {}^r u^t_{k,i}\, {}^r u^t_{k,j} - {}^r u^0_{k,i}\, {}^r u^0_{k,j}\right) \tag{10.59d}$$

with the boundary conditions

$$ {}^0 u^t_i = \hat{u}^t_i\,, \qquad x_i \in \partial\Omega^r_u\,, \tag{10.60a}$$

$$ {}_r\sigma^t_{ij,j} n^r_j = {}_r\hat{t}^t_i\,, \qquad x_i \in \partial\Omega^r_\sigma\,. \tag{10.60b}$$

The form of the function ${}_r U^t({}^0_r\varepsilon^t)$ is assumed known.

Similarly as in the linear case, the system (10.59) can be easily reduced to the system of three equations with three unknowns. It can be seen that the variables ${}^0_r\varepsilon^t_{ij}$ and ${}_r\tilde{\sigma}^t_{ij}$ can be eliminated from the equations so that only Eq. (10.59a) remain in which ${}_r\sigma^t_{ij}$ is treated as a known function of the unknown field ${}^0 u^t_i$. Besides, displacements ${}^r u^0_i$ and ${}^r u^t_i$ appearing in the equations are not additional unknowns—the first is known and the second can be computed as ${}^r u^t_i = {}^r u^0_i + {}^0 u^t_i$.

Consider the set of kinematically admissible displacement fields (trial functions) defined as

$$ \mathcal{P} = \{{}^0\boldsymbol{u}^t(\boldsymbol{x}^r) : \Omega^r \to V^3, \quad {}^0 u^t_i = \hat{u}^t_i \text{ for } \boldsymbol{x}^r \in \partial\Omega^r_u\} \tag{10.61}$$

and the set of their kinematically admissible variations

$$ \mathcal{W} = \{\delta\boldsymbol{u}^t(\boldsymbol{x}^r) : \Omega^r \to V^3, \quad \delta u^t_i = 0 \text{ for } \boldsymbol{x}^r \in \partial\Omega^r_u\,, \quad |\delta\boldsymbol{u}^t| < \epsilon\}. \tag{10.62}$$

If the field ${}^0\boldsymbol{u}^t(\boldsymbol{x}^r)$ is the solution of the strong problem, then ${}^0\boldsymbol{u}^t \in \mathcal{P}$. Besides, for each variation $\delta\boldsymbol{u}^t \in \mathcal{W}$, the following integral equation holds true:

$$ \int_{\Omega^r} ({}_r\sigma^t_{ij,j} + \hat{f}^t_i)\, \delta u^t_i\, \mathrm{d}V^r - \int_{\partial\Omega^r_\sigma} ({}_r\sigma^t_{ij,j} n^r_j - {}_r\hat{t}^t_i)\, \delta\Delta u^t_i\, \mathrm{d}A^r = 0. \tag{10.63}$$

Let us substitute ${}_r\sigma^t_{ij,j}\,\delta u^t_i = ({}_r\sigma^t_{ij}\,\delta u^t_i)_{,j} - {}_r\sigma^t_{ij}\,\delta u^t_{i,j}$ and apply the Gauss–Ostrogradski theorem to the term $({}_r\sigma^t_{ij}\,\delta u^t_i)_{,j}$, taking advantage of the fact that the variation δu^t_i vanishes on the boundary part $\partial\Omega^r_u$. After transformations we arrive at

$$\int_{\Omega^r} {}_r\sigma^t_{ij}\,\delta u^t_{i,j}\,\mathrm{d}V^r = \int_{\Omega^r} \hat{f}^t_i\,\delta u^t_i\,\mathrm{d}V^r + \int_{\partial\Omega^r_\sigma} {}_r\hat{t}^t_i\,\delta u^t_i\,\mathrm{d}A^r. \tag{10.64}$$

Equation (10.64) is the virtual work principle for the problem considered. Let us transform it so that the first Piola–Kirchhoff stress tensor components ${}_r\sigma^t_{ij}$ are replaced by the second Piola–Kirchhoff stress tensor components ${}_r\tilde{\sigma}^t_{ij}$. The following relation holds true:

$${}_r\sigma^t_{ij}\,\delta u^t_{i,j} = (\delta_{ik} + {}^r u^t_{i,k})\,{}_r\tilde{\sigma}^t_{kj}\,\delta u^t_{i,j} = {}_r\tilde{\sigma}^t_{kj}\,(\delta u^t_{k,j} + {}^r u^t_{i,k}\,\delta u^t_{i,j}).$$

Let us denote by

$$\begin{aligned} \delta_r\varepsilon^t_{ij} &= \frac{1}{2}(\delta u^t_{i,j} + \delta u^t_{j,i}) + \frac{1}{2}({}^r u^t_{k,i}\,\delta u^t_{k,j} + {}^r u^t_{k,j}\,\delta u^t_{k,i}) \\ &= \operatorname{sym}\left(\delta u^t_{i,j} + {}^r u^t_{k,i}\,\delta u^t_{k,j}\right) \end{aligned} \tag{10.65}$$

the strain variation corresponding to the displacement variation δu^t_i. Symmetry of the tensor ${}_r\tilde{\sigma}^t_{ij}$ implies that

$${}_r\sigma^t_{ij}\,\delta u^t_{i,j} = {}_r\tilde{\sigma}^t_{kj}\,\delta_r\varepsilon^t_{kj}\,,$$

which allows to finally write Eq. (10.64) in the form

$$\int_{\Omega^r} {}_r\tilde{\sigma}^t_{ij}\,\delta_r\varepsilon^t_{ij}\,\mathrm{d}V^r = \int_{\Omega^r} \hat{f}^t_i\,\delta u^t_i\,\mathrm{d}V^r + \int_{\partial\Omega^r_\sigma} {}_r\hat{t}^t_i\,\delta u^t_i\,\mathrm{d}A^r. \tag{10.66}$$

This is the weak formulation for the considered problem of statics of nonlinear elastic solid.

Let us now try to find—in analogy to the linear problem—the form of the functional $J[{}^0\boldsymbol{u}^t]$ for which the above weak formulation is the stationarity condition. In other words, we look for the functional whose variation δJ (computed for an arbitrary variation $\delta^0\boldsymbol{u}^t$) equals the difference between the left- and right-hand sides of the virtual work equation, i.e. it vanishes at the exact problem solution. This condition appears to be fulfilled if

$$J[{}^0\boldsymbol{u}^t] = \int_{\Omega^r}\left[\int_0^t {}_r\tilde{\sigma}^\tau_{ij}\,\mathrm{d}_r\varepsilon^\tau_{ij}\right]\mathrm{d}V^r - \int_{\Omega^r} \hat{f}^t_i\,{}^0u^t_i\,\mathrm{d}V^r - \int_{\partial\Omega^r_\sigma} {}_r\hat{t}^t_i\,{}^0u^t_i\,\mathrm{d}A^r, \tag{10.67}$$

or, upon substitution of the hyperelastic constitutive equation (10.59c),

$$J[{}^0\boldsymbol{u}^t] = \int_{\Omega^r} {}_rU^t \,\mathrm{d}V^r - \int_{\Omega^r} \hat{f}_i^t \, {}^0u_i^t \,\mathrm{d}V^r - \int_{\partial\Omega^r_\sigma} {}_r\hat{t}_i^t \, {}^0u_i^t \,\mathrm{d}A^r. \tag{10.68}$$

The above functional describes the potential energy for the considered problem of mechanics, given any kinematically admissible displacement field.

While considering the linear problem we were able to show that the stationary point of the potential energy functional was actually its minimum. In other words, the solution of the weak problem was the one minimizing that functional. Let us verify whether this is also true in the case considered now. To do this, let us compute the second variation $\delta^2 J$ (remembering that the second variation of the displacement field is zero). We obtain

$$\delta^2 J = \int_{\Omega^r} \delta^2 {}_rU^t \,\mathrm{d}V^r = \int_{\Omega^r} \left(\frac{\partial^2 {}_rU^t}{\partial {}^0_r\varepsilon^t_{ij} \, \partial {}^0_r\varepsilon^t_{kl}} \, \delta_r\varepsilon^t_{ij} \, \delta_r\varepsilon^t_{kl} + \frac{\partial {}_rU^t}{\partial {}^0_r\varepsilon^t_{ij}} \, \delta^2 {}_r\varepsilon^t_{ij} \right) \mathrm{d}V^r,$$

where the variation $\delta_r\varepsilon^t_{ij}$ is given by Eq. (10.65), while the second variation of strain is

$$\delta^2 {}_r\varepsilon^t_{ij} = \delta u^t_{k,i} \, \delta u^t_{k,j} \,.$$

Denoting

$${}_rC^t_{ijkl} = \frac{\partial^2 {}_rU^t}{\partial {}^0_r\varepsilon^t_{ij} \, \partial {}^0_r\varepsilon^t_{kl}} \,,$$

and substituting Eq. (10.59c), we can rewrite the result as

$$\delta^2 J = \int_{\Omega^r} \left({}_rC^t_{ijkl} \, \delta_r\varepsilon^t_{ij} \, \delta_r\varepsilon^t_{kl} + {}_r\tilde{\sigma}^t_{ij} \, \delta u^t_{k,i} \, \delta u^t_{k,j} \right) \mathrm{d}V. \tag{10.69}$$

Comparing Eq. (10.69) with a similar result for the linear case (10.39), we may observe similarity in the first term which in this case is also a positive definite quadratic form (according to certain thermodynamical limitations on the tangent stiffness tensor component values ${}_rC^t_{ijkl}$). Let us focus, however, on the second term which did not appear in the linear formulation due to vanishing of the second variation of the linearized Cauchy strain $\delta^2\varepsilon_{ij}$. Note that the sign of the expression ${}_r\tilde{\sigma}^t_{ij} \, \delta u^t_{k,i} \, \delta u^t_{k,j}$ depends on the signs of the stress components ${}_r\tilde{\sigma}^t_{ij}$. It may thus appear that, at high compression stress states for instance, the entire integrand is not positive and the existence of the potential energy minimum cannot be guaranteed.

This feature of nonlinear solutions has several undesired consequences. Some solutions may be unstable, i.e. a small perturbation of load may induce unpredictably large changes in the displacements ensuring the system's equilibrium and, in some particular cases, several different solutions may exist for the problem. An example of such an unstable problem is compression of a thin bar. In the small deformation range, it behaves in a way described by solutions presented in the previous sections, i.e. it gets shorter under a homogeneous compression stress state. In the large deformation range, a bending state may additionally appear which—once the load achieves a certain limit value—leads to loss of the solution stability. This feature of the nonlinear solution corresponds well to our experience—buckling observed in some structural elements (say, thin bars or plates) subject to excessive compression loads is the manifestation of such a solution instability.

Instability problems constitute a separate branch of solid mechanics. They will not be discussed in more detail in this book, though.

In analogy to the linear case, the above variational formulation can be easily extended to the case of dynamics in which the equations of motion (10.59a) assume the form

$$ {}_r\sigma^t_{ij,j} + \hat{f}^t_i = \rho^r \ddot{u}^t_i \,, $$

where the density distribution in the reference configuration ρ^r is assumed known. In this case, an additional term appears on the left-hand side of the virtual work equation (10.64) or (10.66) and the equation takes the form

$$ \int_{\Omega^r} {}_r\sigma^t_{ij}\, \delta u^t_{i,j}\, \mathrm{d}V^r + \int_{\Omega^r} \rho^r \ddot{u}^t_i \delta u^t_i\, \mathrm{d}V^r = \int_{\Omega^r} \hat{f}^t_i\, \delta u^t_i\, \mathrm{d}V^r + \int_{\partial\Omega^r_\sigma} {}_r\hat{t}^t_i\, \delta u^t_i\, \mathrm{d}A^r \,. \tag{10.70} $$

The form of the functional J (10.68) is thus extended by the additional term

$$ \int_{\Omega^r} \rho^r \ddot{u}^t_i u^t_i\, \mathrm{d}V^r \,. $$

10.3.2 *Incremental Problem of Nonlinear Mechanics*

The strong formulation for this problem is given by the system of equations (8.6). For the needs of the following derivations it will only be assumed that $\rho\Delta\ddot{u}_i \approx 0$ and that the loads $\hat{f}_i = \rho f_i$ are known (which allows to eliminate the variable ρ).

Let us assume that—without limiting generality of our discussion—the updated Lagrangian description is adopted at a typical time step $[t, t{+}\Delta t]$, i.e. $C^r = C^t$. This allows to simplify the form of equations by removing the displacements ${}^r u^t_i$ and their gradients in the formulae for strain increments. We can also skip for clarity of notation the indices r and t as all quantities appearing in the equations refer to the time instant $\tau = t$ and their increments to the time step $[t, t{+}\Delta t]$. Neglecting additionally the issue of constitutive state parameters which are not important in the following derivations, we rewrite the considered system of equations in the form

$$\Delta\sigma_{ij,j} + \Delta\hat{f}_i = 0, \tag{10.71a}$$
$$\Delta\sigma_{ij} = \Delta\tilde{\sigma}_{ij} + \Delta u_{i,k}(\tilde{\sigma}_{kj} + \Delta\tilde{\sigma}_{kj}), \tag{10.71b}$$
$$\Delta\tilde{\sigma}_{ij} = C_{ijkl}\,\Delta\varepsilon_{kl} + H_{ij}\Delta t, \tag{10.71c}$$
$$\Delta\varepsilon_{ij} = \frac{1}{2}\left(\Delta u_{i,j} + \Delta u_{j,i} + \Delta u_{k,i}\,\Delta u_{k,j}\right), \tag{10.71d}$$

with the boundary conditions

$$\Delta u_i = \Delta\hat{u}_i\,, \qquad x_i \in \partial\Omega_u\,, \tag{10.72a}$$
$$\Delta\sigma_{ij}n_j = \Delta\hat{t}_i\,, \qquad x_i \in \partial\Omega_\sigma\,. \tag{10.72b}$$

The symbols σ_{ij} and $\tilde{\sigma}_{ij}$ in the above formulation stand for the first and the second Piola–Kirchhoff stress tensor components, respectively. Let us recall that these two tensors have equal values at $\tau = t$ but their increments are different.

Similarly as in the previous cases, the system (10.71) can be easily reduced to three equations with three unknowns. It can be seen that $\Delta\varepsilon_{ij}$ and $\Delta\tilde{\sigma}_{ij}$ may be eliminated from the equations in a straightforward way, leaving only Eq. (10.71a) in which $\Delta\sigma_{ij}$ is treated as a known function of the unknown field Δu_i.

Consider the set of kinematically admissible displacement fields (trial functions)

$$\mathcal{P} = \{\Delta\boldsymbol{u}(\boldsymbol{x}) : \Omega \to V^3, \quad \Delta u_i = \Delta\hat{u}_i \text{ for } \boldsymbol{x} \in \partial\Omega_u\} \tag{10.73}$$

and the set of their kinematically admissible variations

$$\mathcal{W} = \{\delta\Delta\boldsymbol{u}(\boldsymbol{x}) : \Omega \to V^3, \quad \delta\Delta u_i = 0 \text{ for } \boldsymbol{x} \in \partial\Omega_u\,, \quad |\delta\Delta\boldsymbol{u}| < \epsilon\}. \tag{10.74}$$

If the field $\Delta\boldsymbol{u}(\boldsymbol{x}) \in \mathcal{P}$ is the solution of the problem (10.71)–(10.72) then for each variation $\delta\Delta\boldsymbol{u} \in \mathcal{W}$ the following integral equation holds true:

$$\int_\Omega (\Delta\sigma_{ij,j} + \Delta\hat{f}_i)\,\delta\Delta u_i\,\mathrm{d}V - \int_{\partial\Omega_\sigma} (\Delta\sigma_{ij}n_j - \Delta\hat{t}_i)\,\delta\Delta u_i\,\mathrm{d}A = 0. \tag{10.75}$$

Let us substitute $\Delta\sigma_{ij,j}\,\delta\Delta u_i = (\Delta\sigma_{ij}\,\delta\Delta u_i)_{,j} - \Delta\sigma_{ij}\,\delta\Delta u_{i,j}$ and apply the Gauss–Ostrogradski theorem to the term $(\Delta\sigma_{ij}\,\delta\Delta u_i)_{,j}$, taking advantage of the fact that the variation $\delta\Delta u_i$ vanishes on the boundary part $\partial\Omega_u$. After transformations we arrive at

$$\int_\Omega \Delta\sigma_{ij}\,\delta\Delta u_{i,j}\,\mathrm{d}V = \int_\Omega \Delta\hat{f}_i\,\delta\Delta u_i\,\mathrm{d}V + \int_{\partial\Omega_\sigma} \Delta\hat{t}_i\,\delta\Delta u_i\,\mathrm{d}A. \tag{10.76}$$

This is the virtual work principle for the problem considered.

Let replace the first Piola–Kirchhoff stress tensor by the second Piola–Kirchhoff stress tensor. The integrand on the left-hand side of the equation can be transformed as follows:

$$\Delta\sigma_{ij}\,\delta\Delta u_{i,j} = \Delta\tilde{\sigma}_{ij}\,\delta\Delta u_{i,j} + (\tilde{\sigma}_{kj} + \Delta\tilde{\sigma}_{kj})\,\Delta u_{i,k}\,\delta\Delta u_{i,j}\,. \tag{10.77}$$

Let us denote, cf. Eq. (4.60),

$$\overline{\Delta\varepsilon_{ij}} = \frac{1}{2}(\Delta u_{i,j} + \Delta u_{j,i}), \qquad \overline{\overline{\Delta\varepsilon_{ij}}} = \frac{1}{2}\,\Delta u_{k,i}\,\Delta u_{k,j}$$

and, by analogy,

$$\delta\overline{\Delta\varepsilon_{ij}} = \frac{1}{2}\,(\delta\Delta u_{i,j} + \delta\Delta u_{j,i}), \qquad \delta\overline{\overline{\Delta\varepsilon_{ij}}} = \frac{1}{2}(\Delta u_{k,i}\,\delta\Delta u_{k,j} + \delta\Delta u_{k,i}\,\Delta u_{k,j})$$

(all the above expressions are symmetric tensors). Symmetry of the tensor $\tilde{\sigma}_{ij}$ implies

$$\Delta\sigma_{ij}\,\delta\Delta u_{i,j} = \Delta\tilde{\sigma}_{ij}\delta\overline{\Delta\varepsilon_{ij}} + (\tilde{\sigma}_{ij} + \Delta\tilde{\sigma}_{ij})\,\delta\overline{\overline{\Delta\varepsilon_{ij}}} = \Delta\tilde{\sigma}_{ij}\,\delta\Delta\varepsilon_{ij} + \tilde{\sigma}_{ij}\,\delta\overline{\overline{\Delta\varepsilon_{ij}}}\,. \tag{10.78}$$

In summary, the virtual work equation (10.76) assumes the form

$$\int_\Omega (\Delta\tilde{\sigma}_{ij}\,\delta\Delta\varepsilon_{ij} + \tilde{\sigma}_{ij}\,\delta\overline{\overline{\Delta\varepsilon_{ij}}})\,\mathrm{d}V = \int_\Omega \Delta\hat{f}_i\,\delta\Delta u_i\,\mathrm{d}V + \int_{\partial\Omega_\sigma} \Delta\hat{t}_i\,\delta\Delta u_i\,\mathrm{d}A\,. \tag{10.79}$$

This is the weak formulation for the incremental problem of statics of nonlinear solid mechanics.

Let us now try to find—in analogy to the linear problem—the form of the functional J for which the above weak formulation is the stationarity condition. In a general case of nonlinear mechanics the answer may not always be positive. In particular, it depends on the form of the constitutive equation for the considered material. In our particular case, however, once the exact form of the constitutive equations has been replaced by their time-integrated incremental form (10.71c), such a functional exists and has the form:

$$\begin{aligned} J[\boldsymbol{u}] = &\int_\Omega \left(\frac{1}{2}C_{ijkl}\Delta\varepsilon_{ij}\Delta\varepsilon_{kl} + \tilde{\sigma}_{ij}\overline{\overline{\Delta\varepsilon_{ij}}} + \Delta t\; H_{ij}\Delta\varepsilon_{ij}\right)\mathrm{d}V \\ &- \int_\Omega \Delta\hat{f}_i\Delta u_i\mathrm{d}V - \int_{\partial\Omega_\sigma} \Delta\hat{t}_i\Delta u_i\,\mathrm{d}A\,. \end{aligned} \tag{10.80}$$

To verify it, let us compute the functional variation. It is important to note that the quantities $\tilde{\sigma}_{ij}$, C_{ijkl} and H_{ij} in the above expression are taken at the time instant $\tau = t$ and thus do not depend on $\Delta\boldsymbol{u}$ and are not subject to variation. The first integral variation reads

$$\delta \int_\Omega \left(\frac{1}{2} C_{ijkl} \Delta\varepsilon_{ij} \Delta\varepsilon_{kl} + \tilde{\sigma}_{ij} \overline{\overline{\Delta\varepsilon}}_{ij} + \Delta t\ H_{ij} \Delta\varepsilon_{ij} \right) \mathrm{d}V$$
$$= \int_\Omega \left(C_{ijkl} \Delta\varepsilon_{kl}\ \delta\Delta\varepsilon_{ij} + \tilde{\sigma}_{ij}\ \delta\overline{\overline{\Delta\varepsilon}}_{ij} + \Delta t\ H_{ij}\ \delta\Delta\varepsilon_{ij} \right) \mathrm{d}V$$
$$= \int_\Omega \left(\Delta\tilde{\sigma}_{ij}\ \delta\Delta\varepsilon_{ij} + \tilde{\sigma}_{ij}\ \delta\overline{\overline{\Delta\varepsilon}}_{ij} \right) \mathrm{d}V.$$

It is seen that it equals the left-hand side of Eq. (10.79). Its right-hand side is equal to the variation of the remaining two integrals in Eq. (10.80) taken with the minus sign. This means that the weak solution of the considered problem ensures stationarity of the functional J defined by Eq. (10.80).

It is also interesting to investigate the character of this stationarity point. To do this, let us compute the second variation of the functional J. Using the expressions for the first variation and the following relations,

$$\delta^2 \Delta\varepsilon_{ij} = \delta^2 \overline{\Delta\varepsilon}_{ij} + \delta^2 \overline{\overline{\Delta\varepsilon}}_{ij}\,, \qquad \delta^2 \overline{\Delta\varepsilon}_{ij} = 0, \qquad \delta^2 \overline{\overline{\Delta\varepsilon}}_{ij} = \delta\Delta u_{k,i} \delta\Delta u_{k,j}\,,$$

one can find

$$\delta^2 J = \int_\Omega \left[C_{ijkl}\ \delta\Delta\varepsilon_{ij}\ \delta\Delta\varepsilon_{kl} + \left(C_{ijkl} \Delta\varepsilon_{kl} + \tilde{\sigma}_{ij} + \Delta t\ H_{ij} \right) \delta^2 \overline{\overline{\Delta\varepsilon}}_{ij} \right] \mathrm{d}V$$
$$= \int_\Omega \left[C_{ijkl}\ \delta\Delta\varepsilon_{ij}\ \delta\Delta\varepsilon_{kl} + (\tilde{\sigma}_{ij} + \Delta\tilde{\sigma}_{ij})\ \delta\Delta u_{k,i} \delta\Delta u_{k,j} \right] \mathrm{d}V. \tag{10.81}$$

The form of this expression does not guarantee its positive sign. Even if the term $C_{ijkl}\ \delta\Delta\varepsilon_{ij}\ \delta\Delta\varepsilon_{kl}$ is a positive definite quadratic form, the second term in the integrand depends on the stress components $\tilde{\sigma}_{ij} + \Delta\tilde{\sigma}_{ij}$ which may be negative. Similarly as in the case of nonlinear elasticity, we conclude that the form of the problem equations allows for non-existence of the potential energy minimum and—in consequence—for the solution instability.

In analogy to the non-incremental case, the variational formulation can be easily extended to the problem of dynamics, i.e. to the case when the equations of motion (10.71a) assume the form

$$\Delta\sigma_{ij,j} + \Delta\hat{f}_i = \rho \Delta\ddot{u}_i\,,$$

where the density distribution at $\tau = t$ is assumed known. In this case, an additional term appears on the left-hand side of the virtual work equation (10.76) or (10.79) and the equation takes the form

$$\int_\Omega \Delta\sigma_{ij}\ \delta\Delta u_{i,j}\ \mathrm{d}V + \int_\Omega \rho\Delta\ddot{u}_i\ \delta\Delta u_i\ \mathrm{d}V$$
$$= \int_\Omega \Delta\hat{f}_i\ \delta\Delta u_i\ \mathrm{d}V + \int_{\partial\Omega_\sigma} \Delta\hat{t}_i\ \delta\Delta u_i\ \mathrm{d}A. \tag{10.82}$$

The form (10.80) of the functional J is then extended by the additional term

$$\int_{\Omega} \rho \Delta \ddot{u}_i \Delta u_i \, \mathrm{d}V.$$

10.4 Variational Formulations for Heat Conduction Problems

Let us start from analyzing the linear, stationary, 1D heat conduction problem. The domain Ω is a segment (a, b). The temperature $\hat{T}$ $(x = a)$ and the inward heat flux $\hat{q}$ $(x = b)$ are imposed on its ends, respectively. Inside the domain there are distributed volumetric heat sources $\hat{g}(x)$, corresponding to ρr in Eq. (9.26), assumed temperature-independent here. The boundary value problem is thus expressed as

$$\lambda T_{,xx} + \hat{g}(x) = 0, \qquad x \in (a, b), \tag{10.83a}$$

$$T(a) = \hat{T}, \tag{10.83b}$$

$$\lambda T_{,x}(b) = \hat{q}. \tag{10.83c}$$

In analogy to the previously discussed mechanical problems, the above formulation is called the strong formulation of the problem.

Let us define the set of trial functions $\mathcal{P}$ which are required to be sufficiently smooth and fulfill the condition (10.83b)

$$\mathcal{P} = \{T(x) : [a, b] \rightarrow R, \;\; T(a) = \hat{T}\}. \tag{10.84}$$

Their variations form the set

$$\mathcal{W} = \{\delta T(x) : [a, b] \rightarrow R, \;\; \delta T(a) = 0, \;\; |\delta T| < \epsilon\}. \tag{10.85}$$

According to the definitions, if $T \in \mathcal{P}$ is the solution of the problem (10.83) then for each variation $\delta T \in \mathcal{W}$ the following equation holds true:

$$\int_a^b (\lambda T_{,xx} + \hat{g}) \, \delta T \, \mathrm{d}x - (\lambda T_{,x}(b) - \hat{q}) \, \delta T(b) = 0. \tag{10.86}$$

Let us transform the first term of the equation making use of the relation $T_{,xx} \, \delta T = (T_{,x} \, \delta T)_{,x} - T_{,x} \, \delta T_{,x}$:

$$\begin{aligned} \int_a^b \lambda T_{,xx} \, \delta T \, \mathrm{d}x &= \left[\lambda T_{,x} \, \delta T\right]_{x=a}^{x=b} - \int_a^b \lambda T_{,x} \, \delta T_{,x} \, \mathrm{d}x \\ &= \lambda T_{,x}(b) \, \delta T(b) - \int_a^b \lambda T_{,x} \, \delta T_{,x} \, \mathrm{d}x. \end{aligned}$$

Substituting this result into Eq. (10.86) we obtain

$$\int_a^b \lambda T_{,x}\, \delta T_{,x}\, \mathrm{d}x - \int_a^b \hat{g}\, \delta T\, \mathrm{d}x - \hat{q}\, \delta T(b) = 0. \tag{10.87}$$

This is the "heat" counterpart for the virtual work principle, i.e. the *virtual temperature principle*.

It is now straightforward to formulate the functional $J[T]$ in the form

$$J[T] = \int_a^b \frac{1}{2}\lambda (T_{,x})^2\, \mathrm{d}x - \int_a^b \hat{g} T\, \mathrm{d}x - \hat{q}\, T(b), \tag{10.88}$$

whose variation δJ is identical to the left-hand side of Eq. (10.87) and thus it must vanish. In other words, the stationarity condition for the functional (10.88) generates Eq. (10.83a) inside the interval $[a, b]$ and Eq. (10.83c) as the natural boundary condition. The condition (10.83b) is the essential boundary condition and must be fulfilled *a priori* by the function $T(x)$.

It is also easy to compute the second variation as

$$\delta^2 J = \int_a^b \lambda (\delta T_{,x})^2\, \mathrm{d}x\,. \tag{10.89}$$

Since the thermal conductivity coefficient λ is always positive, this expression is positive, too. Thus, the stationarity condition for the functional $J[T]$, satisfied by definition by the weak problem solution (10.87), is always the minimum condition.

The weak formulation for the 3D stationary heat conduction problem can be derived in an analogous way. Consider Eq. (9.27), assuming at the moment that $\dot{T} = 0$, together with the boundary conditions (9.33)–(9.34) in which the given temperature distribution is imposed on a part of the body's boundary while on the remaining part a given heat flux and convection heat exchange are defined,

$$(\lambda_{ij} T_{,j})_{,i} + \hat{g} = 0, \qquad \boldsymbol{x} \in \Omega, \tag{10.90a}$$

$$T = \hat{T}, \qquad \boldsymbol{x} \in \partial\Omega_T\,, \tag{10.90b}$$

$$\lambda_{ij} T_{,j} n_i = \hat{q} + \alpha(T_\infty - T), \qquad \boldsymbol{x} \in \partial\Omega_q\,. \tag{10.90c}$$

Defining the trial functions as fields $T(\boldsymbol{x})$ fulfilling the boundary condition (10.90b) and their variations $\delta T(\boldsymbol{x})$ with the homogeneous condition $\delta T = 0$, $\boldsymbol{x} \in \partial\Omega_T$, one can formulate the following integral equation

$$\int_\Omega \left[(\lambda_{ij} T_{,j})_{,i} + \hat{g}\right] \delta T\, \mathrm{d}V - \int_{\partial\Omega_q} \left[\lambda_{ij} T_{,j} n_i - \hat{q} - \alpha(T_\infty - T)\right] \delta T\, \mathrm{d}A = 0, \tag{10.91}$$

which must hold true for each variation δT if T is the solution of the problem (10.90). Making use of the equality

$$(\lambda_{ij}T_{,j})_{,i}\delta T = (\lambda_{ij}T_{,j}\delta T)_{,i} - \lambda_{ij}T_{,j}\delta T_{,i}$$

and applying the Gauss–Ostrogradski theorem to the volume integral of $(\lambda_{ij}T_{,j}\delta T)_{,i}$, one obtains after transformations

$$\int_{\Omega} \lambda_{ij}T_{,j}\delta T_{,i}\,\mathrm{d}V - \int_{\Omega} \hat{g}\,\delta T\,\mathrm{d}V - \int_{\partial\Omega_q} \big[\hat{q} + \alpha(T_\infty - T)\big]\,\delta T\,\mathrm{d}A = 0. \quad (10.92)$$

This is the virtual temperature principle in the 3D formulation, i.e. the weak formulation of the problem (10.90).

The coefficients λ_{ij}, $\hat{g}$, $\hat{q}$, α, appearing in Eq. (10.92) may generally be functions of temperature. The virtual temperature equation is then nonlinear and it is difficult to derive the form of a variational functional corresponding to this formulation. However, if linearity is assumed and the above quantities are treated as material constants, then the functional whose variation is the left-hand side of Eq. (10.92) has the form

$$J[T] = \int_{\Omega} \left(\frac{1}{2}\lambda_{ij}T_{,i}T_{,j} - \hat{g}T\right)\mathrm{d}V - \int_{\partial\Omega_q} \left[\hat{q}T - \frac{1}{2}\alpha(T_\infty - T)^2\right]\mathrm{d}A. \quad (10.93)$$

Its second variation,

$$\delta^2 J = \int_{\Omega} \lambda_{ij}\delta T_{,i}\delta T_{,j}\,\mathrm{d}V + \int_{\partial\Omega_q} \alpha(\delta T)^2\,\mathrm{d}A, \quad (10.94)$$

is always positive due to positive definition of the thermal conductivity tensor λ_{ij} and the convection heat exchange coefficient α. Hence, the temperature field $T(\boldsymbol{x})$ ensuring the thermal equilibrium in the considered system corresponds to the minimum of the functional $J[T]$.

Similarly as in the problem of mechanics, it can be said that Eqs. (10.90a) and (10.90c) are the Euler equations to the functional (10.93). The boundary condition on the heat flux (10.90c) is the natural condition here while the temperature condition (10.90b) is the essential condition.

Taking into account nonlinearity, i.e.

$$\lambda_{ij} = \lambda_{ij}(T), \qquad \hat{g} = \hat{g}(T), \qquad \alpha = \alpha(T), \qquad c = c(T),$$

does not imply significant changes in the above derivations. The variational equation (10.92) remains the same. The only difference is the form of the functional J which cannot usually be expressed in a simple closed form for nonlinear problems. Moreover, it cannot be guaranteed that this functional exists at all for a given nonlinearity type.

In the above derivations, it has been assumed for clarity that the heat flow is stationary, i.e. the term $\rho c\dot{T}$ can be neglected on the right-hand side of Eq. (10.90a). To take nonstationary flows into account, only a minor modification in the above formulation is necessary. Including a non-zero value of $\rho c\dot{T}$ in the right-hand side of Eq. (10.90a), we can see that this equation must be fulfilled at each time instant $\tau \in [0, \bar{t}]$. At each time instant, appropriate boundary conditions can also be formulated. Assuming that the rate $\dot{T}$ is independent of the instantaneous solution T (because its value is extrapolated on the basis of solutions for the previous time instants), we can follow the same transformations as in the stationary case, treating the expression $-\rho c\dot{T}$ as an additional volumetric heat source. In other words, the quantity $\hat{g}$ is replaced by the quantity $\hat{g} - \rho c\dot{T}$ (i.e. $\rho(g - c\dot{T})$). In particular, the virtual temperature equation (10.92) assumes at each time instant the form

$$\int_{\Omega} \lambda_{ij} T_{,j} \delta T_{,i} \, \mathrm{d}V - \int_{\Omega} \rho(g - c\dot{T}) \delta T \, \mathrm{d}V - \int_{\partial\Omega_q} \left[\hat{q} + \alpha(T_\infty - T)\right] \delta T \, \mathrm{d}A = 0. \tag{10.95}$$

Reference

1. Gelfand I.M., Fomin S.W., 1963. *Calculus of Variations*. Prentice-Hall.

Chapter 11
Discrete Formulations in Thermomechanics

In Chap. 10, a general methodology has been presented that is useful in a search for approximate solutions of thermomechanical problems. It consists in replacement of the local (strong) formulation of the problem by the corresponding weak formulation. Contrary to the former one, being a system of partial differential equations with appropriate initial and boundary conditions, the latter takes the form of a scalar integral equation. In this equation, apart from the unknown fields sought in the local differential equations of the problem, additional fields appear, known as variations of the unknown fields. A field that fulfills the mentioned integral equation for all admissible variations is called the solution of the weak formulation of the problem considered.

The following conclusions result from the material presented:

1. The weak problem solution depends on a class of functions (and their corresponding variations) in which the solution is sought. In other words, for each class of functions meeting certain requirements of continuity and differentiability, a corresponding weak problem solution can be found, specific to the class considered. To the contrary, in the case of strong formulation, there is only one solution and any attempts to find it in a class of functions that does not contain it must inevitably fail.
2. If the class of functions in which the weak problem solution is sought contains the exact solution of the strong formulation, then the weak problem solution is the same as the strong problem solution. Otherwise, the weak problem solution is called approximate solution of the strong formulation. It is the best approximate of the latter solution in the considered class of functions.
3. In the weak formulation, the requirements regarding continuity and differentiability of the fields appearing in this formulation are significantly less restrictive than in the corresponding strong formulation. This means that one can search for approximate solutions of the strong formulation in classes of functions that do not meet its requirements of continuity and differentiability. In such a case, it may not be even possible to substitute the approximate solution found into the local

M. Kleiber and P. Kowalczyk, *Introduction to Nonlinear Thermomechanics of Solids*, Lecture Notes on Numerical Methods in Engineering and Sciences,
DOI 10.1007/978-3-319-33455-4_11

differential equations of the strong formulation. In spite of it, it is still called the approximate solution.
4. In several problems of thermomechanics (albeit not in all), it is possible to define a functional on a domain of all fields fulfilling essential boundary conditions which reaches its minimum for the field being the weak problem solution. The value of the functional may thus be a good criterion to compare quality of different approximate solutions of the problem considered.

The conclusions listed above indicate that the weak formulation is a good basis to look for approximate solutions for various problems of thermomechanics of solids. Indeed, widely used computer programs performing analysis of deformation, stresses and temperature distribution in solids and structures, as well as simulations of industrial processes such as plastic forming, contain in their "computational engines" practical implementations of the weak formulations just discussed.

Subsequent chapters of this book are devoted to one of methods in which the weak formulations are applied in a highly automated way. This allows to solve tasks with practically arbitrary geometries and initial-boundary conditions. In this chapter, a methodology of seeking weak problem solutions will be presented in which the class of functions is defined by a certain set of discrete parameters and corresponding set of predefined functions (called shape functions). As will be shown, the weak formulation can be this way transformed to a system of algebraic equations in which the unknown is the vector of discrete values of the considered parameters. In the next chapter, fundamentals of the finite element method will be presented. Nowadays, this is the most popular method of automatized definition of shape functions and building of discrete vector-matrix weak formulations for a truly wide class of problems of solid thermomechanics.

11.1 Discrete Formulations in Heat Conduction Problems

Contrary to the previously introduced order of presentation, let us start here from the heat conduction problems. Clarity of presentation justifies this—the heat conduction problem is described by a differential equation with just one unknown: the scalar temperature field. The reader will thus have the opportunity to first learn the most simple case of scalar field discretization, to proceed then to a more complex case of a vector displacement field.

11.1.1 Linear Problem of Stationary Heat Conduction

Let us start with the linear case of stationary heat conduction. The weak formulation of this problem, for a body occupying the domain Ω bounded by the boundary $\partial\Omega = \partial\Omega_T \cup \partial\Omega_q$, is given by Eq. (10.92), i.e.

$$\int_{\Omega} \lambda_{ij} T_{,j} \delta T_{,i} \, \mathrm{d}V - \int_{\Omega} \hat{g} \, \delta T \, \mathrm{d}V - \int_{\partial \Omega_q} \left[\hat{q} + \alpha (T_\infty - T) \right] \delta T \, \mathrm{d}A = 0. \quad (11.1)$$

The field $T(\boldsymbol{x})$ and its variation $\delta T(\boldsymbol{x})$ must belong, respectively, to the classes $\mathcal{P}$ and $\mathcal{W}$ defined in a way assuring fulfillment of the condition

$$\forall_{T \in \mathcal{P}, \, \delta T \in \mathcal{W}} \; T + \delta T \in \mathcal{P}, \qquad \boldsymbol{x} \in \Omega,$$

and the essential boundary condition

$$T = \hat{T}, \quad \delta T = 0, \qquad \boldsymbol{x} \in \partial \Omega_T \, .$$

Besides, the fields must be continuous and meet certain criteria regarding their differentiability—the form of expressions in Eq. (11.1) implies that they must be first-order differentiable at all points $\boldsymbol{x} \in \Omega$ except for a set of measure zero. The field $T(\boldsymbol{x}) \in \mathcal{P}$ that ensures equality in Eq. (11.1) for each variation $\delta T(\boldsymbol{x}) \in \mathcal{W}$ is the weak problem solution.

Consider the following class of temperature fields,

$$\mathcal{P}^{(N)} = \left\{ \tilde{T}(\boldsymbol{x}) = \hat{\Phi}(\boldsymbol{x}) + \sum_{\alpha=1}^{N} \Phi_\alpha(\boldsymbol{x}) q_\alpha \right\},$$

which is a linear combinations of N predefined linearly independent functions $\Phi_\alpha(\boldsymbol{x})$ (further called *shape functions*) that meet the required criteria of continuity and differentiability, with coefficients q_α (also called *system degrees of freedom*) that form an N-element array (vector) of real numbers. Forms of functions Φ_α are chosen so as to fulfil (exactly or approximately) the condition $\Phi_\alpha = 0$ on $\partial \Omega_T$ while the additional field $\hat{\Phi}(\boldsymbol{x})$ is an arbitrary function such that $\hat{\Phi} = \hat{T}$ on $\partial \Omega_T$. Under such assumptions, the fields $\tilde{T}(\boldsymbol{x})$ fulfil the essential boundary conditions on $\partial \Omega_T$ for arbitrary values of q_α. The tilde $\tilde{}$ above the symbol T emphasizes that we are talking about approximate temperature fields that may in general form only a non-exact solution to the problem considered.

Extending our summation convention to the case of repeating indices $\alpha = 1, \ldots, N$, we can write

$$\begin{aligned} \mathcal{P}^{(N)} &= \left\{ \tilde{T}(\boldsymbol{x}) = \hat{\Phi}(\boldsymbol{x}) + \Phi_\alpha(\boldsymbol{x}) q_\alpha \right\}, \\ \mathcal{W}^{(N)} &= \left\{ \delta \tilde{T}(\boldsymbol{x}) = \Phi_\alpha(\boldsymbol{x}) \, \delta q_\alpha \right\}, \end{aligned} \quad (11.2)$$

where the real coefficients δq_α are sufficiently small to make the temperature variations fit the limit $|\delta \tilde{T}| < \epsilon$, cf. Eq. (10.85). The function form of the variations ensures that $\delta \tilde{T} = 0$ on $\partial \Omega_T$.

Let us now substitute the so-defined temperature fields and their variations into Eq. (11.1). Let us also make use of the fact that the coefficients q_α and δq_α are not functions of $\boldsymbol{x}$ and thus may be moved out from the integrand expressions. We obtain the following equation:

$$\delta q_\alpha \left[q_\beta \int_\Omega \lambda_{ij} \Phi_{\alpha,i} \Phi_{\beta,j} \, \mathrm{d}V + \int_\Omega \left(\lambda_{ij} \Phi_{\alpha,i} \hat{\Phi}_{,j} - \hat{g} \Phi_\alpha \right) \mathrm{d}V \right.$$
$$\left. + \int_{\partial\Omega_q} \left[\alpha(\hat{\Phi} - T_\infty) - \hat{q} \right] \Phi_\alpha \, \mathrm{d}A + q_\beta \int_{\partial\Omega_q} \alpha \Phi_\alpha \Phi_\beta \, \mathrm{d}A \right] = 0.$$

Introducing the notation

$$K_{\alpha\beta} = \int_\Omega \lambda_{ij} \Phi_{\alpha,i} \Phi_{\beta,j} \, \mathrm{d}V + \int_{\partial\Omega_q} \alpha \Phi_\alpha \Phi_\beta \, \mathrm{d}A, \tag{11.3a}$$

$$Q_\alpha = \int_\Omega \left(\hat{g} \Phi_\alpha - \lambda_{ij} \Phi_{\alpha,i} \hat{\Phi}_{,j} \right) \mathrm{d}V + \int_{\partial\Omega_q} \left[\hat{q} + \alpha(T_\infty - \hat{\Phi}) \right] \Phi_\alpha \, \mathrm{d}A, \tag{11.3b}$$

one can rewrite the above equation in the short form as,

$$\delta q_\alpha \left(K_{\alpha\beta} q_\beta - Q_\alpha \right) = 0.$$

This equation must be fulfilled for each variation $\delta\tilde{T}$, i.e. for each set of coefficients δq_α, $\alpha = 1, \ldots, N$. This means that the coefficients q_α must satisfy the following set of equations, written below in both the absolute and index notations as

$$K_{\alpha\beta} q_\beta = Q_\alpha, \qquad \mathbf{K}_{N\times N} \mathbf{q}_{N\times 1} = \mathbf{Q}_{N\times 1}. \tag{11.4}$$

The weak formulation (11.1) has been in this way transformed to assume the form of a system of algebraic equations in which the unknown vector $\mathbf{q}$ contains the sought coefficients q_α of the approximate temperature field (11.2). The coefficient matrix $\mathbf{K}$ is called in this case the heat conductivity matrix, while the right-hand-side vector $\mathbf{Q}$ contains components of "thermal load", i.e. certain heat flow quantities associated with particular degrees of freedom. Hence, approximate solution of the considered problem consists in determining the matrix $\mathbf{K}$ and the vector $\mathbf{Q}$ and then solving the system (11.4) with the help of one of commonly known standard numerical methods. Obviously, all the operations must be preceded by the proper choice of the functions Φ_α, $\alpha = 1, \ldots, N$, whose form is crucial to the quality of the approximate solution, as well as to the computational effort needed to generate the coefficient matrix and the r.h.s. vector of the system.

It is noteworthy that the matrix $\mathbf{K}$ is symmetric and positive definite (because of positive definition of the tensor $\boldsymbol{\lambda}$ and positive value of the convection coefficient α). This fact is of great importance for the issues of existence and stability of the solution, as well as for the choice of an efficient numerical method for the matrix factorization in the solution procedure of the equations system.

It was mentioned in Sect. 10.4 that the weak solution of the problem of stationary heat conduction on hand is the field that minimizes a certain functional $J[T]$ whose form was given in Eq. (10.93). Substituting the discrete approximate of the temperature field $\tilde{T}(\boldsymbol{x})$ (11.2) into this formula, one obtains after transformations

$$J = \frac{1}{2} q_\alpha K_{\alpha\beta} q_\beta - q_\alpha Q_\alpha + C, \qquad C = \frac{1}{2} \int_{\partial\Omega_q} \alpha T_\infty^2 \, \mathrm{d}A. \tag{11.5}$$

As it can be seen from the above formula, the functional $J[T]$ becomes in the present formulation a quadratic polynomial function of N variables $J(q_\alpha)$ whose minimization is a standard algebraic problem, apparently equivalent to solving the system of equations (11.4).

11.1.2 General Form of the Heat Conduction Problem

In the previous Section, a stationary character of heat conduction and linearity of equations were assumed. Let us now consider more general cases, by taking first into account in the equations terms containing the time derivative $\dot{T}$. These terms describe non-stationarity of the phenomena considered.

The weak formulation of this problem is given by Eq. (10.95):

$$\int_\Omega \lambda_{ij} T_{,j} \, \delta T_{,i} \, \mathrm{d}V - \int_\Omega \rho(g - c\dot{T}) \, \delta T \, \mathrm{d}V - \int_{\partial\Omega_q} \left[\hat{q} + \alpha(T_\infty - T)\right] \delta T \, \mathrm{d}A = 0, \tag{11.6}$$

in which δT is an arbitrary variation of the temperature field at a certain time instant $\tau = t$ from the interval $[0, \bar{t}]$, fulfilling the condition $\delta T = 0$ at $\boldsymbol{x} \in \partial\Omega_T$.

Consider classes of functions (11.2) which are to be used for approximation of the sought field T and its variation δT. Let us additionally assume that the coefficients q_α and functions $\hat{\Phi}$ depend on time τ and focus our attention on their values and time derivatives at the particular time instant $\tau = t$. Substituting the so-defined fields $\tilde{T}$ and $\delta\tilde{T}$ into Eq. (11.6) and moving out of the integrand expressions the coefficients q_α, $\dot{q}_\alpha$ and δq_α we arrive at

$$\delta q_\alpha \left[q_\beta \int_\Omega \lambda_{ij} \Phi_{\beta,j} \, \Phi_{\alpha,i} \, \mathrm{d}V + \int_\Omega \lambda_{ij} \hat{\Phi}_{,j} \, \Phi_{\alpha,i} \, \mathrm{d}V - \int_\Omega \rho g \Phi_\alpha \, \mathrm{d}V \right.$$
$$+ \dot{q}_\beta \int_\Omega \rho c \Phi_\beta \, \Phi_\alpha \, \mathrm{d}V + \int_\Omega \rho c \dot{\hat{\Phi}} \, \Phi_\alpha \, \mathrm{d}V$$

$$-\int_{\partial\Omega_q} (\hat{q} + \alpha T_\infty)\, \Phi_\alpha \,\mathrm{d}A + q_\beta \int_{\partial\Omega_q} \alpha \Phi_\beta\, \Phi_\alpha \,\mathrm{d}A + \int_{\partial\Omega_q} \alpha \hat{\Phi} \Phi_\alpha \,\mathrm{d}A \Bigg] = 0.$$

Introducing the notation

$$\begin{aligned} K_{\alpha\beta} &= \int_\Omega \lambda_{ij} \Phi_{\alpha,i} \Phi_{\beta,j} \,\mathrm{d}V + \int_{\partial\Omega_q} \alpha \Phi_\alpha \Phi_\beta \,\mathrm{d}A, \\ C_{\alpha\beta} &= \int_\Omega \rho c \Phi_\alpha \Phi_\beta \,\mathrm{d}V, \\ Q_\alpha &= -\int_\Omega \lambda_{ij} \Phi_{\alpha,i} \hat{\Phi}_{,j} \,\mathrm{d}V + \int_\Omega \rho (g \Phi_\alpha - c \Phi_\alpha \dot{\hat{\Phi}}) \,\mathrm{d}V \\ &\quad + \int_{\partial\Omega_q} [\hat{q} + \alpha(T_\infty - \hat{\Phi})] \Phi_\alpha \,\mathrm{d}A, \end{aligned}$$

one can rewrite the above equation in the following short form,

$$\delta q_\alpha \left(K_{\alpha\beta}\, q_\beta + C_{\alpha\beta}\, \dot{q}_\beta - Q_\alpha \right) = 0.$$

This equation must be fulfilled for each variation $\delta \tilde{T}$, i.e. for each set of coefficients δq_α, $\alpha = 1, \ldots, N$. This means that the coefficients q_α must satisfy the set of equations written below in both the absolute and index notations as

$$K_{\alpha\beta}\, q_\beta + C_{\alpha\beta}\, \dot{q}_\beta = Q_\alpha\,, \qquad \mathbf{K}_{N\times N}\, \mathbf{q}_{N\times 1} + \mathbf{C}_{N\times N}\, \dot{\mathbf{q}}_{N\times 1} = \mathbf{Q}_{N\times 1}. \tag{11.7}$$

The weak formulation (11.6) has thus been transformed to assume the form of a linear system of differential equations with respect to N functions $q_\alpha(\tau)$. It is noteworthy that the matrices **K** and **C** are symmetric. Solution methods of this kind for equations systems will be discussed in Chap. 13.

Let us now consider the nonlinear case of heat conduction. Assume that the quantities λ_{ij}, g, c and α are no more constant but dependent in a known way on temperature T. For simplicity, let us limit ourselves to just a stationary case, although extension of the equations towards the non-stationary formulation is quite easy with the methodology presented above.

Let us start once again from the variational equation (11.1) with the same classes of functions $\mathcal{P}^{(N)}$ and their variations $\mathcal{W}^{(N)}$ as defined in (11.2). Performing the same transformations we arrive at the linear system of algebraic equations identical with Eq. (11.4). The only essential difference lies in the fact that both the coefficient matrix **K** and the right-hand side vector **Q** are now functions of **q**. The system is thus nonlinear and the following way of expressing it appears appropriate:

$$r_\alpha(q_\beta) = 0, \qquad \mathbf{r}_{N\times 1}(\mathbf{q}_{N\times 1}) = \mathbf{0}_{N\times 1}\,, \tag{11.8}$$

where

$$\mathbf{r} = \mathbf{Q}(\mathbf{q}) - \mathbf{K}(\mathbf{q})\,\mathbf{q}.$$

Solution of such a system requires iterative methods that consist in solving a sequence of linear equations sets whose solutions converge to the desired solution $\mathbf{q}$. One of the best known methods is the Newton–Raphson method in which—knowing an approximate solution $\tilde{\mathbf{q}}$—one constructs a set of algebraic equations whose solution is a corrector $\boldsymbol{\delta}$ such that the subsequent solution approximate $\tilde{\mathbf{q}} := \tilde{\mathbf{q}} + \boldsymbol{\delta}$ is better (closer to the exact solution) than the previous one. This set of equations has the form

$$\bar{K}_{\alpha\beta}\,\delta_\beta = r_\alpha\,, \qquad \bar{\mathbf{K}}\boldsymbol{\delta} = \mathbf{r}, \tag{11.9}$$

where $\bar{\mathbf{K}}$ and $\mathbf{r}$ are computed for the current approximate $\tilde{\mathbf{q}}$,

$$\begin{aligned}\bar{K}_{\alpha\beta} = -\frac{\partial r_\alpha(\tilde{q}_\zeta)}{\partial q_\beta} &= K_{\alpha\beta}(\tilde{q}_\zeta) + \frac{\partial K_{\alpha\gamma}(\tilde{q}_\zeta)}{\partial q_\beta}\,\tilde{q}_\gamma - \frac{\partial Q_\alpha(\tilde{q}_\zeta)}{\partial q_\beta}\\ &= K_{\alpha\beta}(\tilde{q}_\zeta) + \left(\frac{\mathrm{d}K_{\alpha\gamma}}{\mathrm{d}T}\,\tilde{q}_\gamma - \frac{\mathrm{d}Q_\alpha}{\mathrm{d}T}\right)\Phi_\beta\,, \qquad (11.10\text{a})\\ r_\alpha = r_\alpha(\tilde{q}_\zeta) &= Q_\alpha(\tilde{q}_\zeta) - K_{\alpha\beta}(\tilde{q}_\zeta)\,\tilde{q}_\beta\,. \qquad (11.10\text{b})\end{aligned}$$

The matrix $\bar{\mathbf{K}}_{N\times N}$ is called *effective (algorithmic) tangent matrix* for the considered problem. The Newton–Raphson method will be discussed in more detail in Chap. 13.

In the case of nonlinear and non-stationary heat conduction problem, the system (11.7) holds true, however, the matrices $\mathbf{K}$ and $\mathbf{C}$ become functions of $\mathbf{q}$ while the r.h.s. vector is additionally a function of time (as it contains the known functions $\hat{q}(\tau)$ and $T_\infty(\tau)$). Hence

$$\begin{aligned}&K_{\alpha\beta}(q_\zeta)\,q_\beta + C_{\alpha\beta}(q_\zeta)\,\dot{q}_\beta = Q_\alpha(\tau, q_\zeta)\,,\\ &\mathbf{K}(\mathbf{q})\,\mathbf{q} + \mathbf{C}(\mathbf{q})\,\dot{\mathbf{q}} = \mathbf{Q}(\tau, \mathbf{q}).\end{aligned} \tag{11.11}$$

11.2 Discrete Formulations in Solid Mechanics Problems

The derivations of discrete formulations for heat conduction problems can be as well applied to problems of solid mechanics. The essential difference consists in the fact that the approximated fields—displacements or their increments—have the vector form. Thus the construction of shape functions and their corresponding unknown coefficients becomes more complex and the matrices and the right-hand side vectors in the derived sets of equations have more elaborated forms.

11.2.1 Linear Problem of Statics

Let us start from the simplest case in which we consider the problem of statics under the assumption of small deformations and a linear elastic constitutive equation. The weak formulation of this problem for a body occupying the domain Ω bounded by the boundary $\partial\Omega = \partial\Omega_u \cup \partial\Omega_\sigma$ is given by Eq. (10.36) which after using the constitutive equation and symmetry of the stiffness tensor assumes the form

$$\int_\Omega C_{ijkl}\, u_{k,l}\, \delta u_{i,j}\, \mathrm{d}V = \int_\Omega \hat{f}_i\, \delta u_i\, \mathrm{d}V + \int_{\partial\Omega_\sigma} \hat{t}_i\, \delta u_i\, \mathrm{d}A. \tag{11.12}$$

The fields $u_i(\boldsymbol{x})$ and their variations $\delta u_i(\boldsymbol{x})$ must belong, respectively, to the classes $\mathcal{P}$ and $\mathcal{W}$ defined in a way assuring fulfillment of the condition

$$\forall_{u_i\in\mathcal{P},\ \delta u_i\in\mathcal{W}}\ u_i + \delta u_i \in \mathcal{P},$$

and the essential boundary conditions

$$u_i = \hat{u}_i, \quad \delta u_i = 0, \qquad \boldsymbol{x} \in \partial\Omega_u\,.$$

Besides, the fields must be continuous and meet certain criteria regarding their differentiability—the form of expressions in Eq. (11.12) implies that they must be first-order differentiable at all points $\boldsymbol{x} \in \Omega$ except for a set of measure zero. The vector field $u_i(\boldsymbol{x}) \in \mathcal{P}$ that ensures equality in Eq. (11.12) for each variation $\delta u_i(\boldsymbol{x}) \in \mathcal{W}$ is the weak problem solution.

Consider the following class of displacement fields and their variations

$$\begin{aligned} \mathcal{P}^{(N)} &= \left\{\tilde{u}_i(\boldsymbol{x}) = \hat{\Phi}_i(\boldsymbol{x}) + \Phi_{i\alpha}(\boldsymbol{x})q_\alpha\right\}, \\ \mathcal{W}^{(N)} &= \{\delta\tilde{u}_i(\boldsymbol{x}) = \Phi_{i\alpha}(\boldsymbol{x})\,\delta q_\alpha\}, \end{aligned} \tag{11.13}$$

which are linear combinations of three N-element sets of predefined linearly independent shape functions $\Phi_{i\alpha}(\boldsymbol{x})$ ($i = 1, 2, 3$, $\alpha = 1, \ldots, N$) that meet the required criteria of continuity and differentiability, with the N-element sets of real coefficients q_α and δq_α, respectively. Function forms of $\Phi_{i\alpha}$ are chosen so as to fulfil (exactly or approximately) the condition $\Phi_{i\alpha} = 0$ on $\partial\Omega_u$ while the additional fields $\hat{\Phi}_i(\boldsymbol{x})$ are arbitrary functions such that $\hat{\Phi}_i = \hat{u}_i$ on $\partial\Omega_u$. With such assumptions, the fields $\tilde{u}_i(\boldsymbol{x})$ fulfil the essential boundary conditions on $\partial\Omega_u$ for any arbitrary values of q_α while variations $\delta\tilde{u}_i(\boldsymbol{x})$ vanish there for all δq_α.

Let us now substitute the so-defined displacement fields and their variations into Eq. (11.12). Noting again that the coefficients q_α and δq_α are not functions of $\boldsymbol{x}$ and thus may be moved out of the integrand expressions, we obtain

$$\delta q_\alpha \left[q_\beta \int_\Omega C_{ijkl}\, \Phi_{i\alpha,j}\, \Phi_{k\beta,l}\, \mathrm{d}V + \int_\Omega C_{ijkl}\, \Phi_{i\alpha,j}\, \hat{\Phi}_{k,l}\, \mathrm{d}V \right.$$
$$\left. - \int_\Omega \hat{f}_i \Phi_{i\alpha}\, \mathrm{d}V - \int_{\partial\Omega_\sigma} \hat{t}_i \Phi_{i\alpha}\, \mathrm{d}A \right] = 0.$$

Introducing the notation

$$K_{\alpha\beta} = \int_\Omega C_{ijkl} \Phi_{i\alpha,j} \Phi_{k\beta,l}\, \mathrm{d}V, \tag{11.14a}$$

$$Q_\alpha = -\int_\Omega C_{ijkl}\, \Phi_{i\alpha,j}\, \hat{\Phi}_{k,l}\, \mathrm{d}V + \int_\Omega \hat{f}_i \Phi_{i\alpha}\, \mathrm{d}V + \int_{\partial\Omega_\sigma} \hat{t}_i \Phi_{i\alpha}\, \mathrm{d}A, \tag{11.14b}$$

one can rewrite the above equation in the following short form,

$$\delta q_\alpha \left(K_{\alpha\beta}\, q_\beta - Q_\alpha \right) = 0.$$

This equation must be fulfilled for each variation $\delta\tilde{u}_i$, i.e. for each set of coefficients δq_α, $\alpha = 1, \ldots, N$. This means that the coefficients q_α must satisfy the set of equations, written below in both the absolute and index notations as

$$K_{\alpha\beta}\, q_\beta = Q_\alpha\,, \qquad \mathbf{K}_{N\times N}\, \mathbf{q}_{N\times 1} = \mathbf{Q}_{N\times 1}. \tag{11.15}$$

This is a linear system of algebraic equations that has apparently the same form as in the case of the linear stationary heat conduction problem (11.4). Totally different are, however, the forms of the coefficient matrix and the right-hand side vector. Their physical meaning is obviously different, too. The matrix **K** in Eq. (11.15) is called the stiffness matrix of the considered mechanical system while **Q** is called the vector of generalized external forces associated with generalized displacements (as the coefficients q_α can be called), i.e. the system degrees of freedom. This issue will be discussed in more detail in Chap. 12.

It is noteworthy that the matrix **K** is symmetric. Besides, it is positive definite, which results from thermodynamical constraints that must be met by the elastic stiffness tensor components C_{ijkl}.

Let us also note that the potential energy functional (10.37), whose minimum is the weak problem solution, assumes for this class of approximate displacement fields (11.13) the following form:

$$J = \frac{1}{2} K_{\alpha\beta}\, q_\alpha\, q_\beta - Q_\alpha\, q_\alpha = \frac{1}{2} \mathbf{q}^{\mathrm{T}} \mathbf{K} \mathbf{q} - \mathbf{q}^{\mathrm{T}} \mathbf{Q}. \tag{11.16}$$

It is thus no more a functional but rather a quadratic polynomial function of N variables $J(q_\alpha)$ whose minimization is equivalent to solving the system of equations (11.15).

11.2.2 Linear Problem of Dynamics

Let us now consider a more general case in which dynamic effects are additionally included. A weak formulation of this problem is given by the variational equation (10.42), which only differs from Eq. (11.12) by the presence of the volume integral of the quantity $\rho\ddot{u}_i\delta u_i$.

In this case we proceed similarly as in the statics case, noting only that the assumed approximation of displacements by a combination of the shape functions $\Phi_{i\alpha}(\boldsymbol{x})$ holds true at each time instant (any time changes of the displacement field $\tilde{u}_i$ can be described by appropriate time changes of the coefficients q_α, which become functions of τ now). This can be expressed as

$$\tilde{u}_i(\boldsymbol{x},\tau) = \hat{\Phi}_i(\boldsymbol{x},\tau) + \Phi_{i\alpha}(\boldsymbol{x})\, q_\alpha(\tau)$$

which implies

$$\dot{\tilde{u}}_i^t = \dot{\hat{\Phi}}(\boldsymbol{x},\tau) + \Phi_{i\alpha}(\boldsymbol{x})\,\dot{q}_\alpha^t\,, \qquad \ddot{\tilde{u}}_i^t = \ddot{\hat{\Phi}}(\boldsymbol{x},\tau) + \Phi_{i\alpha}(\boldsymbol{x})\,\ddot{q}_\alpha^t\,.$$

In view of these conclusions Eq. (10.42) assumes the following form:

$$\begin{aligned}\delta q_\alpha \Bigg[q_\beta \int_\Omega C_{ijkl}\,\Phi_{i\alpha,j}\,\Phi_{k\beta,l}\,\mathrm{d}V + \int_\Omega C_{ijkl}\,\Phi_{i\alpha,j}\,\hat{\Phi}_{k,l}\,\mathrm{d}V \\ + \ddot{q}_\beta \int_\Omega \rho\Phi_{i\alpha}\Phi_{i\beta}\,\mathrm{d}V + \int_\Omega \rho\Phi_{i\alpha}\ddot{\hat{\Phi}}_i\,\mathrm{d}V \\ - \int_\Omega \hat{f}_i\Phi_{i\alpha}\,\mathrm{d}V - \int_{\partial\Omega_\sigma} \hat{t}_i\Phi_{i\alpha}\,\mathrm{d}A \Bigg] = 0.\end{aligned}$$

The above equation must be fulfilled for each set of coefficients δq_α which requires the expression in brackets to vanish for each α. Introducing the notation

$$M_{\alpha\beta} = \int_\Omega \rho\Phi_{i\alpha}\Phi_{i\beta}\,\mathrm{d}V \tag{11.17}$$

(the so-called mass matrix) and adding one term to the previous definition of Q_α,

$$Q_\alpha = -\int_\Omega C_{ijkl}\,\Phi_{i\alpha,j}\,\hat{\Phi}_{k,l}\,\mathrm{d}V + \int_\Omega (\hat{f}_i - \rho\ddot{\hat{\Phi}}_i)\Phi_{i\alpha}\,\mathrm{d}V + \int_{\partial\Omega_\sigma} \hat{t}_i\Phi_{i\alpha}\,\mathrm{d}A,$$

we arrive at the following system of equations for the coefficients q_α:

$$M_{\alpha\beta}\,\ddot{q}_\beta + K_{\alpha\beta}\,q_\beta = Q_\alpha\,, \qquad \mathbf{M}\,\ddot{\mathbf{q}} + \mathbf{K}\,\mathbf{q} = \mathbf{Q}. \tag{11.18}$$

This is a linear system of differential equations with respect to N unknown functions $q_\alpha(\tau)$ which must be fulfilled at each time instant τ. Solution methods for this kind of equations systems will be discussed in Chap. 13.

11.2.3 Nonlinear Elastic Problem with Large Deformations

Let us now proceed to the problem of nonlinear elasticity, starting from the case of statics. The base for the discrete formulation is in this case the variational virtual work equation (10.66) which is rewritten below for the reference configuration coincident with the initial configuration, $C^r = C^0$. This assumption does not limit generality of our discussion but it simplifies the form of the subsequent derivations. For simplicity of notation, let us also skip the index 0 wherever it denotes the reference configuration (but not the initial time instant for defining displacements and strains):

$$\int_\Omega \tilde{\sigma}^t_{ij}\, \delta\varepsilon^t_{ij}\, \mathrm{d}V = \int_\Omega \hat{f}^t_i\, \delta u^t_i\, \mathrm{d}V + \int_{\partial\Omega_\sigma} \hat{t}^t_i\, \delta u^t_i\, \mathrm{d}A. \tag{11.19}$$

Additionally, the following equations hold true,

$$\begin{aligned} {}^0\varepsilon^t_{ij} &= \operatorname{sym}\left({}^0u^t_{i,j}\right) + \frac{1}{2}\,{}^0u^t_{k,i}\,{}^0u^t_{k,j}\,, \\ \delta\varepsilon^t_{ij} &= \operatorname{sym}\left(\delta u^t_{i,j} + \delta u^t_{k,i}\,{}^0u^t_{k,j}\right) \end{aligned}$$

and, according to the hyperelastic constitutive law,

$$\tilde{\sigma}^t_{ij} = \frac{\partial U^t}{\partial\,{}^0\varepsilon^t_{ij}} = \bar{C}^t_{ijkl}\,{}^0\varepsilon^t_{kl}\,,$$

where $\bar{C}^t_{ijkl}$ is the secant material stiffness tensor.

Similarly as in the linear case, let us postulate the approximate solution ${}^0\tilde{u}^t_i$ and its variation $\delta\tilde{u}^t_i$ in the form (11.13),

$$\begin{aligned} \mathcal{P}^{(N)} &= \left\{{}^0\tilde{u}^t_i(\boldsymbol{x}) = \hat{\Phi}_i(\boldsymbol{x}) + \Phi_{i\alpha}(\boldsymbol{x})q^t_\alpha\right\}, \\ \mathcal{W}^{(N)} &= \left\{\delta\tilde{u}^t_i(\boldsymbol{x}) = \Phi_{i\alpha}(\boldsymbol{x})\,\delta q^t_\alpha\right\}, \end{aligned} \tag{11.20}$$

Equation (11.19) may be thus transformed to the following form:

$$\delta q^t_\alpha \left[\int_\Omega \tilde{\sigma}^t_{ij}\left(\Phi_{i\alpha,j} + \Phi_{k\alpha,i}\,{}^0\tilde{u}^t_{k,j}\right)\mathrm{d}V - \int_\Omega \hat{f}^t_i\,\Phi_{i\alpha}\,\mathrm{d}V - \int_{\partial\Omega_\sigma} \hat{t}^t_i\,\Phi_{i\alpha}\,\mathrm{d}A\right] = 0.$$

To fulfil it for each variation, i.e. for each set of values of the coefficients δq^t_α, it is necessary that the expression in brackets vanishes for each $\alpha = 1, \dots, N$. Denoting

$$F^t_\alpha = \int_\Omega \tilde{\sigma}^t_{ij} \left(\Phi_{i\alpha,j} + \Phi_{k\alpha,i}\, {}^0\tilde{u}^t_{k,j}\right) \mathrm{d}V,$$

$$Q^t_\alpha = \int_\Omega \hat{f}^t_i \Phi_{i\alpha} \,\mathrm{d}V + \int_{\partial\Omega_\sigma} \hat{t}^t_i \Phi_{i\alpha} \,\mathrm{d}A,$$

and noting that F^t_α depends on displacements (and thus also on q^t_ζ), one can express this condition as

$$F^t_\alpha(q^t_\zeta) = Q^t_\alpha\,, \qquad \mathbf{F}^t(\mathbf{q}^t) = \mathbf{Q}^t, \tag{11.21}$$

or shorter

$$r_\alpha(q^t_\zeta) = 0, \qquad \mathbf{r}(\mathbf{q}^t) = \mathbf{0}, \tag{11.22}$$

where $\mathbf{r} = \mathbf{Q}^t - \mathbf{F}^t$.

Equation (11.22) is a nonlinear system of N algebraic equations for q^t_α. Nonlinearity is due to nonlinear dependence of stress $\tilde{\sigma}^t_{ij}$ on strain ${}^0\varepsilon^t_{ij}$ as well as of the latter on the displacement field ${}^0u^t_i$.

Let us apply the Newton–Raphson procedure, mentioned Sect. 11.1.2, to solve this system (this method will be discussed in more detail in Chap. 13). In this procedure, linear systems of equations are sequentially solved to compute subsequent correctors δ_α to the solution q^t_α. A system built for each subsequent solution approximate $\tilde{q}^t_\alpha$ assumes the form

$$\bar{K}_{\alpha\beta}\, \delta_\beta = r_\alpha\,, \qquad \bar{\mathbf{K}}\boldsymbol{\delta} = \mathbf{r},$$

where $\bar{\mathbf{K}}$ and $\mathbf{r}$ are computed for the current solution approximate $\tilde{\mathbf{q}}^t$,

$$\bar{K}_{\alpha\beta} = -\frac{\partial r_\alpha(\tilde{q}^t_\zeta)}{\partial q_\beta}\,, \qquad r_\alpha = r_\alpha(\tilde{q}^t_\zeta) = Q^t_\alpha - F^t_\alpha(\tilde{q}^t_\zeta).$$

The next approximate is obtained by adding the corrector to the previous one, $\tilde{q}^t_\alpha := \tilde{q}^t_\alpha + \delta_\alpha$. In view of the properties

$$\frac{\partial \tilde{\sigma}^t_{ij}}{\partial\, {}^0\varepsilon^t_{kl}} = C^t_{ijkl}\,, \qquad \frac{\partial\, {}^0\varepsilon^t_{kl}}{\partial q^t_\beta} = \operatorname{sym}\left(\Phi_{k\beta,l} + \Phi_{n\beta,k}\, {}^0\tilde{u}^t_{n,l}\right), \tag{11.23}$$

one can write after transformations

$$\bar{K}_{\alpha\beta} = \bar{K}^{(c)}_{\alpha\beta} + \bar{K}^{(\sigma)}_{\alpha\beta}\,,$$

$$\bar{K}^{(c)}_{\alpha\beta} = \int_\Omega C^t_{ijkl} \left(\Phi_{i\alpha,j} + \Phi_{m\alpha,i}\, {}^0\tilde{u}^t_{m,j}\right) \left(\Phi_{k\beta,l} + \Phi_{n\beta,k}\, {}^0\tilde{u}^t_{n,l}\right) \mathrm{d}V,$$

$$\bar{K}^{(\sigma)}_{\alpha\beta} = \int_{\Omega} \tilde{\sigma}^t_{ij}\, \Phi_{k\alpha,i}\Phi_{k\beta,j}\, \mathrm{d}V.$$

The matrix $\bar{\mathbf{K}}_{N\times N}$, called the effective tangent matrix of the considered problem, is a sum of the so-called constitutive tangent matrix $\bar{\mathbf{K}}^{(c)}$ and the initial stress matrix $\bar{\mathbf{K}}^{(\sigma)}$. Both the matrices are symmetric, however, positive definition of $\bar{\mathbf{K}}$ is not obvious. Although thermodynamical constraints imposed on the stiffness tensor components C^t_{ijkl} guarantee this property for the constitutive matrix $\bar{\mathbf{K}}^{(c)}$, stress $\tilde{\sigma}^t_{ij}$ in the other component matrix may have positive or negative sign. This fact may result in instability of the nonlinear solution, which was already mentioned in Sect. 10.3.1.

It is easy to extend the above formulation to the case of dynamics. Proceeding in the same way as in the linear case, one arrives at a nonlinear system of differential equations in the form

$$M_{\alpha\beta}\ddot{q}^t_{\beta} + F^t_{\alpha}(q^t_{\zeta}) = Q^t_{\alpha}, \qquad \mathbf{M}\ddot{\mathbf{q}}^t + \mathbf{F}^t(\mathbf{q}^t) = \mathbf{Q}^t, \tag{11.24}$$

where the mass matrix components $M_{\alpha\beta}$ are given by the same Eq. (11.17) as in the linear case.

11.2.4 Incremental Form of Nonlinear Mechanics Problem

In this section discrete formulation for the general nonlinear problem of solid mechanics will be presented in the incremental form. A weak formulation for this problem, for a typical time increment (step) $[t, t+\Delta t]$ and with the updated Lagrangian description employed ($C^r = C^t$), is given by Eq. (10.79) in the form

$$\int_{\Omega}(\Delta\tilde{\sigma}_{ij}\,\delta\Delta\varepsilon_{ij} + \tilde{\sigma}_{ij}\,\delta\overline{\overline{\Delta\varepsilon}}_{ij})\,\mathrm{d}V = \int_{\Omega}\Delta\hat{f}_i\,\delta\Delta u_i\,\mathrm{d}V + \int_{\partial\Omega_{\sigma}}\Delta\hat{t}_i\,\delta\Delta u_i\,\mathrm{d}A. \tag{11.25}$$

Let us recall the general incremental form of the constitutive equation,

$$\Delta\tilde{\sigma}_{ij} = C_{ijkl}\,\Delta\varepsilon_{kl} + H_{ij}\Delta t$$

and incremental relations between linear and quadratic components of strain and its variation and the displacement gradients,

$$\begin{aligned}
\overline{\Delta\varepsilon}_{ij} &= \mathrm{sym}(\Delta u_{i,j}), & \delta\overline{\Delta\varepsilon}_{ij} &= \mathrm{sym}(\delta\Delta u_{i,j}),\\
\overline{\overline{\Delta\varepsilon}}_{ij} &= \tfrac{1}{2}\,\Delta u_{k,i}\,\Delta u_{k,j}\,, & \delta\overline{\overline{\Delta\varepsilon}}_{ij} &= \mathrm{sym}(\delta\Delta u_{k,i}\,\Delta u_{k,j}),\\
\Delta\varepsilon_{ij} &= \overline{\Delta\varepsilon}_{ij} + \overline{\overline{\Delta\varepsilon}}_{ij}\,, & \delta\Delta\varepsilon_{ij} &= \delta\overline{\Delta\varepsilon}_{ij} + \delta\overline{\overline{\Delta\varepsilon}}_{ij}\,.
\end{aligned}$$

Let us now consider the following classes of incremental displacement fields and their variations:

$$\mathcal{P}^{(N)} = \left\{ \Delta\tilde{u}_i = \Delta\hat{\Phi}_i(\boldsymbol{x}) + \Phi_{i\alpha}(\boldsymbol{x})\Delta q_\alpha \right\},$$
$$\mathcal{W}^{(N)} = \{ \delta\Delta\tilde{u}_i = \Phi_{i\alpha}(\boldsymbol{x})\,\delta\Delta q_\alpha \}, \tag{11.26}$$

which are linear combinations of three N-element sets of predefined linearly independent shape functions $\Phi_{i\alpha}(\boldsymbol{x})$ ($i = 1, 2, 3$, $\alpha = 1, \ldots, N$) with the N-element sets of real coefficients Δq_α and $\delta\Delta q_\alpha$, respectively. The functions $\Phi_{i\alpha}$ vanish on $\partial\Omega_u$, while the functions $\hat{\Phi}_i(\boldsymbol{x})$ equal $\hat{u}_i$ on this part of the boundary. Under such assumptions the fields $\Delta\tilde{u}_i(\boldsymbol{x})$ fulfil the essential boundary conditions on $\partial\Omega_u$ for any arbitrary values of Δq_α while variations $\delta\Delta\tilde{u}_i(\boldsymbol{x})$ vanish there for all $\delta\Delta q_\alpha$.

Let us now substitute the so-defined approximate fields of displacement increments and their variations into Eq. (11.25). We arrive at the following equation:

$$\delta\Delta q_\alpha \left[\int_\Omega \left[\Delta\tilde{\sigma}_{ij}\,\Phi_{i\alpha,j} + (\tilde{\sigma}_{ij} + \Delta\tilde{\sigma}_{ij})\,\Phi_{k\alpha,i}\,\Delta\tilde{u}_{k,j} \right] \mathrm{d}V - \int_\Omega \Delta\hat{f}_i \Phi_{i\alpha}\,\mathrm{d}V - \int_{\partial\Omega_\sigma} \Delta\hat{t}_i \Phi_{i\alpha}\,\mathrm{d}A \right] = 0.$$

This equation must be fulfilled for each variation $\delta\Delta\tilde{u}_i$, i.e. for each set of coefficients $\delta\Delta q_\alpha$, $\alpha = 1, \ldots, N$. This means that the expression in brackets must vanish for each $\alpha = 1, \ldots, N$. Introducing the notation

$$\Delta F_\alpha = \int_\Omega \left[\Delta\tilde{\sigma}_{ij}\,\Phi_{i\alpha,j} + (\tilde{\sigma}_{ij} + \Delta\tilde{\sigma}_{ij})\,\Phi_{k\alpha,i}\,\Delta\tilde{u}_{k,j} \right] \mathrm{d}V,$$
$$\Delta Q_\alpha = \int_\Omega \Delta\hat{f}_i \Phi_{i\alpha}\,\mathrm{d}V + \int_{\partial\Omega_\sigma} \Delta\hat{t}_i \Phi_{i\alpha}\,\mathrm{d}A,$$

one obtains, similarly as in the nonlinear problem of elasticity, a nonlinear system of N algebraic equations for the unknown incremental coefficients Δq_α,

$$\Delta F_\alpha(\Delta q_\zeta) = \Delta Q_\alpha\,, \qquad \Delta\mathbf{F}(\Delta\mathbf{q}) = \Delta\mathbf{Q}. \tag{11.27}$$

This system may as well be expressed in a non-incremental form. Defining at the time instant t and for the reference configuration C^t

$$F^t_\alpha = \int_\Omega \tilde{\sigma}^t_{ij} \Phi_{i\alpha,j}\,\mathrm{d}V,$$
$$Q^t_\alpha = \int_\Omega \hat{f}^t_i \Phi_{i\alpha}\,\mathrm{d}V + \int_{\partial\Omega_\sigma} \hat{t}^t_i \Phi_{i\alpha}\,\mathrm{d}A$$

and noting that having solved the problem for the previous time step we have $F^t_\alpha = Q^t_\alpha$, one can write

$$F^{t+\Delta t}_\alpha(\Delta q_\zeta) = Q^{t+\Delta t}_\alpha , \qquad \mathbf{F}^{t+\Delta t}(\Delta \mathbf{q}) = \mathbf{Q}^{t+\Delta t} , \tag{11.28}$$

where $(\cdot)^{t+\Delta t} = (\cdot)^t + \Delta(\cdot)$. Introducing further the notation

$$\mathbf{r} = \Delta\mathbf{Q} - \Delta\mathbf{F} = \mathbf{Q}^{t+\Delta t} - \mathbf{F}^{t+\Delta t}$$

one can express it shortly as

$$r_\alpha(\Delta q_\zeta) = 0, \qquad \mathbf{r}(\Delta\mathbf{q}) = \mathbf{0}. \tag{11.29}$$

Again, this nonlinear system of equations can be solved with the Newton–Raphson procedure. Forming the sequence of systems of equations for the subsequent approximate solution correctors δ_α

$$\bar{K}_{\alpha\beta}\,\delta_\alpha = r_\alpha ,$$

where

$$\bar{K}_{\alpha\beta} = -\frac{\partial r_\alpha(\Delta\tilde{q}^t_\zeta)}{\partial \Delta q_\beta} , \qquad r_\alpha = r_\alpha(\Delta\tilde{q}^t_\zeta),$$

while the subsequent solution approximates are updated using the scheme $\Delta\tilde{q}_\alpha := \Delta\tilde{q}_\alpha + \delta_\alpha$, we can derive, after transformations, the following formulae for the effective tangent matrix:

$$\begin{aligned}
\bar{K}_{\alpha\beta} &= \bar{K}^{(c)}_{\alpha\beta} + \bar{K}^{(\sigma)}_{\alpha\beta} ,\\
\bar{K}^{(c)}_{\alpha\beta} &= \int_\Omega C_{ijkl}\,\left(\Phi_{i\alpha,j} + \Phi_{m\alpha,i}\,\Delta\tilde{u}_{m,j}\right)\left(\Phi_{k\beta,l} + \Phi_{n\beta,k}\,\Delta\tilde{u}_{n,l}\right)\mathrm{d}V,\\
\bar{K}^{(\sigma)}_{\alpha\beta} &= \int_\Omega (\tilde{\sigma}_{ij} + \Delta\tilde{\sigma}_{ij})\,\Phi_{k\alpha,i}\Phi_{k\beta,j}\,\mathrm{d}V.
\end{aligned}$$

In analogy to the previously considered cases the above formulation can be extended to include dynamic effects. These are described by an additional term that appears in Eq. (11.25), equal to the volume integral of the quantity $\rho\Delta\ddot{u}_i\delta u_i$. Appropriate transformations, similar to those in the linear formulation, lead to the following nonlinear system of differential equations for the functions Δq_α:

$$M_{\alpha\beta}\Delta\ddot{q}_\beta + \Delta F_\alpha(\Delta q_\zeta) = \Delta Q_\alpha , \qquad \mathbf{M}\Delta\ddot{\mathbf{q}} + \Delta\mathbf{F}(\Delta\mathbf{q}) = \Delta\mathbf{Q} \tag{11.30}$$

or

$$M_{\alpha\beta}\ddot{q}^{t+\Delta t}_\beta + F^{t+\Delta t}_\alpha(\Delta q_\zeta) = Q^{t+\Delta t}_\alpha , \tag{11.31}$$

$$\mathbf{M}\ddot{\mathbf{q}}^{t+\Delta t} + \mathbf{F}^{t+\Delta t}(\Delta\mathbf{q}) = \mathbf{Q}^{t+\Delta t},$$

where the mass matrix $M_{\alpha\beta}$ is given by Eq. (11.17).

11.3 Weighted Residual Method

In Chap. 10, methods of derivation of weak formulations for thermomechanical problems were presented in which application of the variational calculus principles played the crucial role. Let us recall that a weak formulation has the form of an integral equation in which—apart from the sought fields of displacement or temperature—additional fields appear, called variations of the unknown fields. Besides, we have shown in Sects. 11.1 and 11.2 that—with a certain function form postulated for the approximate problem solution—the weak formulation equation may be transformed to become a system of equations for a certain N-dimensional unknown vector of real coefficients (or, in time-dependent problems, real time functions) that define the approximate solution. The resulting systems of equations are discrete forms of the weak formulations presented in Chap. 10.

In this Section, we are going to discuss an alternative method of discrete form derivation for weak formulations of thermomechanical problems. In the *weighted residual method*, as it is called, there is no need to employ the "function variation" notion. The advantage of this method is its more intuitive character. Its weakness, compared to variational methods, is its disability to construct a functional minimized by the weak problem solution (as the notion of function variation is necessary to formulate the problem of functional minimization).

Consider a linear system of M partial differential equations in the form

$$A_{ij}(\boldsymbol{x})\,u_j(\boldsymbol{x}) + B_i(\boldsymbol{x}) = 0, \qquad \boldsymbol{x} \in \Omega,\;\; i, j = 1, 2, \ldots, M \tag{11.32a}$$

with the following boundary conditions

$$\bar{A}_{rj}(\boldsymbol{x})\,u_j(\boldsymbol{x}) + \bar{B}_r(\boldsymbol{x}) = 0, \qquad \boldsymbol{x} \in \partial\Omega,\;\; r = 1, 2, \ldots, \bar{M}. \tag{11.32b}$$

Here, A_{ij} and $\bar{A}_{rj}$ denote linear differential operators while the unknown functions $u_i(\boldsymbol{x})$ are assumed to be sufficiently smooth. Following the summation convention, expressions containing doubled index j in the above equations are to be summed up. The system (11.32) may be expressed in the matrix form as

$$\mathbf{A}(\boldsymbol{x})\,\mathbf{u}(\boldsymbol{x}) + \mathbf{B}(\boldsymbol{x}) = \mathbf{0}, \tag{11.33a}$$

$$\bar{\mathbf{A}}(\boldsymbol{x})\,\mathbf{u}(\boldsymbol{x}) + \bar{\mathbf{B}}(\boldsymbol{x}) = \mathbf{0}. \tag{11.33b}$$

Let us assume the following, known from Sect. 11.2, approximation of the fields $u_i(\boldsymbol{x})$:

$$u_i(\boldsymbol{x}) \approx \tilde{u}_i(\boldsymbol{x}) = \phi_{i\alpha}(\boldsymbol{x})\, q_\alpha\,, \qquad \alpha = 1, 2, \ldots, N, \tag{11.34a}$$

i.e.

$$\mathbf{u}_{M\times 1}(\boldsymbol{x}) \approx \tilde{\mathbf{u}}_{M\times 1}(\boldsymbol{x}) = \boldsymbol{\phi}_{M\times N}(\boldsymbol{x})\, \mathbf{q}_{N\times 1}\,, \tag{11.34b}$$

where, for each $i = 1, 2, \ldots, M$, the *test functions* $\phi_{i1}\,,\ \phi_{i2}\,,\ \ldots\,,\ \phi_{iN}$ are chosen to form a set of linearly independent functions of $\boldsymbol{x}$ while $q_1\,,\ q_2\,,\ \ldots\,,\ q_N$ are scalar parameters whose values are to be determined in the procedure of approximate solution of the boundary value problem (11.32). Substituting Eq. (11.34) into Eq. (11.33) one obtains

$$\mathbf{A}(\boldsymbol{x})\, \tilde{\mathbf{u}}(\boldsymbol{x}) + \mathbf{B}(\boldsymbol{x}) = \mathbf{R}(\boldsymbol{x}), \tag{11.35a}$$

$$\bar{\mathbf{A}}(\boldsymbol{x})\, \tilde{\mathbf{u}}(\boldsymbol{x}) + \bar{\mathbf{B}}(\boldsymbol{x}) = \bar{\mathbf{R}}(\boldsymbol{x}), \tag{11.35b}$$

where $\mathbf{R} = \{R_i(\boldsymbol{x})\}$ and $\bar{\mathbf{R}} = \{\bar{R}_r(\boldsymbol{x})\}$ are arrays (vectors) of residual functions, generally different from zero due to approximate character of the assumed solution form (11.34).

In order to formulate relations necessary to compute the parameters $q_1\,,\ q_2\,,\ \ldots\,,\ q_N\,,$ let us define a "weighted sum" of the residual functions and require it to vanish:

$$\int_\Omega \psi_i(\boldsymbol{x})\, R_i(\boldsymbol{x})\,\mathrm{d}V + \int_{\partial\Omega} \bar{\psi}_r(\boldsymbol{x})\, \bar{R}_r(\boldsymbol{x})\,\mathrm{d}A = 0. \tag{11.36}$$

The functions ψ_i and $\bar{\psi}_r$ form an arbitrary set of $(M+\bar{M})$ *weight functions*. Satisfaction of Eq. (11.37) does not obviously guarantee that the residual functions vanish but it is intuitively clear that if Eq. (11.37) is satisfied for a sufficiently large number of N sets of appropriately selected (in particular: linearly independent) weight functions, i.e.

$$\int_\Omega \psi_{i\alpha}(\boldsymbol{x})\, R_i(\boldsymbol{x})\,\mathrm{d}V + \int_{\partial\Omega} \bar{\psi}_{r\alpha}(\boldsymbol{x})\, \bar{R}_r(\boldsymbol{x})\,\mathrm{d}A = 0, \tag{11.37a}$$

$$\int_\Omega \boldsymbol{\psi}_{M\times N}(\boldsymbol{x})\, \mathbf{R}_{N\times 1}(\boldsymbol{x})\,\mathrm{d}V + \int_{\partial\Omega} \bar{\boldsymbol{\psi}}_{\bar{M}\times N}(\boldsymbol{x})\, \bar{\mathbf{R}}_{N\times 1}(\boldsymbol{x})\,\mathrm{d}A = \mathbf{0}, \tag{11.37b}$$

then the functions $\tilde{u}_i$ in Eq. (11.34) should be good approximates of the unknown solution u_i.

Substituting Eq. (11.34) into Eq. (11.35) and employing the residual vectors computed by Eq. (11.35) in Eq. (11.37) one obtains

$$\left[\int_\Omega \psi_{i\alpha} A_{ij} \phi_{j\beta}\,\mathrm{d}V + \int_{\partial\Omega} \bar{\psi}_{r\alpha} \bar{A}_{rj} \phi_{j\beta}\,\mathrm{d}A\right] q_\beta$$
$$+ \int_\Omega \psi_{i\alpha} B_i\,\mathrm{d}V + \int_{\partial\Omega} \bar{\psi}_{r\alpha} \bar{B}_r\,\mathrm{d}A = 0. \tag{11.38}$$

This equation may be rewritten in a more compact vector-matrix notation:

$$K_{\alpha\beta}\, q_\beta = Q_\alpha\,, \qquad \mathbf{K}_{N\times N}\, \mathbf{q}_{N\times 1} = \mathbf{Q}_{N\times 1}\,, \tag{11.39}$$

where

$$K_{\alpha\beta} = \int_\Omega \psi_{i\alpha} A_{ij} \phi_{j\beta}\, \mathrm{d}V + \int_{\partial\Omega} \bar{\psi}_{r\alpha} \bar{A}_{rj} \phi_{j\beta}\, \mathrm{d}A, \tag{11.40}$$

$$Q_\alpha = -\int_\Omega \psi_{i\alpha} B_i\, \mathrm{d}V - \int_{\partial\Omega} \bar{\psi}_{r\alpha} \bar{B}_r\, \mathrm{d}A. \tag{11.41}$$

If the matrix $\boldsymbol{K}$ is non-singular, the system (11.39), being a linear set of algebraic equations, can be solved for the unknown parameters $q_1\,,\ q_2\,,\ \ldots\,,\ q_N$. Employing Eq. (11.34), we come at the approximate solution to the boundary value problem considered.

The above described solution method for systems of differential equations is called the weighted residual method. Its accuracy and efficiency depend obviously on the selection of the test functions $\phi_{i\alpha}$ ($\alpha = 1, 2, \ldots, N$) as well as the weight functions $\psi_{i\alpha}$ ($i = 1, 2, \ldots, M$) and $\bar{\psi}_{r\alpha}$ ($r = 1, 2, \ldots, \bar{M}$).

Considering the test functions $\phi_{i\alpha}$, one can naturally distinguish between the following cases:

1. $\phi_{i\alpha}$ do not guarantee automatic fulfillment of Eqs. (11.32) by the combination (11.34), i.e. $R_i \neq 0$ and $\bar{R}_r \neq 0$.
2. $\phi_{i\alpha}$ guarantee that the combination (11.34) satisfies automatically (i.e. independently of the values of q_α) the differential equations (11.32a) in Ω, i.e. $R_i = 0$, $\bar{R}_r \neq 0$; this version of the weighted residual method is called the boundary method and its particular case is the *boundary element method.*
3. $\phi_{i\alpha}$ guarantee that the combination (11.34) automatically satisfies the boundary conditions (11.32b) on $\partial\Omega$, or at least on its selected part, so that $R_i \neq 0$ and e.g. $\bar{R}_r = 0$ at $\boldsymbol{x} \in \partial\Omega^{(1)}$ and $\bar{R}_r \neq 0$ at $\boldsymbol{x} \in \partial\Omega^{(2)}$ (with $\partial\Omega = \partial\Omega^{(1)} \cup \partial\Omega^{(2)}$).

The latter case is the most frequent choice in applications of the weighted residual method in the problems of thermomechanics considered. It is convenient then to slightly modify the above derivations by assuming that

$$\tilde{u}_i(\boldsymbol{x}) = \chi_i(\boldsymbol{x}) + \phi_{i\alpha}(\boldsymbol{x})\, q_\alpha\,, \tag{11.42}$$

where the functions $\phi_{i\alpha}(\boldsymbol{x})$ are selected so as to satisfy for arbitrary $\alpha = 1, 2, \ldots, N$ the conditions

$$\bar{A}_{rj}\phi_{j\alpha} = 0, \qquad \boldsymbol{x} \in \partial\Omega^{(1)}, \tag{11.43}$$

while the functions $\chi_i(\boldsymbol{x})$ satisfy the conditions

$$\bar{A}_{rj}\chi_j + \bar{B}_r = 0, \qquad \boldsymbol{x} \in \partial\Omega^{(1)}, \qquad r = 1, \ldots, \bar{M}, \quad j = 1, \ldots, M. \tag{11.44}$$

This way the boundary conditions are automatically (i.e. independently of the values of q_α, $\alpha = 1, 2, \ldots, N$) satisfied on $\partial\Omega^{(1)}$.

Under such assumptions the residuals R_i and $\bar{R}_i$ are expressed as

$$R_i = A_{ij}(\chi_j + \phi_{j\alpha}q_\alpha) + B_i\,, \tag{11.45a}$$

$$\bar{R}_r = \begin{cases} 0, & \boldsymbol{x} \in \partial\Omega^{(1)}, \\ \bar{A}_{rj}(\chi_j + \phi_{j\alpha}q_\alpha) + \bar{B}_i\,, & \boldsymbol{x} \in \partial\Omega^{(2)}, \end{cases} \tag{11.45b}$$

which, when substituted into Eq. (11.37) yields the equations system (11.39) in which

$$K_{\alpha\beta} = \int_\Omega \psi_{i\alpha}A_{ij}\phi_{j\beta}\,\mathrm{d}V + \int_{\partial\Omega^{(2)}} \bar{\psi}_{r\alpha}\bar{A}_{rj}\phi_{j\beta}\,\mathrm{d}A, \tag{11.46}$$

$$Q_\alpha = -\int_\Omega \psi_{i\alpha}(A_{ij}\chi_j + B_i)\,\mathrm{d}V - \int_{\partial\Omega^{(2)}} \bar{\psi}_{r\alpha}(\bar{A}_{rj}\chi_j + \bar{B}_r)\,\mathrm{d}A. \tag{11.47}$$

Let us now present an application example of the weighted residual method for the simplest case of a differential equations system considered in this book, i.e. the linear problem of stationary heat conduction. The equations have the form

$$(\lambda_{kl}T_{,l})_{,k} + \rho r = 0, \tag{11.48a}$$

$$T = \hat{T}, \qquad \boldsymbol{x} \in \partial\Omega_T\,, \tag{11.48b}$$

$$\lambda_{kl}T_{,l}n_k = \hat{q} + \alpha(T_\infty - T), \qquad \boldsymbol{x} \in \partial\Omega_q\,. \tag{11.48c}$$

The unknown field is thus the scalar distribution of temperature $T(\boldsymbol{x})$. The operator A and the field B (no indices are necessary as $M = 1$ here) are defined as

$$A(\boldsymbol{x}) = \frac{\partial}{\partial x_k}\left(\lambda_{kl}\frac{\partial}{\partial x_l}\right), \qquad B(\boldsymbol{x}) = -\rho r.$$

Besides, $\bar{M} = 1$ (although there are two boundary conditions, only one of them is imposed at each boundary point). Thus, the operator $\bar{A}$ and the field $\bar{B}$ assume different forms at different boundary parts:

$$\begin{array}{lll} \bar{A}(\boldsymbol{x}) = 1, & \bar{B}(\boldsymbol{x}) = \hat{T}, & \boldsymbol{x} \in \partial\Omega_T\,, \\ \bar{A}(\boldsymbol{x}) = n_k\lambda_{kl}\frac{\partial}{\partial x_l} + \alpha\,, & \bar{B}(\boldsymbol{x}) = -\hat{q} - \alpha T_\infty\,, & \boldsymbol{x} \in \partial\Omega_q\,. \end{array}$$

The first step is the choice of appropriate test functions $\phi_\alpha(\boldsymbol{x})$. Let us do it so that the approximate solution $\tilde{T}(\boldsymbol{x}) = \phi_\alpha(\boldsymbol{x})q_\alpha$ satisfies the boundary conditions on $\partial\Omega_T$ for arbitrary coefficients q_α. Equations (11.42)–(11.43) are a good choice in this case, with the boundary parts $\partial\Omega^{(1)}$ and $\partial\Omega^{(2)}$ replaced by $\partial\Omega_T$ and $\partial\Omega_q$, respectively. Hence, the approximate temperature field is assumed as

$$\tilde{T}(\boldsymbol{x}) = \chi(\boldsymbol{x}) + \phi_\alpha(\boldsymbol{x})\, q_\alpha\,,$$

where the functions $\chi(\boldsymbol{x})$ and $\phi(\boldsymbol{x})$ satisfy the conditions

$$\chi = \hat{T}, \qquad \phi_\alpha = 0, \qquad \boldsymbol{x} \in \partial\Omega_T.$$

This leads to the following expressions for the residual fields R and $\bar{R}$:

$$\begin{aligned} R_i &= (\lambda_{kl}\phi_{\beta,l})_{,k}\, q_\beta + (\lambda_{kl}\chi_{,l})_{,k} + \rho r, \\ \bar{R}_r &= \begin{cases} 0, & \boldsymbol{x} \in \partial\Omega_T\,, \\ n_k\lambda_{kl}\left(\chi_{,l} + \phi_{\beta,l}q_\beta\right) + \alpha(\chi + \phi_\beta q_\beta) - \hat{q} - \alpha T_\infty\,, & \boldsymbol{x} \in \partial\Omega_q\,. \end{cases} \end{aligned}$$

Upon substitution of the above expressions into the weighted residual equation, which in this case assumes the form

$$\int_\Omega \psi_\alpha(\boldsymbol{x})\, R(\boldsymbol{x})\,\mathrm{d}V + \int_{\partial\Omega} \bar{\psi}_\alpha(\boldsymbol{x})\, \bar{R}(\boldsymbol{x})\,\mathrm{d}A = 0, \tag{11.49}$$

one obtains a system of equations in the form (11.39),

$$K_{\alpha\beta}\, q_\beta = Q_\alpha\,, \tag{11.50}$$

where

$$\begin{aligned} K_{\alpha\beta} &= \int_\Omega \psi_\alpha(\lambda_{kl}\phi_{\beta,l})_{,k}\,\mathrm{d}V \\ &\quad + \int_{\partial\Omega_q} \bar{\psi}_\alpha n_k\lambda_{kl}\phi_{\beta,l}\,\mathrm{d}A + \int_{\partial\Omega_q} \bar{\psi}_\alpha\alpha\phi_\beta\,\mathrm{d}A, \end{aligned} \tag{11.51a}$$

$$\begin{aligned} Q_\alpha &= -\int_\Omega \psi_\alpha\left[\rho r + (\lambda_{kl}\chi_{,l})_{,k}\right]\mathrm{d}V \\ &\quad + \int_{\partial\Omega_q} \bar{\psi}_\alpha\left[\hat{q} - \alpha(\chi - T_\infty)\right]\mathrm{d}A - \int_{\partial\Omega_q} \bar{\psi}_\alpha\, n_k\lambda_{kl}\chi_{,l}\,\mathrm{d}A. \end{aligned} \tag{11.51b}$$

Let us now focus on the weight functions $\psi_\alpha(\boldsymbol{x})$ and $\bar{\psi}_\alpha(\boldsymbol{x})$. First, let us note that if the boundary condition on $\partial\Omega_T$ is automatically satisfied, values of $\bar{\psi}_\alpha$ on this boundary part do not matter and we may, for instance, require that they are zero. Hence, the integrals over $\partial\Omega_q$ in the above equations may be replaced by integrals over the entire boundary $\partial\Omega$. It is thus allowed to make use of the Gauss–Ostrogradski theorem to those of boundary integrals in Eq. (11.51) that contain n_k in the integrand expression. This leads to the following transformations:

$$\begin{aligned} \int_{\partial\Omega_q} \bar{\psi}_\alpha n_k\lambda_{kl}\phi_{\beta,l}\,\mathrm{d}A &= \int_\Omega (\bar{\psi}_\alpha\lambda_{kl}\phi_{\beta,l})_{,k}\,\mathrm{d}V \\ &= \int_\Omega \left[\bar{\psi}_{\alpha,k}\lambda_{kl}\phi_{\beta,l} + \bar{\psi}_\alpha(\lambda_{kl}\phi_{\beta,l})_{,k}\right]\mathrm{d}V\,, \end{aligned}$$

$$\int_{\partial\Omega_q} \bar{\psi}_\alpha n_k \lambda_{kl} \chi_{,l}\, \mathrm{d}A = \int_\Omega (\bar{\psi}_\alpha \lambda_{kl} \chi_{,l})_{,k}\, \mathrm{d}V$$
$$= \int_\Omega \left[\bar{\psi}_{\alpha,k} \lambda_{kl} \chi_{,l} + \bar{\psi}_\alpha (\lambda_{kl} \chi_{,l})_{,k} \right] \mathrm{d}V\,,$$

which allow to rewrite the formulae (11.51) in the form:

$$K_{\alpha\beta} = \int_\Omega \left[(\psi_\alpha + \bar{\psi}_\alpha)(\lambda_{kl}\phi_{\beta,l})_{,k} + \bar{\psi}_{\alpha,k}\lambda_{kl}\phi_{\beta,l} \right] \mathrm{d}V$$
$$+ \int_{\partial\Omega_q} \bar{\psi}_\alpha \alpha \phi_\beta\, \mathrm{d}A, \tag{11.52a}$$
$$Q_\alpha = -\int_\Omega \left[\psi_\alpha \rho r + (\psi_\alpha + \bar{\psi}_\alpha)(\lambda_{kl}\chi_{,l})_{,k} + \bar{\psi}_{\alpha,k}\lambda_{kl}\chi_{,l} \right] \mathrm{d}V$$
$$+ \int_{\partial\Omega_q} \bar{\psi}_\alpha \left[\hat{q} - \alpha(\chi - T_\infty) \right] \mathrm{d}A. \tag{11.52b}$$

The matrix $K_{\alpha\beta}$ (with the square size $N \times N$) is generally non-symmetric.

Let us now note that if the weight functions $\psi(\boldsymbol{x})$ and $\bar{\psi}(\boldsymbol{x})$ may be chosen arbitrarily, there are no obstacles to particularly assume that

$$\psi_\alpha(\boldsymbol{x}) = -\phi_\alpha(\boldsymbol{x}), \qquad \bar{\psi}_\alpha(\boldsymbol{x}) = -\psi_\alpha(\boldsymbol{x}). \tag{11.53}$$

This choice does not violate the assumptions made for these functions due to the boundary condition on the boundary part $\partial\Omega_T$—let us recall that to satisfy this condition we required that $\phi_\alpha = 0$ and $\bar{\psi}_\alpha = 0$ on $\partial\Omega_T$. Upon such a choice the coefficient matrix and the right-hand side vector in the system (11.50) assume much simpler forms; besides, the matrix $K_{\alpha\beta}$ appears to be symmetric:

$$K_{\alpha\beta} = \int_\Omega \phi_{\alpha,k}\lambda_{kl}\phi_{\beta,l}\, \mathrm{d}V + \int_{\partial\Omega_q} \phi_\alpha \alpha \phi_\beta\, \mathrm{d}A\,, \tag{11.54a}$$
$$Q_\alpha = \int_\Omega \left[\phi_\alpha \rho r - \phi_{\alpha,k}\lambda_{kl}\chi_{,l} \right] \mathrm{d}V + \int_{\partial\Omega_q} \phi_\alpha \left[\hat{q} - \alpha(\chi - T_\infty) \right] \mathrm{d}A. \tag{11.54b}$$

Let us now compare the resulting system of equations (11.50), (11.54) with the system (11.3), (11.4) derived for the same problem with the use of the variational approach. It is clear that they are identical as long as the functions ϕ_α and χ are the same as the functions Φ_α and $\hat{\Phi}$, respectively. We can thus conclude that the method applied here leads to the same discrete formulation of the linear heat conduction problem as the discretization of an appropriate variational formulation.

It is easy to ascertain that the weighted residual method may be employed in an analogous manner to derive discrete systems of equations corresponding to all the other—both "thermal" and "mechanical"—problems described in Chap. 10. In order to obtain systems with symmetric matrices **K**, **C** and **M**, i.e. systems of the same forms as those derived from variational formulations, one must assure the same forms of weight and test functions. This particular case of the method is called in the literature the *Galerkin method*. In summary, the Galerkin method is an alternative approach to derive approximate discrete formulations for problems of solid thermomechanics.

Remark. Application of the Galerkin method in a search for an approximate solution of an arbitrary system of differential equations (i.e. assuming the weight and the test functions in the same form) does not guarantee symmetry of the coefficient matrix in the resulting discretized formulation. It is also necessary that the differential operator in Eq. (11.32a) is symmetric, i.e. that for arbitrary, sufficiently smooth fields $v_i^1(\boldsymbol{x})$ and $v_i^2(\boldsymbol{x})$ the condition

$$\int_\Omega v_i^1 A_{ij} v_j^2 \, \mathrm{d}V = \int_\Omega v_i^2 A_{ij} v_j^1 \, \mathrm{d}V. \tag{11.55}$$

is satisfied. As it is easy to show, Eq. (11.55) is fulfilled in all the considered problems of thermomechanics.

Chapter 12
Fundamentals of Finite Element Method

12.1 Introduction

The main advantage of variational formulations of boundary value problems is that they make it possible to search for approximate solutions of problems in a unified, rational way. In the approximate methods, the solution space of an infinite order is replaced by a finite order one. The fundamental difficulty in application of such methods to problems of practical importance is the choice of function classes that approximate the solution in the considered domain, i.e. the global shape functions Φ_α. The functions must enable adequate representation of the solution by taking into account the specific geometry of the domain, inhomogeneous material properties, as well as all other characteristic features of the problem (like distribution of mass forces or heat sources, functions imposed on the boundary, etc.). Efficient generation of approximating functions defined in the entire domain (global shape functions) is therefore impossible in general.

Along with progress in computer methods, an approach based on approximation of solution with localized functions have become possible. Its name is the *finite element method*, further referred to as FEM. The method, being now the basis for almost all effective systems for numerical analysis of solid thermomechanics problems, has been widely discussed in the literature, e.g. [1–5]. In this chapter, fundamentals of its formulations will only be shortly presented.

In this approach, the domain Ω is understood as divided into a finite number of disjoint subdomains called finite elements. The way the elements are defined should ensure their simple and convenient geometrical shape so that approximating functions can be defined in a relatively straightforward way within each of the elements. For a 2D domain, this could be e.g. triangles or tetragons while for a 3D domain one may consider e.g. tetrahedra or other simple polyhedra able to completely fill the domain Ω.

M. Kleiber and P. Kowalczyk, *Introduction to Nonlinear Thermomechanics of Solids*, Lecture Notes on Numerical Methods in Engineering and Sciences,
DOI 10.1007/978-3-319-33455-4_12

Sufficiently fine domain discretization (i.e. its division into elements) ensures sufficiently accurate description of the real problem defined in the domain Ω. The solution within particular finite elements has then a simple form which makes it possible to efficiently approximate it with generalized coordinates (degrees of freedom), defined in the so-called elements' nodal points, and with analytically simple, smooth functions defined inside each element. The set of all possible values of all generalized coordinates defines thus the set of trial functions within each of the elements (and thus within the entire domain Ω) while the particular values of the generalized coordinate obtained as the solution of the discrete problem defines the approximate solution of the problem in the domain Ω. Increasing the number of degrees of freedom in the system, by increasing the number of elements or the number of degrees of freedom in each element, usually leads to a better accuracy of the solution. Similar geometrical shape of the elements used and application of simple functions interpolating the solution between the nodes allow to effectively algorithmize the whole computational process, which makes the task ideally suitable for computer applications.

Let us thus divide the domain Ω into E finite elements Ω_e, $e = 1, 2, \ldots, E$, being disjoint open sets, $\Omega_e \cap \Omega_f = \emptyset$ for $e \neq f$. The boundary of the eth element is denoted by $\partial\Omega_e$, the common boundary of two neighbouring elements e and f by $\partial\Omega_{ef} = \partial\Omega_e \cap \partial\Omega_f$ while the common part of the eth element boundary and the boundary of the entire domain Ω by $\partial\Omega_{\bar{e}} = \partial\Omega \cap \partial\Omega_e$. The following relation holds true,

$$\partial\Omega_e = \left(\bigcup_{f=1}^{E_{(e)}} \partial\Omega_{ef} \right) \cup \partial\Omega_{\bar{e}} \,,$$

in which $E_{(e)}$ is the number of elements neighbouring the eth one, i.e. such elements Ω_f for which $\partial\Omega_e \cap \partial\Omega_f \neq \emptyset$.

Upon such assumptions integrals over Ω appearing in the discrete formulation equations discussed in Chap. 11 can be expressed as sums of integrals over particular element domains,

$$\int_{\Omega} (\cdot)\,\mathrm{d}V = \sum_{e=1}^{E} \int_{\Omega_e} (\cdot)\,\mathrm{d}V. \tag{12.1}$$

Due to assumed simple geometry of elements and simple forms of the approximating functions (which implies simple forms of the integrands) within the element domains, efficient and algorithmized evaluation of this kind of expressions is possible for entire discretized domains with arbitrary complex shapes.

Integrals over the boundary $\partial\Omega$ or its parts can also be expressed as sums

$$\int_{\partial\Omega} (\cdot)\,\mathrm{d}A = \sum_{\bar{e}\in\{E_{\partial\Omega}\}} \int_{\partial\Omega_{\bar{e}}} (\cdot)\,\mathrm{d}A, \tag{12.2}$$

where $\{E_{\partial\Omega}\}$ denotes the set of all elements for which $\partial\Omega_{\bar{e}}$ is non-empty.

Remark. The assumed simplicity of finite elements may make it impossible to exactly describe the domain Ω, e.g. curvatures of its boundary. In such a case Eqs. (12.1) and (12.2) are only approximate. In further derivations, for clarity of the discussion, this issue will not be addressed and we will assume that the error of geometrical mapping is negligible.

An important advantage of the FEM approximation is an intuitive meaning assigned to the coefficients q_α in the approximate solutions sought. In classical formulations the approximating functions $\Phi_\alpha(\boldsymbol{x})$ are associated with particular points like element vertices, for instance. These points, called *nodes*, are usually common to several neighbouring elements. In the case of a scalar field $u(\boldsymbol{x})$ (this may be temperature or a displacement component) whose approximate is sought in the form of the combination

$$\tilde{u}(\boldsymbol{x}) = \Phi_\alpha(\boldsymbol{x})\, q_\alpha, \qquad \alpha = 1, 2, \ldots, N, \tag{12.3}$$

one may assume that N is the number of nodes $\boldsymbol{x}_\alpha \in \Omega$ in the FEM model and require that the coefficients q_α equal values of the approximated field at corresponding nodes, respectively, i.e.

$$\tilde{u}(\boldsymbol{x}_\alpha) = q_\alpha\,. \tag{12.4}$$

The so-defined coefficients q_α are usually called *nodal parameters.*

It appears to be quite straightforward to define shape functions so as to satisfy the condition (12.4). It is sufficient that their values at all nodes are

$$\Phi_\alpha(\boldsymbol{x}_\beta) = \delta_{\alpha\beta} \qquad \alpha, \beta = 1, 2, \ldots, N\,. \tag{12.5}$$

In other words, the shape function $\Phi_\alpha(\boldsymbol{x})$ associated with the node α should assume the unit value at this node and vanish at all remaining nodes. Considering a particular finite element e spanned on N_e nodes we should consequently require that Eq. (12.5) is true for all N_e shape functions associated with the element nodes. All other shape functions (associated with nodes that do not belong to our element e) must vanish within this element.

An additional property typically required from the shape functions $\Phi_\alpha(\boldsymbol{x})$ is the condition

$$\sum_{\alpha=1}^{N} \Phi_\alpha(\boldsymbol{x}) = 1. \tag{12.6}$$

It implies that if all coefficients q_α have the same value $\bar{u}$ (i.e. the approximated function has the same value at all nodes) then $\tilde{u}(\boldsymbol{x}) = \bar{u} \equiv \text{const.}$

Figure 12.1 presents an example of such a discretization (with the use of triangular elements) for a 2D domain and an example of the shape function associated

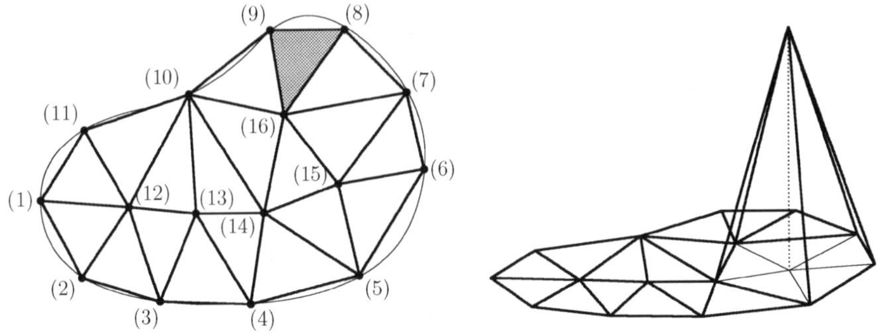

Fig. 12.1 Example of a 2D domain discretization and a shape function form for the node $\alpha = 15$

with a selected node. Shape functions are linear within each element and continuous (although non-differentiable) on the element boundaries. This does not violate the requirements regarding trial functions appearing in weak problem formulations discussed in Chap. 10.

Remark. In some formulations, higher requirements regarding differentiability of shape functions may be necessary (e.g. continuity of the gradient ∇u at element boundaries). Shape functions with higher number of degrees of freedom per node are then utilized—nodal parameters may then be taken as the nodal values u and derivatives $u_{,i}$ of the function. The forms of such functions corresponding to the so-defined parameters must be selected appropriately. This issue will further be addressed in Sect. 12.2.

The so-defined shape functions are thus localized functions i.e. (i) each of them assumes non-zero values within the domain of only a few elements neighbouring the associated node and (ii) for each element, there are only a few shape functions that are non-zero within its domain.

Let us now present examples of FEM formulations for two simple problems for which discrete equations have been derived in Chap. 11. These are the linear heat conduction problem and the linear elasticity problem at small deformations.

12.1.1 FEM Formulation for Linear Heat Conduction Problem

The discrete form of this problem has been derived in Sect. 11.1. Let us neglect for simplicity convection heat exchange on the boundary. The only boundary conditions are the imposed temperature $\hat{T}$ on $\partial\Omega_T$ and the heat flux $\hat{q}$ on $\partial\Omega_q$. Assume the FEM approximation of the temperature field in the form (12.3), i.e.

$$\tilde{T}(\boldsymbol{x}) = \Phi_\alpha(\boldsymbol{x})\, q_\alpha\,, \tag{12.7}$$

where q_α denote the temperature values at nodes $\boldsymbol{x}_\alpha$. Assuming the number of nodes with unknown temperatures to be N, one may rewrite Eq. (12.7) in the matrix form as

$$\tilde{T}(\boldsymbol{x}) = \mathbf{\Phi}^{\mathrm{T}}_{1\times N}\, \mathbf{q}_{N\times 1}\,.$$

The required fulfillment of the essential boundary condition on $\partial\Omega_T$ implies modification of the above formula, i.e.

$$\tilde{T}(\boldsymbol{x}) = \Phi_\alpha(\boldsymbol{x})\, q_\alpha + \hat{\Phi}_{\hat{\beta}}(\boldsymbol{x})\, \hat{q}_{\hat{\beta}}\,, \tag{12.8}$$

where the index $\hat{\beta} = 1, \ldots, N_b$ runs over the nodal points located on the boundary part $\partial\Omega_T$ while $\hat{q}_{\hat{\beta}}$ are the temperature values imposed at these nodes (the total number of nodes in the model is thus $\bar{N} = N + N_b$ where N denotes the number of all nodes at which temperature is unknown). The assumed approximation (12.8) is thus compatible with Eq. (11.2) provided $\hat{\Phi}(\boldsymbol{x}) = \hat{\Phi}_{\hat{\beta}}(\boldsymbol{x})\, \hat{q}_{\hat{\beta}}$.

Upon these assumptions, the following system of algebraic equations is obtained, c.f. Eq. (11.4),

$$\mathbf{K}_{N\times N}\, \mathbf{q}_{N\times 1} = \mathbf{Q}_{N\times 1}\,, \tag{12.9}$$

in which unknown is the vector of nodal temperatures $\mathbf{q}$. Components of the matrix $\mathbf{K}$ and the right-hand side vector $\mathbf{Q}$, cf. Eq. (11.3), are expressed as

$$K_{\alpha\beta} = \int_\Omega \lambda_{ij}\, \Phi_{\alpha,i}\, \Phi_{\beta,j}\, \mathrm{d}V, \tag{12.10a}$$

$$\begin{aligned} Q_\alpha &= \int_\Omega \hat{g}\, \Phi_\alpha\, \mathrm{d}V + \int_{\partial\Omega_q} \hat{q}\, \Phi_\alpha\, \mathrm{d}A - \int_\Omega \lambda_{ij}\, \Phi_{\alpha,i}\, \hat{\Phi}_{\hat{\beta},j}\, \hat{q}_{\hat{\beta}}\, \mathrm{d}V \\ &= \int_\Omega \hat{g}\, \Phi_\alpha\, \mathrm{d}V + \int_{\partial\Omega_q} \hat{q}\, \Phi_\alpha\, \mathrm{d}A - \hat{K}_{\alpha\hat{\beta}}\, \hat{q}_{\hat{\beta}}\,, \end{aligned} \tag{12.10b}$$

where the components of the matrix $\hat{\mathbf{K}}_{N\times N_b}$ are computed in the same way as those of $\mathbf{K}_{N\times N}$, i.e. from Eq. (12.10a). Defining the matrix $\mathbf{B}_{3\times N}$ as

$$B_{i\alpha} = \Phi_{\alpha,i}$$

one may rewrite the above formulae in the following matrix form:

$$\mathbf{K}_{N\times N} = \int_\Omega \mathbf{B}^{\mathrm{T}}_{N\times 3}\, \boldsymbol{\lambda}_{3\times 3}\, \mathbf{B}_{3\times N}\, \mathrm{d}V, \tag{12.11a}$$

$$\mathbf{Q}_{N\times 1} = \int_\Omega \hat{g}\, \mathbf{\Phi}_{N\times 1}\, \mathrm{d}V + \int_{\partial\Omega_q} \hat{q}\, \mathbf{\Phi}_{N\times 1}\, \mathrm{d}A - \hat{\mathbf{K}}_{N\times N_b}\, \hat{\mathbf{q}}_{N_b\times 1}\,. \tag{12.11b}$$

Remark. Generally, to ensure convergence of approximate solutions with decreasing element sizes (i.e. with finer FEM discretization) in a problem described by a functional containing derivatives up to mth order of the unknown fields, the following conditions have to be fulfilled:

- consistency conditions—the $(m-1)$th derivatives of the shape functions must be continuous on the element boundaries,
- differentiability conditions—the mth derivatives of the shape functions must be continuous within elements.

In other words, the functions sought must be differentiable in Ω, possibly except for a set of inter-element surfaces, up to the order required by the variational formulation. In the considered formulation (12.10) we have $m = 1$, hence the functions must be continuous in the entire domain and differentiable within elements. It is easy to verify that the assumed form of the approximation formulae meets these requirements.

In view of the properties (12.1) and (12.2), the integrals in Eqs. (12.10) and (12.11) may be computed as sums of integrals over particular element domains Ω_e or their boundaries $\partial\Omega_{\bar{e}}$.

Let us now present a practical method of computing this kind of integrals on an example of the conductivity matrix $\mathbf{K}$. In view of Eq. (12.1) we have

$$K_{\alpha\beta} = \sum_{e=1}^{E} K_{\alpha\beta}^{(e)}, \qquad K_{\alpha\beta}^{(e)} = \int_{\Omega_e} \lambda_{ij} \Phi_{\alpha,i} \Phi_{\beta,j} \, \mathrm{d}V \, . \tag{12.12}$$

Since the shape functions are localized, each matrix $K_{\alpha\beta}^{(e)}$ has usually a large amount of zero elements—non-zero terms appear only in those rows α and columns β for which the shape functions are non-zero, i.e. for only those values of α, β that correspond to nodes that belong to the element e.

Let us first consider an element e whose nodes do not belong to $\partial\Omega_T$, i.e. all its nodal parameters are unknown and belong to $\mathbf{q}$. If $N_e \leq N$, then the local nodal parameter vector $\mathbf{q}_{N_e\times 1}^{(e)}$ can be defined as containing only these nodal parameters from the vector $\mathbf{q}_{N\times 1}$ that are associated with the nodes of the element e. Then,

$$q_a^{(e)} = A_{a\alpha}^{(e)} q_\alpha, \qquad a = 1, \dots, N_e, \quad \alpha = 1, \dots, N. \tag{12.13}$$

where $\mathbf{A}_{N_e\times N}^{(e)}$ is an appropriate Boolean matrix. For example, for the element highlighted in Fig. 12.1, $N_e = 3$, the nodes associated with the element are 16, 8 and 9 while the matrix $\mathbf{A}^{(e)}$ is defined as

$$
\left[\mathbf{A}^{(e)}_{3\times 16}\right] = \begin{array}{c} \begin{matrix} 1 & & 7 & 8 & 9 & 10 & & 15 & 16 \end{matrix} \\ \begin{bmatrix} 0 & \cdots & 0 & 0 & 0 & 0 & \cdots & 0 & 1 \\ 0 & \cdots & 0 & 1 & 0 & 0 & \cdots & 0 & 0 \\ 0 & \cdots & 0 & 0 & 1 & 0 & \cdots & 0 & 0 \end{bmatrix} \end{array} .
$$

In this notation, the local conductivity matrix $\mathbf{k}^{(e)}_{N_e\times N_e}$ can be defined for the element e as

$$
k^{(e)}_{ab} = \int_{\Omega_e} \lambda_{ij}\, \Phi^{(e)}_{a,i} \Phi^{(e)}_{b,j}\, \mathrm{d}V , \qquad K^{(e)}_{\alpha\beta} = A^{(e)}_{a\alpha}\, k^{(e)}_{ab}\, A^{(e)}_{b\beta} , \tag{12.14}
$$

where the local shape functions $\Phi^{(e)}_a$, $a = 1, \ldots, N_e$ are an appropriate subset of the global set

$$
\Phi^{(e)}_a = A^{(e)}_{a\alpha}\, \Phi_\alpha .
$$

For the element highlighted in Fig. 12.1, the local and global matrices have the form, for instance,

$$
\left[\mathbf{k}^{(e)}_{3\times 3}\right] = \begin{bmatrix} k^{(e)}_{11} & k^{(e)}_{12} & k^{(e)}_{13} \\ k^{(e)}_{21} & k^{(e)}_{22} & k^{(e)}_{23} \\ k^{(e)}_{31} & k^{(e)}_{32} & k^{(e)}_{33} \end{bmatrix} ,
$$

$$
\left[\mathbf{K}^{(e)}_{16\times 16}\right] = \begin{array}{c} \begin{matrix} 1 & & 7 & 8 & 9 & 10 & & 15 & 16 \end{matrix} \\ \begin{bmatrix} 0 & \cdots & 0 & 0 & 0 & 0 & \cdots & 0 & 0 \\ \vdots & & \vdots & \vdots & \vdots & \vdots & & \vdots & \vdots \\ 0 & \cdots & 0 & 0 & 0 & 0 & \cdots & 0 & 0 \\ 0 & \cdots & 0 & k^{(e)}_{22} & k^{(e)}_{23} & 0 & \cdots & 0 & k^{(e)}_{21} \\ 0 & \cdots & 0 & k^{(e)}_{32} & k^{(e)}_{33} & 0 & \cdots & 0 & k^{(e)}_{31} \\ 0 & \cdots & 0 & 0 & 0 & 0 & \cdots & 0 & 0 \\ \vdots & & \vdots & \vdots & \vdots & \vdots & & \vdots & \vdots \\ 0 & \cdots & 0 & 0 & 0 & 0 & \cdots & 0 & 0 \\ 0 & \cdots & 0 & k^{(e)}_{12} & k^{(e)}_{13} & 0 & \cdots & 0 & k^{(e)}_{11} \end{bmatrix} \end{array} .
$$

The above notation allows to use at the element level reduced-size local matrices and vectors, their size corresponding to the actual number of element degrees of freedom. Building the global matrix $\mathbf{K}$ consists then in computing local matrices $\mathbf{k}^{(e)}$ for each element and then "assembling" their components at appropriate positions in the matrix $\mathbf{K}$, indicated by the transformation formula (12.14).

The right-hand side vector is defined as

$$
Q_\alpha = \sum_{e=1}^{E} Q^{(e)}_\alpha , \qquad Q^{(e)}_\alpha = p^{(e)}_a A^{(e)}_{a\alpha} , \tag{12.15}
$$

where the local vector components $p_a^{(e)}$ are determined by computing appropriate integrals (cf. Eq. (12.10b)) over the element domain Ω_e or a part of its boundary $\partial\Omega_{\bar{e}}$ (if only it belongs to $\partial\Omega_q$). Also in this case only these shape functions or their gradients are processed in the integrand expressions that assume non-zero values within the element domain.

If the element e contains nodes at which temperature values are imposed, then Eq. (12.13) should be extended to the form

$$q_a^{(e)} = A_{a\alpha}^{(e)} q_\alpha + \hat{A}_{a\hat{\alpha}}^{(e)} \hat{q}_{\hat{\alpha}}, \qquad a = 1, \ldots, N_e, \quad \begin{array}{l} \alpha = 1, \ldots, N, \\ \hat{\alpha} = 1, \ldots, N_b, \end{array} \tag{12.16}$$

where the matrix $\hat{\mathbf{A}}^{(e)}_{N_e \times N_b}$ is a Boolean matrix.

Upon these modifications, the formulae for the local and global conductivity matrix (12.14) remain the same while the matrix $\hat{K}_{\alpha\hat{\beta}}$ in Eq. (12.10b) may be assembled from local matrices $k_{ab}^{(e)}$ as

$$\hat{K}_{\alpha\hat{\beta}} = A_{a\alpha}^{(e)} k_{ab}^{(e)} \hat{A}_{b\hat{\beta}}^{(e)} . \tag{12.17}$$

Remark. From the point of view of general procedures of building and solving the system of FEM equations it is frequently troublesome to explicitly separate from the set of all nodes those nodes at which the values of the unknown field are predefined. Sometimes it is easier to assume that the field sought is unknown at all nodes of the model (i.e. that $\mathbf{q}$ contains nodal parameters associated with all nodes) and then, by additional algebraic transformations of the system of equations, to "enforce" the solution to approximately fulfill the conditions $q_\alpha = \hat{q}_\alpha$ at nodes at which such conditions are imposed. This issue will be discussed in more detail in Sect. 13.1.

12.1.2 FEM Formulation for Linear Static Elasticity Problem

The discrete form of this problem has been derived in Sect. 11.2. Assume the FEM approximation of the displacement field in the form (12.3), i.e.

$$\tilde{u}_i(\boldsymbol{x}) = \Phi_{i\alpha}(\boldsymbol{x})\, q_\alpha , \tag{12.18}$$

where q_α denote the displacement values at nodes.

Let us note an essential difference between the above formulation and the formulation (12.7) for the scalar temperature field—here, three components of the displacement field are approximated. Thus, if the nodal parameters q_α are supposed to have the physical sense of displacement values at nodal points then three nodal parameters are needed at each node. If the number of nodes is N, the index α in Eq. (12.18) should thus run over the range $\alpha = 1, 2, \ldots, 3N$. Equation (12.18) may be written in the following matrix form:

$$\tilde{\mathbf{u}}_{3\times 1}(\boldsymbol{x}) = \boldsymbol{\Phi}^{\mathrm{T}}_{3\times 3N}(\boldsymbol{x})\, \mathbf{q}_{3N\times 1}\,.$$

It seems natural to divide the vector $\mathbf{q}_{3N\times 1}$ into N three-component vectors containing displacement component values at particular nodes:

$$\{\mathbf{q}_{3N\times 1}\} = \left\{\begin{matrix} \mathbf{q}^{\langle 1\rangle}_{3\times 1} \\ \mathbf{q}^{\langle 2\rangle}_{3\times 1} \\ \vdots \\ \mathbf{q}^{\langle N\rangle}_{3\times 1} \end{matrix}\right\},$$

The matrix $\boldsymbol{\Phi}$ may be then written as

$$\left[\boldsymbol{\Phi}^{\mathrm{T}}_{3\times 3N}\right] = \begin{bmatrix} \Phi^{\langle 1\rangle} & & & \Phi^{\langle 2\rangle} & & & & \Phi^{\langle N\rangle} & & \\ & \Phi^{\langle 1\rangle} & & & \Phi^{\langle 2\rangle} & & \cdots & & \Phi^{\langle N\rangle} & \\ & & \Phi^{\langle 1\rangle} & & & \Phi^{\langle 2\rangle} & & & & \Phi^{\langle N\rangle} \end{bmatrix},$$

while the functions $\Phi^{\langle 1\rangle}(\boldsymbol{x}),\ \Phi^{\langle 2\rangle}(\boldsymbol{x}), \ldots, \Phi^{\langle N\rangle}(\boldsymbol{x})$ are the nodal shape functions defined in Eqs. (12.3)–(12.5).

The required fulfillment of the essential boundary condition $u_i = \hat{u}_i$ on $\partial\Omega_u$ implies modification of Eq. (12.18) as

$$\tilde{u}_i(\boldsymbol{x}) = \Phi_{i\alpha}(\boldsymbol{x})\, q_\alpha + \hat{\Phi}_{i\hat{\beta}}(\boldsymbol{x})\, \hat{q}_{\hat{\beta}}\,, \tag{12.19}$$

where the index $\hat{\beta} = 1, \ldots, N_b$ runs over the set of those nodal displacement components on $\partial\Omega_u$ whose values are imposed as the boundary conditions (and collected in the vector $\hat{q}_{\hat{\beta}}$). The assumed approximation (12.19) is thus compatible with Eq. (11.13) provided $\hat{\Phi}_i(\boldsymbol{x}) = \hat{\Phi}_{i\hat{\beta}}(\boldsymbol{x})\, \hat{q}_{\hat{\beta}}$. Since, as it was mentioned at the end of Sect. 12.1.1, boundary conditions of this kind are frequently taken into account in an implicit way in the computational practice (i.e. without explicit separation of predefined parameters $\hat{\mathbf{q}}$ from the vector of all nodal parameters $\mathbf{q}$), this modification will be skipped in the following derivations and the general form (12.18) of the solution will be assumed with all $3N$ displacement components at N nodes treated as unknown.

Upon these assumptions, the following system of algebraic equations is obtained, c.f. Eq. (11.15),

$$\mathbf{K}_{3N\times 3N}\, \mathbf{q}_{3N\times 1} = \mathbf{Q}_{3N\times 1}\,, \tag{12.20}$$

in which unknown is the vector of nodal displacement components $\mathbf{q}$. Components of the matrix $\mathbf{K}$ (called the stiffness matrix) and the right-hand side vector $\mathbf{Q}$ (containing the resultant external force components at nodes), cf. Eq. (11.14), are expressed as

$$K_{\alpha\beta} = \int_{\Omega} C_{ijkl} \Phi_{i\alpha,j} \Phi_{k\beta,l} \, \mathrm{d}V, \tag{12.21a}$$

$$Q_{\alpha} = \int_{\Omega} \hat{f}_i \Phi_{i\alpha} \, \mathrm{d}V + \int_{\partial\Omega_\sigma} \hat{t}_i \Phi_{i\alpha} \, \mathrm{d}A. \tag{12.21b}$$

Using the symmetry properties of the stiffness tensor C_{ijkl} (which allow to replace the terms $\Phi_{i\alpha,j}$ and $\Phi_{k\beta,l}$ in Eq. (12.21a) by their symmetric parts with respect to the pairs of indices ij and kl, respectively) and recalling the notation convention adopted in Chap. 7 that allows to express symmetric tensors of the second and fourth order as 6×1 vectors and 6×6 matrices, respectively, cf. Eq. (7.5), we may rewrite the above expressions in the matrix form as

$$\mathbf{K}_{3N\times 3N} = \int_{\Omega} \mathbf{B}^{\mathrm{T}}_{3N\times 6} \, \mathbf{C}_{6\times 6} \, \mathbf{B}_{6\times 3N} \, \mathrm{d}V, \tag{12.22a}$$

$$\mathbf{Q}_{3N\times 1} = \int_{\Omega} \boldsymbol{\Phi}_{3N\times 3} \, \hat{\mathbf{f}}_{3\times 1} \, \mathrm{d}V + \int_{\partial\Omega_\sigma} \boldsymbol{\Phi}_{3N\times 3} \, \hat{\mathbf{t}}_{3\times 1} \, \mathrm{d}A, \tag{12.22b}$$

where the so-called geometric matrix is defined as

$$\left[\mathbf{B}_{6\times 3N}\right] = \left[\, \mathbf{B}^{\langle 1\rangle}_{6\times 3} \; \mathbf{B}^{\langle 2\rangle}_{6\times 3} \; \cdots \; \mathbf{B}^{\langle N\rangle}_{6\times 3} \,\right]$$

$$\left[\mathbf{B}^{\langle n\rangle}_{6\times 3}\right] = \begin{bmatrix} \Phi^{\langle n\rangle}_{,1} & & \\ & \Phi^{\langle n\rangle}_{,2} & \\ & & \Phi^{\langle n\rangle}_{,3} \\ \Phi^{\langle n\rangle}_{,2} & \Phi^{\langle n\rangle}_{,1} & \\ & \Phi^{\langle n\rangle}_{,3} & \Phi^{\langle n\rangle}_{,2} \\ \Phi^{\langle n\rangle}_{,3} & & \Phi^{\langle n\rangle}_{,1} \end{bmatrix}.$$

Note that this matrix relates the strain components to the nodal parameters according to the formula

$$\boldsymbol{\varepsilon}_{6\times 1} = \mathbf{B}_{6\times 3N} \, \mathbf{q}_{3N\times 1} \,. \tag{12.23}$$

In view of the properties (12.1) and (12.2), the integrals in Eqs. (12.21) and (12.22) may be computed as sums of integrals over particular element domains Ω_e or their boundaries $\partial\Omega_{\bar{e}}$. Similarly as in the heat conduction problem, the element contributions to the stiffness matrix and the nodal load vector can be expressed in their local reduced-dimension forms. If the number of nodes in the element e is N_e then the local number of nodal parameters is $3N_e$ and this is also the dimension of the element matrix $\mathbf{k}^{(e)}$ and the element vectors $\mathbf{q}^{(e)}$, $\mathbf{p}^{(e)}$. The following relations hold true,

$$K_{\alpha\beta} = \sum_{e=1}^{E} K_{\alpha\beta}^{(e)}, \qquad K_{\alpha\beta}^{(e)} = A_{a\alpha}^{(e)} \, k_{ab}^{(e)} \, A_{b\beta}^{(e)}, \tag{12.24a}$$

$$Q_{\alpha} = \sum_{e=1}^{E} Q_{\alpha}^{(e)}, \qquad Q_{\alpha}^{(e)} = p_{a}^{(e)} A_{a\alpha}^{(e)}, \tag{12.24b}$$

in which, as before, the matrix $\mathbf{A}_{3N_e \times 3N}^{(e)}$ is a Boolean matrix that assigns local nodal parameters of the element e to the global parameters of the system, cf. Eq. (12.13),

$$q_{a}^{(e)} = A_{a\alpha}^{(e)} q_{\alpha}, \qquad a = 1, \ldots, 3N_e, \quad \alpha = 1, \ldots, 3N, \tag{12.25}$$

while the matrix $\mathbf{k}_{3N_e \times 3N_e}^{(e)}$ and the vector $\mathbf{p}_{3N_e \times 1}^{(e)}$ are computed from Eq. (12.21) in which local shape functions (assuming non-zero values within the element domain Ω_e) are only processed.

One more aspect of assembling local element vectors and matrices into their global counterparts is worth mentioning here (this issue is not important in heat conduction problems but is frequently crucial in the context of displacement nodal parameters, as in the problem considered now). It happens that various local quantities considered at the element level can be defined in a very convenient way in a local, element-specific coordinate system, aligned with one of the element's edges or planes, for instance. In such a case the local nodal parameter vector $\mathbf{q}_{3N_e \times 1}^{(e)}$ is not just a simple "extract" of the global vector $\mathbf{q}_{3N \times 1}$ but it contains displacement components being a result of a certain transformation (usually rotation) of corresponding components of the vector $\mathbf{q}$. The following relation holds true then for an arbitrary node n which in the local numbering in its neighbouring element e has the counter l,

$$\begin{Bmatrix} q_{3l-2}^{(e)} \\ q_{3l-1}^{(e)} \\ q_{3l}^{(e)} \end{Bmatrix} = \begin{bmatrix} R_{11}^{(e)} & R_{12}^{(e)} & R_{13}^{(e)} \\ R_{21}^{(e)} & R_{22}^{(e)} & R_{23}^{(e)} \\ R_{31}^{(e)} & R_{32}^{(e)} & R_{33}^{(e)} \end{bmatrix} \begin{Bmatrix} q_{3n-2} \\ q_{3n-1} \\ q_{3n} \end{Bmatrix},$$

where $\mathbf{R}_{3\times 3}^{(e)}$ is the rotation matrix of the local coordinate system. Defining the matrix

$$\left[\mathbf{T}_{3N_e \times 3N_e}^{(e)}\right] = \begin{bmatrix} \mathbf{R}_{3\times 3}^{(e)} & & & \\ & \mathbf{R}_{3\times 3}^{(e)} & & \\ & & \ddots & \\ & & & \mathbf{R}_{3\times 3}^{(e)} \end{bmatrix},$$

one may write Eq. (12.24) in the modified form as

$$K_{\alpha\beta}^{(e)} = A_{a\alpha}^{(e)} \, T_{ba}^{(e)} \, k_{bc}^{(e)} \, T_{cd}^{(e)} \, A_{d\beta}^{(e)}, \tag{12.26a}$$

$$Q_{\alpha}^{(e)} = p_{a}^{(e)} \, T_{ab}^{(e)} \, A_{b\alpha}^{(e)}, \tag{12.26b}$$

convenient for algorithmized computation for various elements

The two FEM formulations presented above have been used as instructive examples—they may help the readers to figure out the methodology of deriving analogous formulations for more complex nonlinear and non-stationary cases. They follow essentially the same lines in all cases. It must be only remembered that in non-linear problems the coefficient matrix **K** depends on the unknown vector **q** while in non-stationary cases the vector **q** is a function of time whose derivatives (of the first and second order) appear in the system of equations and make it necessary to apply an appropriate time integration scheme. Let us stress once again that the great advantage of the presented method of solution approximation and domain discretization is easy algorithmization of determining the matrix and vectors in the system of equations for arbitrarily complex geometries of Ω. This makes it particularly attractive from the point of view of practical use in many engineering applications.

12.2 FEM Approximation at the Element Level

The scope of this book excludes a detailed presentation of element-level solution approximation techniques, i.e. appropriate selection of shape function forms for particular element types. Fundamental aspects of this issue will only be briefly discussed now to provide the readers with some basic information on the essence of constructing approximate solutions at the element level and enable their further individual studying the issue.

12.2.1 Simple One-Dimensional Elements

Let us start our discussion from presentation of basic rules governing element approximation of functions in the one-dimensional (1D) case.

Consider a scalar function of a scalar argument $u = u(x)$. The symbol u may denote a displacement component, temperature, etc., while x is the location of a point to which a value of u is assigned.

The essence of FEM is approximation of the function u by the following expression, known from Chap. 11,

$$u(x) \approx \tilde{u}(x) = \Phi_\alpha(x) q_\alpha = \mathbf{\Phi}^{\mathrm{T}}(x)\,\mathbf{q}, \qquad \alpha = 1, 2, \ldots, N, \tag{12.27}$$

where $\mathbf{\Phi}(x)$ and $\mathbf{q}$ are N-dimensional vectors containing subsequent shape functions $\Phi_\alpha(x)$ and their coefficients q_α, respectively. As has been mentioned in the introduction to this chapter, the characteristic feature of FEM is association of particular shape functions with their particular corresponding points x_α, $\alpha = 1, 2, \ldots, N$, called *nodes*. Standard shape functions are defined in a way ensuring fulfillment of the conditions (12.5) and (12.6), i.e. in the 1D case

$$\Phi_\alpha(x_\beta) = \delta_{\alpha\beta}, \qquad \alpha, \beta = 1, 2, \ldots, N, \tag{12.28}$$

$$\sum_{\alpha=1}^{N} \Phi_\alpha(x) = 1. \tag{12.29}$$

These shape function properties imply the following equality at each nodal point

$$\tilde{u}(x_\alpha) = q_\alpha\,, \tag{12.30}$$

i.e. the coefficients q_α called *nodal parameters* or *system degrees of freedom* have clear physical sense—they are the solution values at the nodal points.

An example of the above approximation is shown in Fig. 12.2. This is a first order polynomial approximation in the interval $x \in [a, b]$ for $N = 4$. Note that the particular shape functions are piece-wise linear between nodal points and each of them assumes non-zero values in the segments neighbouring its corresponding node. The segments will be called elements—their number in our example is $E = 3$. Also note that the proposed approximate solution form is continuous and only piece-wise differentiable (i.e. except for the nodal points). In typical thermomechanics problems this assumption appears to be sufficient. In further part of this section discretizations preserving differentiability at nodal points will also be discussed.

Consider a single 1D linear (i.e. with a first order polynomial approximation) finite element with two nodes and two degrees of freedom (Fig. 12.3). If the element length is l then

$$\tilde{u}(x) = q_1 + \frac{x}{l}(q_2 - q_1) = \begin{bmatrix} \Phi_1(x) \; \Phi_2(x) \end{bmatrix} \begin{bmatrix} q_1 \\ q_2 \end{bmatrix}, \tag{12.31}$$

where

$$\Phi_1(x) = \frac{l - x}{l}, \qquad \Phi_2(x) = \frac{x}{l}. \tag{12.32}$$

The index (e) indicating matrices and vectors related to a particular finite element is skipped for clarity in the above as well as in the subsequent formulae.

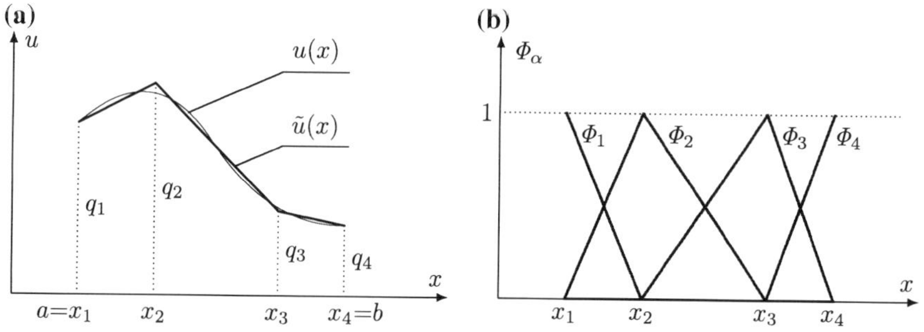

Fig. 12.2 Linear (first order) approximation in the 1D case

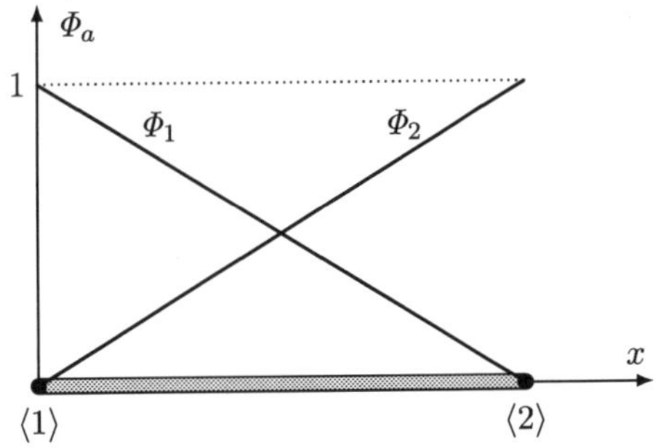

Fig. 12.3 1D element with the first order approximation

In order to determine at the element level the matrices and vectors appearing in discrete formulations of thermomechanics, certain expression must be integrated over the element domain Ω_e. Assuming that the element's cross-section area is A one may define the integral over the domain Ω_e as a definite integral in the interval $x \in [0, l]$ multiplied by A. In particular, the stiffness matrix in the linear elastic formulation, cf. Eq. (12.21a), has in this case the form

$$\left[\mathbf{k}_{2\times 2}\right] = EA \begin{bmatrix} \int_0^l \frac{\mathrm{d}\Phi_1}{\mathrm{d}x}\frac{\mathrm{d}\Phi_1}{\mathrm{d}x}\,\mathrm{d}x & \int_0^l \frac{\mathrm{d}\Phi_1}{\mathrm{d}x}\frac{\mathrm{d}\Phi}{\mathrm{d}x}\,\mathrm{d}x \\ \int_0^l \frac{\mathrm{d}\Phi_2}{\mathrm{d}x}\frac{\mathrm{d}\Phi}{\mathrm{d}x}\,\mathrm{d}x & \int_0^l \frac{\mathrm{d}\Phi_2}{\mathrm{d}x}\frac{\mathrm{d}\Phi_2}{\mathrm{d}x}\,\mathrm{d}x \end{bmatrix} = \frac{EA}{l}\begin{bmatrix} 1 & -1 \\ -1 & 1 \end{bmatrix}, \tag{12.33}$$

where E is the Young modulus of the bar's material. The conductivity matrix in the heat conduction problem has the same form in which the Young modulus E must only be replaced by the heat conductivity coefficient λ. Formulae for local right-hand side vectors $\mathbf{p}_{2\times 1}$ may be determined in a similar simple way.

Let us now analyse the forms of shape functions for higher order 1D elements. An element with three degrees of freedom related to three nodes, respectively, and with shape functions being second order polynomials is shown in Fig. 12.4a. Let us

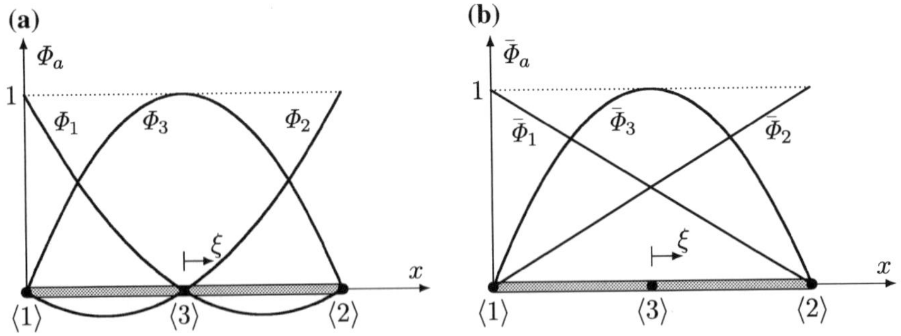

Fig. 12.4 1D element with second order approximation: **a** basic formulation, **b** hierarchical formulation

introduce a dimensionless coordinate $\xi = 2x/l$ assuming values -1, 1 and 0 at the nodes 1, 2 and 3, respectively. Approximation of displacement u is assumed in the form

$$\tilde{u}(\xi) = \begin{bmatrix} \Phi_1(\xi) & \Phi_2(\xi) & \Phi_3(\xi) \end{bmatrix} \begin{bmatrix} q_1 \\ q_2 \\ q_3 \end{bmatrix}, \tag{12.34}$$

where

$$\Phi_1(\xi) = -\frac{1}{2}\xi(1-\xi), \qquad \Phi_2(\xi) = \frac{1}{2}\xi(1+\xi), \qquad \Phi_3(\xi) = (1-\xi^2). \tag{12.35}$$

The following conditions are obviously fulfilled,

$$\tilde{u} = \begin{cases} q_1 & \text{for } \xi = -1 \text{ (node 1)} \\ q_2 & \text{for } \xi = 1 \text{ (node 2)} \\ q_3 & \text{for } \xi = 0 \text{ (node 3)} \end{cases} \tag{12.36}$$

and

$$\sum_{\alpha=1}^{3} \Phi_\alpha(\xi) = 1. \tag{12.37}$$

Integrals over the domain Ω_e may be again computed as definite integrals over $x \in [0, l]$ multiplied by A or—utilizing the dimensionless coordinate ξ—as definite integrals over $\xi \in [-1, 1]$ multiplied by $\frac{1}{2}Al$. Considering the linear elastic problem, one may compute the geometric matrix $\mathbf{B}$ (of the size 1×3 in this case) as

$$\left[\mathbf{B}_{1\times 3}\right] = \begin{bmatrix} \dfrac{\mathrm{d}\Phi_1}{\mathrm{d}x} & \dfrac{\mathrm{d}\Phi_2}{\mathrm{d}x} & \dfrac{\mathrm{d}\Phi_3}{\mathrm{d}x} \end{bmatrix} = \frac{1}{l}\,[\; 2\xi{-}1 \;\; 2\xi{+}1 \;\; -4\xi \;], \tag{12.38}$$

and then determine the element stiffness matrix as

$$\left[\mathbf{k}_{3\times 3}\right] = \frac{EAl}{2}\left[\int_{-1}^{1} \mathbf{B}^{\mathrm{T}}\mathbf{B}\,\mathrm{d}\xi\right] = \frac{EA}{3l}\begin{bmatrix} 7 & 1 & -8 \\ 1 & 7 & -8 \\ -8 & -8 & 15 \end{bmatrix}. \tag{12.39}$$

Note that the 2×2 submatrix of the matrix $\mathbf{k}_{3\times 3}$ corresponding to the first two degrees of freedom is different from the linear element stiffness matrix (12.33).

Let us now define the nodal degrees of freedom in the considered element in another way:

$$\begin{bmatrix} \bar{q}_1 \\ \bar{q}_2 \\ \bar{q}_3 \end{bmatrix} = \begin{bmatrix} q_1 \\ q_2 \\ q_3 - \frac{1}{2}(q_1 + q_2) \end{bmatrix}. \tag{12.40}$$

Such a choice of the parameters $\bar{q}_a$, $a = 1, 2, 3$, implies the following form of the shape functions (Fig. 12.4b):

$$\bar{\Phi}_1(\xi) = \frac{1}{2}(1-\xi), \qquad \bar{\Phi}_2(\xi) = \frac{1}{2}(1+\xi), \qquad \bar{\Phi}_3(\xi) = (1-\xi^2). \tag{12.41}$$

Computing the matrix $\bar{\mathbf{B}}$ according to Eq. (12.38) and the stiffness matrix according to Eq. (12.39), one obtains

$$\bar{\mathbf{k}}_{3\times 3} = \frac{EA}{3l}\begin{bmatrix} 3 & -3 & 0 \\ -3 & 3 & 0 \\ 0 & 0 & 16 \end{bmatrix}. \tag{12.42}$$

Note the following two advantages of this result:

1. The matrix $\mathbf{k}$, previously full, has now zero components (which is for obvious reasons favourable from the computational point of view).
2. The matrix $\bar{\mathbf{k}}_{3\times 3}$ contains (as a submatrix) the matrix $\mathbf{k}_{2\times 2}$ of the linear element—this is no surprise as the shape functions $(12.41)_{1,2}$ are identical as those used previously for the linear element (12.32) (in the translated coordinate system).

The latter feature has great practical importance—it allows to "enrich" the linear element mesh by making nodes more dense and the degree of approximating polynomials higher without modifying the stiffness matrix components corresponding to the original mesh nodes. Since the added nodes have "internal" character, this property extends to the global stiffness matrix and allows—with an appropriate block system solution algorithm—to utilize the solution of the system with the sparse mesh to solve the system with the fine mesh.

The formulations characterized by the shape functions (12.35) and (12.41) are called *standard* and *hierarchical*, respectively.

Up to now, 1D finite elements whose generalized coordinates are nodal values of the unknown function (e.g. displacement) have been discussed. Such elements ensure continuity—but not smoothness—of the approximation function at element boundaries. Such elements are called *Lagrange elements*. Let us now shortly present *Hermite elements*, i.e. elements that preserve continuity of not only the function but also its selected derivatives at the boundaries.

Let us consider as an example the element shown in Fig. 12.5. It has two nodes and two degrees of freedom at each of them: the value of the function sought and the value of its derivative at the node. The displacement approximation has the form

$$\tilde{u}(x) = \begin{bmatrix} \Phi_1(x) & \Phi_2(x) & \Phi_3(x) & \Phi_4(x) \end{bmatrix} \begin{bmatrix} q_1 \\ q_2 \\ q_3 \\ q_4 \end{bmatrix}, \tag{12.43}$$

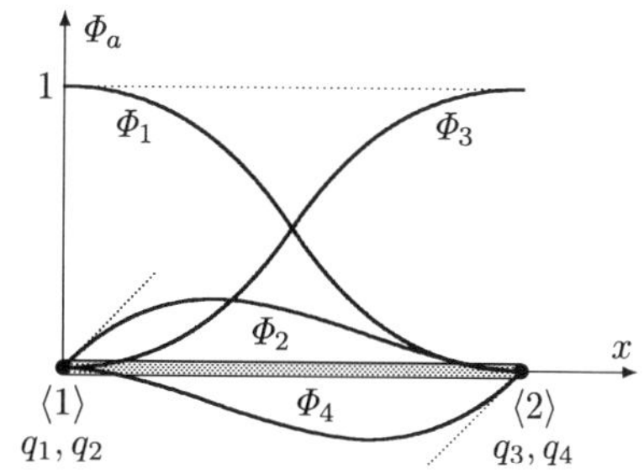

Fig. 12.5 1D element with shape functions ensuring continuity of their first order derivatives at element boundaries

where

$$q_1 = \tilde{u}_1 = \tilde{u}(0), \qquad q_2 = \tilde{u}'_1 = \left.\frac{\mathrm{d}\tilde{u}}{\mathrm{d}x}\right|_{x=0},$$

$$q_3 = \tilde{u}_2 = \tilde{u}(l), \qquad q_4 = \tilde{u}'_2 = \left.\frac{\mathrm{d}\tilde{u}}{\mathrm{d}x}\right|_{x=l},$$

and the shape functions are third order polynomials that upon introduction of a dimensionless variable $\eta = x/l$ are expressed as

$$\begin{aligned}
\Phi_1(x) &= 2\eta^3 - 3\eta^2 + 1, \\
\Phi_2(x) &= l\left(\eta^3 - 2\eta^2 + \eta\right), \\
\Phi_3(x) &= -2\eta^3 + 3\eta^2, \\
\Phi_4(x) &= l\left(\eta^3 - \eta^2\right).
\end{aligned}$$

The functions do not fulfill the conditions (12.28) and (12.29) but they meet their generalized form. The functions Φ_1 and Φ_3, related to nodes 1 and 2, respectively, assume values $\Phi_1 = 1$, $\Phi_3 = 0$ at node 1 and $\Phi_1 = 0$, $\Phi_3 = 1$ at node 2 and sum up to unity for $x \in (0, l)$. Besides, derivatives of the functions Φ_2 and Φ_4, also related to nodes 1 and 2, respectively, assume values $\Phi_{2,x} = 1$, $\Phi_{4,x} = 0$ at node 1 and $\Phi_{2,x} = 0$, $\Phi_{4,x} = 1$ at node 2.

Finite element approximation of the solution ensuring continuity of higher order derivatives can be constructed in a similar way.

12.2.2 Constant Strain Elements

The simplest examples of 2D and 3D finite element formulations are constant strain elements (or, more generally, elements with constant shape function gradients). These are triangular (2D) or tetrahedral (3D) elements with shape functions being first order polynomials with respect to the coordinates x_i (the first order approximation). Let us briefly present both the element types.

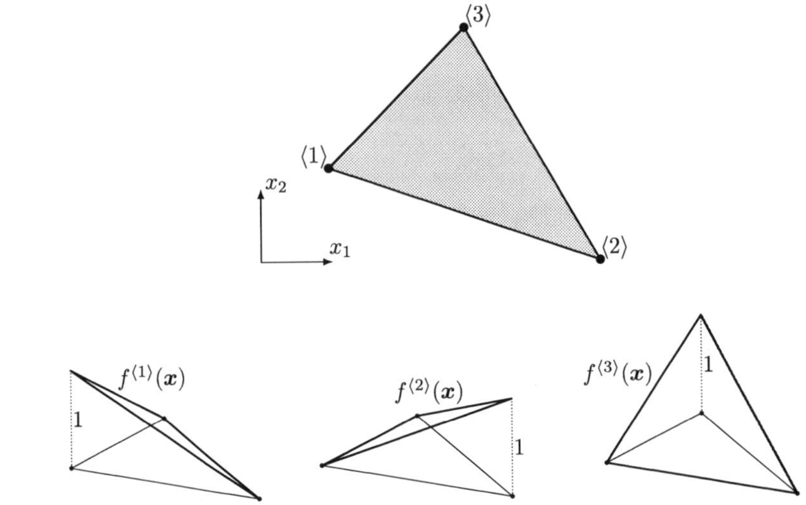

Fig. 12.6 Triangular element with first order approximation

A triangular constant strain element is shown in Fig. 12.6. Assume the local coordinate system $\{x_1^{(e)} x_2^{(e)}\}$ to coincide with the global system $\{x_1 x_2\}$. There are two degrees of freedom corresponding to two displacement components at each of the element nodes 1, 2 and 3. Skipping again for clarity the index (e), one may define the nodal displacement vector for the element as

$$\{\mathbf{q}^{\mathrm{T}}_{1\times 6}\} = \{\, q_1\; q_2\; \ldots\; q_6 \,\} = \left\{\, u_1^{\langle 1\rangle}\; u_2^{\langle 1\rangle}\; u_1^{\langle 2\rangle}\; u_2^{\langle 2\rangle}\; u_1^{\langle 3\rangle}\; u_2^{\langle 3\rangle} \,\right\}. \tag{12.44}$$

The displacement fields $u_1(\boldsymbol{x}),\ u_2(\boldsymbol{x})$ of the element points $\boldsymbol{x} = \{x_1\, x_2\}$ are now expressed by the vector $\mathbf{q}$ and shape functions in the general form

$$\tilde{u}_1(\boldsymbol{x}) = a + bx_1 + cx_2, \qquad \tilde{u}_2(\boldsymbol{x}) = d + ex_1 + fx_2\,. \tag{12.45}$$

It has thus been assumed that the shape functions are linear with respect to $x_1,\ x_2$. Making use in Eq. (12.45) of the conditions

$$\tilde{u}_1(\boldsymbol{x}^{\langle 1\rangle}) = q_1,\ \ \tilde{u}_2(\boldsymbol{x}^{\langle 1\rangle}) = q_2,\ \ldots,\ \ \tilde{u}_2(\boldsymbol{x}^{\langle 3\rangle}) = q_6\,, \tag{12.46}$$

one may determine from the above six equations the six parameters $a, b, \ldots, f$ as functions of the nodal parameters $q_1,\ \ldots,\ q_6$. Substituting them into Eq. (12.45), one obtains

$$\begin{aligned} \tilde{u}_1(\boldsymbol{x}) &= q_1 f^{\langle 1\rangle}(\boldsymbol{x}) + q_3 f^{\langle 2\rangle}(\boldsymbol{x}) + q_5 f^{\langle 3\rangle}(\boldsymbol{x}), \\ \tilde{u}_2(\boldsymbol{x}) &= q_2 f^{\langle 2\rangle}(\boldsymbol{x}) + q_4 f^{\langle 2\rangle}(\boldsymbol{x}) + q_6 f^{\langle 3\rangle}(\boldsymbol{x}), \end{aligned} \tag{12.47}$$

where the nodal shape functions $f^{\langle i\rangle}(\boldsymbol{x})$, $i = 1, 2, 3$ (cf. Fig. 12.6), are given as

$$f^{\langle i\rangle}(\boldsymbol{x}) = \frac{\alpha_i + \beta_i x_1 + \gamma_i x_2}{2\Delta}. \tag{12.48}$$

In the above equation, Δ denotes the element area,

$$\Delta = \frac{1}{2}\begin{vmatrix} 1 & 1 & 1 \\ x_1^{\langle 1\rangle} & x_1^{\langle 2\rangle} & x_1^{\langle 3\rangle} \\ x_2^{\langle 1\rangle} & x_2^{\langle 2\rangle} & x_2^{\langle 3\rangle} \end{vmatrix}, \tag{12.49}$$

while the coefficients α_i, β_i and γ_i are defined as

$$\begin{aligned} \alpha_1 &= x_1^{\langle 2\rangle} x_2^{\langle 3\rangle} - x_1^{\langle 3\rangle} x_2^{\langle 2\rangle}, \quad & \beta_1 &= x_2^{\langle 2\rangle} - x_2^{\langle 3\rangle}, \quad & \gamma_1 &= x_1^{\langle 3\rangle} - x_1^{\langle 2\rangle}, \\ \alpha_2 &= x_1^{\langle 3\rangle} x_2^{\langle 1\rangle} - x_1^{\langle 1\rangle} x_2^{\langle 3\rangle}, \quad & \beta_2 &= x_2^{\langle 3\rangle} - x_2^{\langle 1\rangle}, \quad & \gamma_2 &= x_1^{\langle 1\rangle} - x_1^{\langle 3\rangle}, \\ \alpha_3 &= x_1^{\langle 1\rangle} x_2^{\langle 2\rangle} - x_1^{\langle 2\rangle} x_2^{\langle 1\rangle}, \quad & \beta_3 &= x_2^{\langle 1\rangle} - x_2^{\langle 2\rangle}, \quad & \gamma_3 &= x_1^{\langle 2\rangle} - x_1^{\langle 1\rangle}. \end{aligned} \tag{12.50}$$

It is easy to verify that $f^{\langle i\rangle}(\boldsymbol{x}^{\langle j\rangle}) = \delta_{ij}$.

Equation (12.47) can be written in the matrix form as

$$\tilde{\mathbf{u}}_{2\times 1}(\boldsymbol{x}) = \begin{Bmatrix} \tilde{u}_1(x_1, x_2) \\ \tilde{u}_2(x_1, x_2) \end{Bmatrix} = \boldsymbol{\Phi}_{2\times 6}(\boldsymbol{x})\, \mathbf{q}_{6\times 1}, \tag{12.51}$$

where

$$\boldsymbol{\Phi}_{2\times 6} = \begin{bmatrix} f^{\langle 1\rangle} & 0 & f^{\langle 2\rangle} & 0 & f^{\langle 3\rangle} & 0 \\ 0 & f^{\langle 1\rangle} & 0 & f^{\langle 2\rangle} & 0 & f^{\langle 3\rangle} \end{bmatrix}, \tag{12.52}$$

and the vector $\mathbf{q}$ is given by Eq. (12.44). The strain vector $\tilde{\boldsymbol{\varepsilon}}$ can be expressed as

$$\tilde{\boldsymbol{\varepsilon}}_{3\times 1} = \begin{Bmatrix} \tilde{\varepsilon}_{11} \\ \tilde{\varepsilon}_{22} \\ 2\tilde{\varepsilon}_{12} \end{Bmatrix} = \begin{bmatrix} \frac{\partial}{\partial x_1} & 0 \\ 0 & \frac{\partial}{\partial x_2} \\ \frac{\partial}{\partial x_2} & \frac{\partial}{\partial x_1} \end{bmatrix} \tilde{\mathbf{u}}_{2\times 1} = \mathbf{B}_{3\times 6}\, \mathbf{q}_{6\times 1}, \tag{12.53}$$

where

$$\mathbf{B}_{3\times 6} = \frac{1}{2\Delta}\begin{bmatrix} \beta_1 & 0 & \beta_2 & 0 & \beta_3 & 0 \\ 0 & \gamma_1 & 0 & \gamma_2 & 0 & \gamma_3 \\ \gamma_1 & \beta_1 & \gamma_2 & \beta_2 & \gamma_3 & \beta_3 \end{bmatrix}. \tag{12.54}$$

Note that the matrix $\boldsymbol{B}$ is independent of $\boldsymbol{x}$—this is why the element is called "constant strain". Since in the linear theory of elasticity the material properties are also constant, it is easy to derive the form of the element stiffness matrix—the integrand in Eq. (12.22a) is constant. Assuming the plane stress state, cf. Eq. (7.33),

$$[\mathbf{C}_{3\times3}] = \frac{E}{1-\nu^2}\begin{bmatrix} 1 & \nu & \\ \nu & 1 & \\ & & \frac{1-\nu}{2} \end{bmatrix},$$

one may write down Eq. (12.22a) in the form reduced to two dimensions and $N = 3$ nodes as

$$\mathbf{k}_{6\times6} = \Delta\left[\mathbf{B}^{\mathrm{T}}_{6\times3}\mathbf{C}_{3\times3}\mathbf{B}_{3\times6}\right] = \frac{Et\Delta}{1-\nu^2}\left[\mathbf{k}^{*}_{6\times6} + \frac{1-\nu}{2}\mathbf{k}^{**}_{6\times6}\right],$$

$$\mathbf{k}^{*}_{6\times6} = \begin{bmatrix} \beta_1^2 & \nu\beta_1\gamma_1 & \beta_1\beta_2 & \nu\beta_1\gamma_2 & \beta_1\beta_3 & \nu\beta_1\gamma_3 \\ & \gamma_1^2 & \nu\gamma_1\beta_2 & \gamma_1\gamma_2 & \nu\gamma_1\beta_3 & \gamma_1\gamma_3 \\ & & \beta_2^2 & \nu\beta_2\gamma_2 & \beta_2\beta_3 & \nu\beta_2\gamma_3 \\ & & & \gamma_2^2 & \nu\gamma_2\beta_3 & \gamma_2\gamma_3 \\ & \text{symm.} & & & \beta_3^2 & \nu\beta_3\gamma_3 \\ & & & & & \gamma_3^2 \end{bmatrix},$$

$$\mathbf{k}^{**}_{6\times6} = \begin{bmatrix} \gamma_1^2 & \gamma_1\beta_1 & \gamma_1\gamma_2 & \gamma_1\beta_2 & \gamma_1\gamma_3 & \gamma_1\beta_3 \\ & \beta_1^2 & \beta_1\gamma_2 & \beta_1\beta_2 & \beta_1\gamma_3 & \beta_1\beta_3 \\ & & \gamma_2^2 & \gamma_2\beta_2 & \gamma_2\gamma_3 & \gamma_2\beta_3 \\ & & & \beta_2^2 & \beta_2\gamma_3 & \beta_2\beta_3 \\ & \text{symm.} & & & \gamma_3^2 & \gamma_3\beta_3 \\ & & & & & \beta_3^2 \end{bmatrix},$$

where t denotes the element thickness.

In analogy, the first order polynomial approximation can be defined for a 3D tetrahedral element (Fig. 12.7). This element has four nodes with three degrees of freedom at each of them, equal to displacement component values at these nodes, respectively. This makes 12 nodal parameters q_a altogether:

$$\left\{\mathbf{q}^{\mathrm{T}}_{1\times12}\right\} = \{\, q_1\ q_2\ \cdots\ q_{12}\,\} = \left\{\, u_1^{\langle1\rangle}\ u_2^{\langle1\rangle}\ u_3^{\langle1\rangle}\ \cdots\ u_1^{\langle4\rangle}\ u_2^{\langle4\rangle}\ u_3^{\langle4\rangle}\,\right\}. \quad (12.55)$$

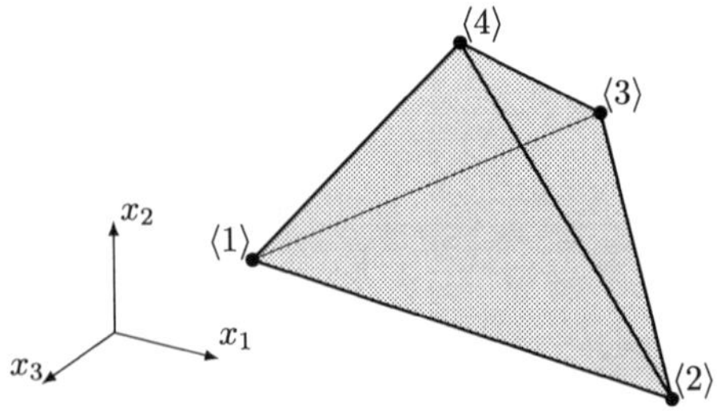

Fig. 12.7 Tetrahedral element with first order approximation

The approximate displacement field at the element points is expressed as

$$\tilde{\mathbf{u}}_{3\times1}(\boldsymbol{x}) = \begin{Bmatrix} \tilde{u}_1(x_1, x_2, x_3) \\ \tilde{u}_2(x_1, x_2, x_3) \\ \tilde{u}_3(x_1, x_2, x_3) \end{Bmatrix} = \mathbf{\Phi}_{3\times12}(\boldsymbol{x})\, \mathbf{q}_{12\times1}\,, \tag{12.56}$$

where

$$\mathbf{\Phi}_{3\times12} = \begin{bmatrix} f^{\langle1\rangle} & 0 & 0 & f^{\langle2\rangle} & 0 & 0 & f^{\langle3\rangle} & 0 & 0 & f^{\langle4\rangle} & 0 & 0 \\ 0 & f^{\langle1\rangle} & 0 & 0 & f^{\langle2\rangle} & 0 & 0 & f^{\langle3\rangle} & 0 & 0 & f^{\langle4\rangle} & 0 \\ 0 & 0 & f^{\langle1\rangle} & 0 & 0 & f^{\langle2\rangle} & 0 & 0 & f^{\langle3\rangle} & 0 & 0 & f^{\langle4\rangle} \end{bmatrix}, \tag{12.57}$$

and $f^{\langle i\rangle}$ are the shape functions corresponding to the nodes $i = 1, 2, 3, 4$. The functions are first order polynomials with respect to x_1, x_2, x_3 and are expressed as

$$f^{\langle i\rangle}(\boldsymbol{x}) = \frac{\alpha_i + \beta_i x_1 + \gamma_i x_2 + \delta_i x_3}{6V}, \tag{12.58}$$

where V denotes the element volume

$$V = \frac{1}{6}\begin{vmatrix} 1 & 1 & 1 & 1 \\ x_1^{\langle1\rangle} & x_1^{\langle2\rangle} & x_1^{\langle3\rangle} & x_1^{\langle4\rangle} \\ x_2^{\langle1\rangle} & x_2^{\langle2\rangle} & x_2^{\langle3\rangle} & x_2^{\langle4\rangle} \\ x_3^{\langle1\rangle} & x_3^{\langle2\rangle} & x_3^{\langle3\rangle} & x_3^{\langle4\rangle} \end{vmatrix}, \tag{12.59}$$

while the coefficients α_i, β_i, γ_i and δ_i are defined as

$$\begin{aligned} \alpha_i &= \begin{vmatrix} x_1^{\langle j\rangle} & x_1^{\langle k\rangle} & x_1^{\langle l\rangle} \\ x_2^{\langle j\rangle} & x_2^{\langle k\rangle} & x_2^{\langle l\rangle} \\ x_3^{\langle j\rangle} & x_3^{\langle k\rangle} & x_3^{\langle l\rangle} \end{vmatrix}, \qquad \beta_i = -\begin{vmatrix} 1 & 1 & 1 \\ x_2^{\langle j\rangle} & x_2^{\langle k\rangle} & x_2^{\langle l\rangle} \\ x_3^{\langle j\rangle} & x_3^{\langle k\rangle} & x_3^{\langle l\rangle} \end{vmatrix}, \\ \gamma_i &= \begin{vmatrix} x_1^{\langle j\rangle} & x_1^{\langle k\rangle} & x_1^{\langle l\rangle} \\ 1 & 1 & 1 \\ x_3^{\langle j\rangle} & x_3^{\langle k\rangle} & x_3^{\langle l\rangle} \end{vmatrix}, \qquad \delta_i = \begin{vmatrix} x_1^{\langle j\rangle} & x_1^{\langle k\rangle} & x_1^{\langle l\rangle} \\ x_2^{\langle j\rangle} & x_2^{\langle k\rangle} & x_2^{\langle l\rangle} \\ 1 & 1 & 1 \end{vmatrix}. \end{aligned} \tag{12.60}$$

Hence, if the strain vector $\tilde{\boldsymbol{\varepsilon}}$ is expressed as

$$\tilde{\boldsymbol{\varepsilon}}_{6\times1} = \begin{Bmatrix} \tilde{\varepsilon}_{11} \\ \tilde{\varepsilon}_{22} \\ \tilde{\varepsilon}_{33} \\ 2\tilde{\varepsilon}_{12} \\ 2\tilde{\varepsilon}_{23} \\ 2\tilde{\varepsilon}_{31} \end{Bmatrix} = \begin{bmatrix} \frac{\partial}{\partial x_1} & 0 & 0 \\ 0 & \frac{\partial}{\partial x_2} & 0 \\ 0 & 0 & \frac{\partial}{\partial x_3} \\ \frac{\partial}{\partial x_2} & \frac{\partial}{\partial x_1} & 0 \\ 0 & \frac{\partial}{\partial x_3} & \frac{\partial}{\partial x_2} \\ \frac{\partial}{\partial x_3} & 0 & \frac{\partial}{\partial x_1} \end{bmatrix} \tilde{\mathbf{u}}_{3\times1} = \mathbf{B}_{6\times12}\, \mathbf{q}_{12\times1}\,, \tag{12.61}$$

then the geometric matrix has the following form,

$$\mathbf{B}_{6\times 12} = \frac{1}{6V}\begin{bmatrix} \beta_1 & 0 & 0 & \beta_2 & 0 & 0 & \beta_3 & 0 & 0 & \beta_4 & 0 & 0 \\ 0 & \gamma_1 & 0 & 0 & \gamma_2 & 0 & 0 & \gamma_3 & 0 & 0 & \gamma_4 & 0 \\ 0 & 0 & \delta_1 & 0 & 0 & \delta_2 & 0 & 0 & \delta_3 & 0 & 0 & \delta_4 \\ \gamma_1 & \beta_1 & 0 & \gamma_2 & \beta_2 & 0 & \gamma_3 & \beta_3 & 0 & \gamma_4 & \beta_4 & 0 \\ 0 & \delta_1 & \gamma_1 & 0 & \delta_2 & \gamma_2 & 0 & \delta_3 & \gamma_3 & 0 & \delta_4 & \gamma_4 \\ \delta_1 & 0 & \beta_1 & \delta_2 & 0 & \beta_2 & \delta_3 & 0 & \beta_3 & \delta_4 & 0 & \beta_4 \end{bmatrix}. \tag{12.62}$$

In this case, again, the matrix $\mathbf{B}$ is independent of $\boldsymbol{x}$, which allows to easily derive closed-form formulae for the stiffness matrix components. We leave this task to the readers as an exercise.

12.2.3 Isoparametric Elements

In the previous section, the methodology of stiffness matrix derivation for simplest finite element formulations has been shown. Looking at the presented formulae, it is easy to figure out that application of such derivation schemes to more complex element types (quadrilateral or triangular with higher order approximation, for instance) will not lead us to closed-form formulae. To the contrary, one should rather expect that integrands in the formulae as well as the domains over which they are to be integrated will be much more complex. Below, we shall briefly present the methodology of stiffness matrix derivation for an important class of finite elements called the isoparametric elements. Our discussion is limited to plane problems only—its extension to 3D cases is straightforward.

The isoparametric elements are finite elements in which geometry is approximated with the same functions as the generalized coordinate fields (say, displacements). If geometry is approximated with lower or higher order functions than generalized coordinates, then elements are called sub- or superparametric, respectively.

Consider a plane quadrilateral element with generally curvilinear edges. Construction of the isoparametric element formulation consists in mapping of the so-called base element, i.e. a predefined rectangle (usually it is a unit square) with a predefined number of nodes, onto the corresponding curvilinear element with the same number of nodes. According to this definition, geometry of an element defined by shape functions being n-order polynomials is described by the same n-order polynomial functions. This allows to automatically fulfill compatibility conditions for generalized strain fields at element boundaries.

Let us illustrate this by considering the simplest case of a four-node quadrilateral element Ω^e with straight edges. Define the base element $\hat{\Omega}$ in a certain coordinate system (ξ_1, ξ_2), as a square whose vertices have the coordinates $(-1, -1)$, $(1, -1)$, $(1, 1)$ and $(-1, 1)$, respectively, see Fig. 12.8. Note that the element $\hat{\Omega}$ may be a common base element for several different quadrilateral elements of various geometric proportions that may appear in any finite element mesh.

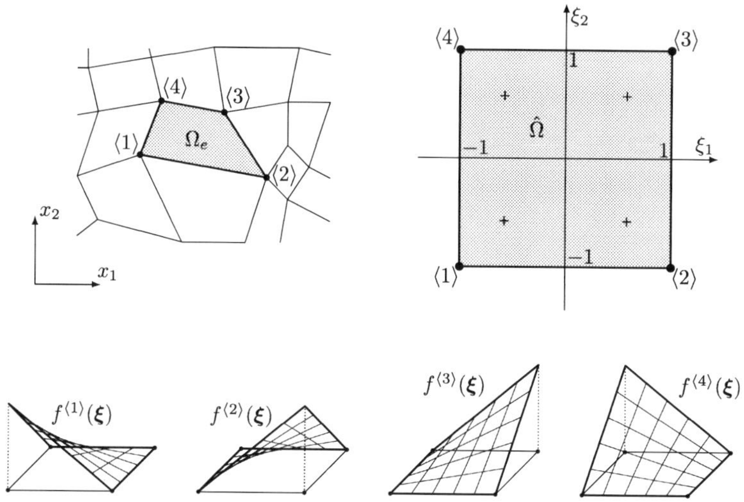

Fig. 12.8 Quadrilateral four-node isoparametric element

Let us define a unique mapping of the base coordinates $\boldsymbol{\xi}$ onto the global ones $\boldsymbol{x}$,

$$\boldsymbol{x}(\boldsymbol{\xi}) = f^{\langle i\rangle}(\boldsymbol{\xi})\,\boldsymbol{X}^{\langle i\rangle}, \tag{12.63}$$

where $\boldsymbol{X}^{\langle i\rangle} = (X_1^{\langle i\rangle},\, X_2^{\langle i\rangle})$ denote coordinates of the ith nodal point, $i = 1, 2, 3, 4$, and $f^{\langle i\rangle}(\boldsymbol{\xi})$ are differentiable shape functions, such that they (i) fulfill the known condition $f^{\langle i\rangle}(\boldsymbol{\xi}^{\langle j\rangle}) = \delta_{ij}$ and (ii) transform the sides of $\hat{\Omega}$ onto straight segments (i.e. the transformation image of the square $\hat{\Omega}$ is a tetragon). Figure 12.8 presents graphs of functions that meet these requirements—they are expressed by the following bilinear (i.e. linear with respect to each of the coordinates ξ_1, ξ_2 separately) formulae:

$$\begin{aligned} f^{\langle 1\rangle} &= \frac{1}{4}(1-\xi_1)(1-\xi_2), \qquad & f^{\langle 2\rangle} &= \frac{1}{4}(1+\xi_1)(1-\xi_2), \\ f^{\langle 3\rangle} &= \frac{1}{4}(1+\xi_1)(1+\xi_2), \qquad & f^{\langle 4\rangle} &= \frac{1}{4}(1-\xi_1)(1+\xi_2). \end{aligned} \tag{12.64}$$

Equation (12.63) may be also written in the matrix form as

$$\mathbf{x}_{2\times 1}(\boldsymbol{\xi}) = \boldsymbol{\Phi}_{2\times 8}(\boldsymbol{\xi})\,\mathbf{X}_{8\times 1}\,, \tag{12.65}$$

where

$$\begin{aligned} \{\mathbf{x}^{\mathrm{T}}\} &= \{\, x_1 \; x_2 \,\}, \\ \{\mathbf{X}^{\mathrm{T}}\} &= \left\{\, X_1^{\langle 1\rangle} \; X_2^{\langle 1\rangle} \; X_1^{\langle 2\rangle} \; X_2^{\langle 2\rangle} \; X_1^{\langle 3\rangle} \; X_2^{\langle 3\rangle} \; X_1^{\langle 4\rangle} \; X_2^{\langle 4\rangle} \,\right\}, \end{aligned}$$

$$[\boldsymbol{\Phi}] = \begin{bmatrix} f^{\langle 1\rangle} & 0 & f^{\langle 2\rangle} & 0 & f^{\langle 3\rangle} & 0 & f^{\langle 4\rangle} & 0 \\ 0 & f^{\langle 1\rangle} & 0 & f^{\langle 2\rangle} & 0 & f^{\langle 3\rangle} & 0 & f^{\langle 4\rangle} \end{bmatrix}.$$

Defining the nodal parameter vector $\mathbf{q}_{8\times 1}$ as

$$\{\mathbf{q}^{\mathrm{T}}\} = \left\{ u_1^{\langle 1\rangle}\ u_2^{\langle 1\rangle}\ u_1^{\langle 2\rangle}\ u_2^{\langle 2\rangle}\ u_1^{\langle 3\rangle}\ u_2^{\langle 3\rangle}\ u_1^{\langle 4\rangle}\ u_2^{\langle 4\rangle} \right\}, \tag{12.66}$$

one may apply an identical transformation to approximate displacements inside the element:

$$\tilde{\mathbf{u}}_{2\times 1}(\boldsymbol{\xi}) = \boldsymbol{\Phi}_{2\times 8}(\boldsymbol{\xi})\,\mathbf{q}_{8\times 1}\,. \tag{12.67}$$

The difference between the above approximation and that used in the previous formulations, e.g. Eq. (12.51), consists in the fact that here displacements are not defined at a point with given coordinates $\boldsymbol{x}$ but rather at its counterimage in the base element configuration with given coordinates $\boldsymbol{\xi}$.

To use the base element $\hat{\Omega}$ to derive the stiffness matrix of the element Ω_e one must transform all functions of (x_1, x_2) describing the element Ω_e with the mapping (12.63). The Jacobian of this mapping has the form

$$\left[\mathbf{J}_{2\times 2}\right] = \begin{bmatrix} \frac{\partial x_1}{\partial \xi_1} & \frac{\partial x_2}{\partial \xi_1} \\ \frac{\partial x_1}{\partial \xi_2} & \frac{\partial x_2}{\partial \xi_2} \end{bmatrix} = \begin{bmatrix} \frac{\partial f^{\langle i\rangle}}{\partial \xi_1} X_1^{\langle i\rangle} & \frac{\partial f^{\langle i\rangle}}{\partial \xi_1} X_2^{\langle i\rangle} \\ \frac{\partial f^{\langle i\rangle}}{\partial \xi_2} X_1^{\langle i\rangle} & \frac{\partial f^{\langle i\rangle}}{\partial \xi_2} X_1^{\langle i\rangle} \end{bmatrix} \tag{12.68}$$

and its determinant $j = \det \mathbf{J}$ is non-zero in the entire domain $\hat{\Omega}$, which is implied by the assumed uniqueness of the mapping (12.63).

For an arbitrary field F defined on the domain $\hat{\Omega}$ one may write

$$\frac{\partial F}{\partial x_1} = \frac{\partial F}{\partial \xi_1}\frac{\partial \xi_1}{\partial x_1} + \frac{\partial F}{\partial \xi_2}\frac{\partial \xi_2}{\partial x_1}\,, \qquad \frac{\partial F}{\partial x_2} = \frac{\partial F}{\partial \xi_1}\frac{\partial \xi_1}{\partial x_2} + \frac{\partial F}{\partial \xi_2}\frac{\partial \xi_2}{\partial x_2}\,, \tag{12.69}$$

which can also be expressed in the matrix notation as

$$\begin{bmatrix} \frac{\partial F}{\partial x_1} \\ \frac{\partial F}{\partial x_2} \end{bmatrix} = \begin{bmatrix} \frac{\partial \xi_1}{\partial x_1} & \frac{\partial \xi_2}{\partial x_1} \\ \frac{\partial \xi_1}{\partial x_2} & \frac{\partial \xi_2}{\partial x_2} \end{bmatrix} \begin{bmatrix} \frac{\partial F}{\partial \xi_1} \\ \frac{\partial F}{\partial \xi_2} \end{bmatrix} = \left[\mathbf{J}_{2\times 2}^{-1}\right] \begin{bmatrix} \frac{\partial F}{\partial \xi_1} \\ \frac{\partial F}{\partial \xi_2} \end{bmatrix}. \tag{12.70}$$

The above formula refers in particular to the shape functions $f^{\langle i\rangle}$ for which we have:

$$\frac{\partial f^{\langle 1\rangle}}{\partial \xi_1} = -\frac{1}{4}(1-\xi_2), \qquad \frac{\partial f^{\langle 1\rangle}}{\partial \xi_2} = -\frac{1}{4}(1-\xi_1),$$

$$\frac{\partial f^{\langle 2\rangle}}{\partial \xi_1} = \frac{1}{4}(1-\xi_2), \qquad \frac{\partial f^{\langle 2\rangle}}{\partial \xi_2} = -\frac{1}{4}(1+\xi_1),$$

$$\frac{\partial f^{\langle 3\rangle}}{\partial \xi_1} = \frac{1}{4}(1+\xi_2), \qquad \frac{\partial f^{\langle 3\rangle}}{\partial \xi_2} = \frac{1}{4}(1+\xi_1),$$
$$\frac{\partial f^{\langle 4\rangle}}{\partial \xi_1} = -\frac{1}{4}(1+\xi_2), \qquad \frac{\partial f^{\langle 4\rangle}}{\partial \xi_2} = \frac{1}{4}(1-\xi_1).$$

Given the above relations, it is easy to determine the geometric matrix **B** which assumes here the following form:

$$\mathbf{B}_{3\times 8} = \begin{bmatrix} \frac{\partial f^{\langle 1\rangle}}{\partial x_1} & 0 & \frac{\partial f^{\langle 2\rangle}}{\partial x_1} & 0 & \frac{\partial f^{\langle 3\rangle}}{\partial x_1} & 0 & \frac{\partial f^{\langle 4\rangle}}{\partial x_1} & 0 \\ 0 & \frac{\partial f^{\langle 1\rangle}}{\partial x_2} & 0 & \frac{\partial f^{\langle 2\rangle}}{\partial x_2} & 0 & \frac{\partial f^{\langle 3\rangle}}{\partial x_2} & 0 & \frac{\partial f^{\langle 4\rangle}}{\partial x_2} \\ \frac{\partial f^{\langle 1\rangle}}{\partial x_2} & \frac{\partial f^{\langle 1\rangle}}{\partial x_1} & \frac{\partial f^{\langle 2\rangle}}{\partial x_2} & \frac{\partial f^{\langle 2\rangle}}{\partial x_1} & \frac{\partial f^{\langle 3\rangle}}{\partial x_2} & \frac{\partial f^{\langle 3\rangle}}{\partial x_1} & \frac{\partial f^{\langle 4\rangle}}{\partial x_2} & \frac{\partial f^{\langle 4\rangle}}{\partial x_1} \end{bmatrix}. \tag{12.71}$$

The stiffness matrix **k** of the element Ω_e with the thickness t is thus expressed as

$$\mathbf{k}_{8\times 8} = \int_{\Omega_e} t\mathbf{B}^{\mathrm{T}}_{8\times 3}\,\mathbf{C}_{3\times 3}\,\mathbf{B}_{3\times 8}\,\mathrm{d}V. \tag{12.72}$$

Note that the integrand is not constant here and thus computation of the above integral is not trivial (as it was for the linear triangular element). Also note that the matrix components in the integrand are defined as functions of the base coordinates ξ_1, ξ_2 so that it would be more convenient to perform the integration over the base element domain $\hat{\Omega}$ rather than over Ω_e, especially that $\hat{\Omega}$ has a more regular, simple shape. We therefore observe that since

$$\mathrm{d}V = j\mathrm{d}\hat{V},$$

integration of any function over Ω_e may be replaced by integration over $\hat{\Omega}$ according to the formula

$$\int_{\Omega_e} (\cdot)\,\mathrm{d}V = \int_{\hat{\Omega}} (\cdot)\,j\mathrm{d}\hat{V} = \int_{-1}^{1}\int_{-1}^{1} (\cdot)\,j\mathrm{d}\xi_1\mathrm{d}\xi_2\,.$$

In particular,

$$\mathbf{k}_{8\times 8} = \int_{-1}^{1}\int_{-1}^{1} t\boldsymbol{B}^{\mathrm{T}}_{8\times 3}\,\mathbf{C}_{3\times 3}\,\mathbf{B}_{3\times 8}\,j\,\mathrm{d}\xi_1\mathrm{d}\xi_2\,. \tag{12.73}$$

Instead of the inconvenient integral over an irregular domain Ω_e we may thus perform integration over the base element domain $\hat{\Omega}$.

It is noteworthy that a similar methodology can be employed for elements with more complex approximation forms. For example, placing additional four nodes at central points of the base square edges and using shape functions in the form of second order polynomials with respect to each of the coordinates ξ_1, ξ_2 (see the 1D example in Fig. 12.4) allows to define curvilinear second order quadrilateral elements. Higher order triangular elements can be defined in a similar way. Figure 12.9 presents examples of such elements and their corresponding base configurations. In all these

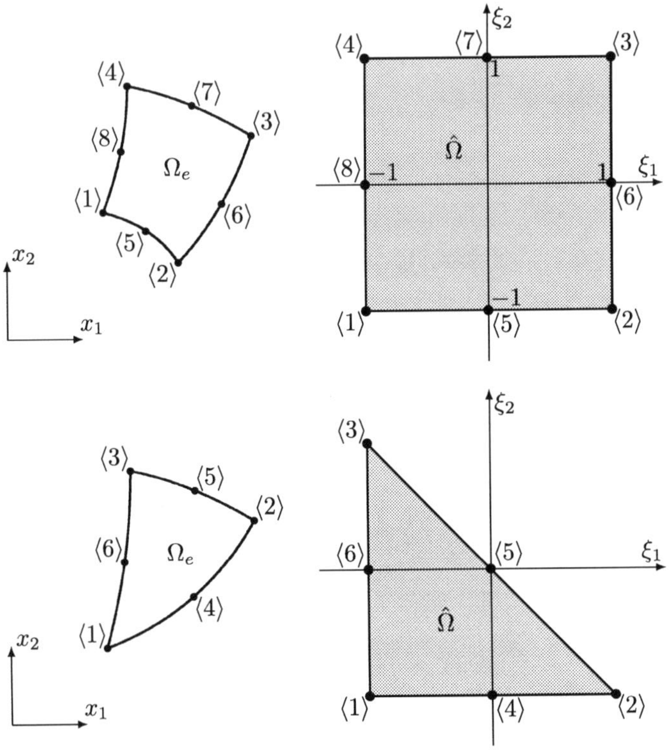

Fig. 12.9 Isoparametric elements with second order approximation

cases, the presented transformations can be used as well which allows to express the element stiffness matrix as an integral over a simple base element geometric domain. The above discussion can easily be generalized to the 3D problems.

Let us finally mention the issue of numerical integration of expressions like that in Eq. (12.73). There are several computational methods that can be employed here, of which the most popular is the so-called Gauss quadrature. It consists in replacing the integral over a, say, 2D domain $\hat{\Omega}$ by a double sum in the form

$$\int_{\hat{\Omega}} F(\xi_1, \xi_2)\,\mathrm{d}\xi_1\mathrm{d}\xi_2 = \sum_{i=1}^{M}\sum_{j=1}^{N} F(\xi_1^{(k)}, \xi_2^{(l)}) W_k W_l \,, \tag{12.74}$$

in which M and N denote the numbers of quadrature points in the directions ξ_1 and ξ_2, respectively, while W_k are called the Gauss weights. The choice of the number of integration points depends on the order of polynomials appearing in F and on the required accuracy of the procedure. Generally, an accurate integral value of a single variable polynomial of the order $2p$ can be computed with $K = p + 1$ Gauss integration points. In most cases, interpolation functions have the same order with respect to both their arguments ξ_1 and ξ_2, so that $M = N$ can usually be assumed. In the above discussed case of bilinear functions, Eq. (12.64), a 2×2 quadrature is sufficient with the Gauss point coordinates $\xi_i^{(k)} = \pm 1/\sqrt{3}$ (depicted by crosses in Fig. 12.8) and weights $W_1 = W_2 = 1$.

References

1. Bathe K.J., 1996. *Finite Element Procedures*. Prentice-Hall.
2. Belytschko T., Liu W.K., Moran B., 2000. *Nonlinear Finite Elements for Continua and Structures*. John Wiley & Sons.
3. Kleiber M., 1989. *Incremental Finite Element Modelling in Non-linear Solid Mechanics*. Ellis Horwood.
4. Kleiber M., Antúnez H., Hien T.D., Kowalczyk P., 1997. *Parameter Sensitivity in Nonlinear Mechanics; Theory and Finite Element Computations*. John Wiley & Sons.
5. Zienkiewicz O.C., Taylor R.L., 2000. *The Finite Element Method, 5th edition*. Butterworth-Heinemann.

Chapter 13
Solution of FEM Equation Systems

13.1 Introduction

In previous chapters, equations for various thermomechanical problems derived within the FEM approximation methodology have been presented. Here, methods of their solution are going to be briefly described.

Let us first consider the way the displacement boundary conditions are taken into account in the global system of algebraic equations. It is convenient to discuss this issue on a simple example of the following system of equations:

$$\begin{bmatrix} K_{11} & K_{12} & 0 & 0 & 0 \\ K_{12} & K_{22} & K_{23} & 0 & 0 \\ 0 & K_{23} & K_{33} & K_{34} & 0 \\ 0 & 0 & K_{34} & K_{44} & K_{45} \\ 0 & 0 & 0 & K_{45} & K_{55} \end{bmatrix} \begin{bmatrix} q_1 \\ q_2 \\ q_3 \\ q_4 \\ q_5 \end{bmatrix} = \begin{bmatrix} Q_1 \\ Q_2 \\ Q_3 \\ Q_4 \\ Q_5 \end{bmatrix}. \tag{13.1}$$

This system's structure is typical of FEM, i.e. the coefficient matrix **K** is symmetric and banded.

Consider an exemplary boundary condition in the form $q_4 = \hat{q}_4$. In this case, the external load component Q_4 becomes an unknown reaction force at the corresponding mesh node. Let us remove the fourth equation from the system (13.1) and rewrite it as

$$\begin{bmatrix} K_{11} & K_{12} & 0 & 0 \\ K_{21} & K_{22} & K_{23} & 0 \\ 0 & K_{23} & K_{33} & 0 \\ 0 & 0 & 0 & K_{55} \end{bmatrix} \begin{bmatrix} q_1 \\ q_2 \\ q_3 \\ q_5 \end{bmatrix} = \begin{bmatrix} Q_1 \\ Q_2 \\ Q_3 - K_{34}\hat{q}_4 \\ Q_5 - K_{45}\hat{q}_4 \end{bmatrix}. \tag{13.2}$$

The unknown q_4 has been eliminated and the corresponding terms containing $q_4 = \hat{q}_4$ have been moved to the right-hand side. The solution of the system (13.2), i.e. the values of q_1, q_2, q_3 and q_5, is the desired solution of the system (13.1) fulfilling the imposed boundary condition $q_4 = \hat{q}_4$.

M. Kleiber and P. Kowalczyk, *Introduction to Nonlinear Thermomechanics of Solids*, Lecture Notes on Numerical Methods in Engineering and Sciences,
DOI 10.1007/978-3-319-33455-4_13

The above approach has a disadvantage that makes programming of a solution algorithm inconvenient—it changes the original problem dimension, i.e. the size of the system of equations. Let us thus employ another approach in which we assume in Eq. (13.1) $K_{44} = 1$, $K_{34} = 0$, $K_{45} = 0$ and replace Q_4 by $\hat{q}_4$. The resulting system is

$$\begin{bmatrix} K_{11} & K_{12} & 0 & 0 & 0 \\ K_{12} & K_{22} & K_{33} & 0 & 0 \\ 0 & K_{23} & K_{33} & K_{34} & 0 \\ 0 & 0 & 0 & 1 & 0 \\ 0 & 0 & 0 & K_{45} & K_{55} \end{bmatrix} \begin{bmatrix} q_1 \\ q_2 \\ q_3 \\ q_4 \\ q_5 \end{bmatrix} = \begin{bmatrix} Q_1 \\ Q_2 \\ Q_3 \\ \hat{q}_4 \\ Q_5 \end{bmatrix} . \tag{13.3}$$

This solution meets the required condition $q_4 = \hat{q}_4$ while the size of the matrix **K** remains unchanged. However, the matrix symmetry has been lost, which is again inconvenient from the computational point of view.

To fix it, let us write down the system of Eq. (13.3) in the form combining the essential features of the formulations (13.2) and (13.3):

$$\begin{bmatrix} K_{11} & K_{12} & 0 & 0 & 0 \\ K_{12} & K_{22} & K_{23} & 0 & 0 \\ 0 & K_{23} & K_{33} & 0 & 0 \\ 0 & 0 & 0 & 1 & 0 \\ 0 & 0 & 0 & 0 & K_{55} \end{bmatrix} \begin{bmatrix} q_1 \\ q_2 \\ q_3 \\ q_4 \\ q_5 \end{bmatrix} = \begin{bmatrix} Q_1 \\ Q_2 \\ Q_3 - K_{34}\hat{q}_4 \\ \hat{q}_4 \\ Q_5 - K_{45}\hat{q}_4 \end{bmatrix} . \tag{13.4}$$

The resulting matrix has unchanged size, is symmetric and banded, and the solution of the system (13.4) fulfills the boundary condition $q_4 = \hat{q}_4$. The unknown reaction Q_4 can be computed from the equation

$$Q_n = K_{34}q_3 + K_{44}\hat{q}_4 + K_{45}q_5 , \tag{13.5}$$

cf. the fourth equation of the system (13.1).

An alternative, numerically convenient method of including boundary conditions in the FEM system consists in multiplication of the diagonal stiffness matrix components that correspond to the imposed parameter values by a certain large number κ and modification of the load vector as follows,

$$\begin{bmatrix} K_{11} & K_{12} & 0 & 0 & 0 \\ K_{12} & K_{22} & K_{23} & 0 & 0 \\ 0 & K_{23} & K_{33} & K_{34} & 0 \\ 0 & 0 & K_{34} & K_{44} \cdot \kappa & K_{45} \\ 0 & 0 & 0 & K_{45} & K_{55} \end{bmatrix} \begin{bmatrix} q_1 \\ q_2 \\ q_3 \\ q_4 \\ q_5 \end{bmatrix} = \begin{bmatrix} Q_1 \\ Q_2 \\ Q_3 \\ K_{44} \cdot \kappa\hat{q}_4 \\ Q_5 \end{bmatrix} . \tag{13.6}$$

This system ensures that the boundary condition is approximately fulfilled since for $\kappa \to \infty$ the fourth equation becomes equivalent to $q_4 = \hat{q}_4$. In practice, the value of κ cannot be too large as this would lead to ill conditioning of the matrix **K** and eventually result in a significant inaccuracy in the numerical solution of the system (13.6).

Let us stress that both the above methods preserve symmetry and banded structure of the stiffness matrix and they do not change its size. The are also both valid if $\hat{q}_4 = 0$.

Let us now discuss the methods of introducing the displacement boundary conditions to the FEM system of equations in a more general manner. Recall that the FEM equations result from minimization of a functional that expresses the potential energy of the system,

$$J[\mathbf{q}] = \frac{1}{2}\mathbf{q}^{\mathrm{T}}\mathbf{K}\mathbf{q} - \mathbf{q}^{\mathrm{T}}\mathbf{Q}, \tag{13.7}$$

by making its first variation vanish,

$$\delta J = \frac{\partial J}{\partial q_\alpha}\,\delta q_\alpha = 0,$$

which leads to

$$\mathbf{K}\mathbf{q} = \mathbf{Q}. \tag{13.8}$$

Assume that the solution $\mathbf{q} = \{q_1, q_2, \ldots, q_N\}$ is additionally required to fulfill the condition

$$q_{\hat{\alpha}} = \hat{q}_{\hat{\alpha}} \tag{13.9}$$

where $\hat{\alpha}$ is a predefined value of the index α. We shall use the Lagrange multiplier method and define

$$J^*[\mathbf{q}, \lambda] = \frac{1}{2}\mathbf{q}^{\mathrm{T}}\mathbf{K}\mathbf{q} - \mathbf{q}^{\mathrm{T}}\mathbf{Q} + \lambda(q_{\hat{\alpha}} - \hat{q}_{\hat{\alpha}}). \tag{13.10}$$

Introducing the notation

$$q_{\hat{\alpha}} = \mathbf{e}_{\hat{\alpha}}^{\mathrm{T}}\mathbf{q} = \mathbf{q}^{\mathrm{T}}\mathbf{e}_{\hat{\alpha}}\,,$$

where

$$\mathbf{e}_{\hat{\alpha}} = \{0\ldots 0\underbrace{\ 1\ }_{\hat{\alpha}}0\ldots 0\},$$

we obtain

$$\delta J^* = \frac{\partial J^*}{\partial q_\alpha}\,\delta q_\alpha + \frac{\partial J^*}{\partial \lambda}\,\delta\lambda = 0$$

i.e., since q_α and λ are mutually independent,

$$\begin{aligned} &\mathbf{K}\mathbf{q} - \mathbf{Q} + \lambda\mathbf{e}_{\hat{\alpha}} = 0 \qquad && (N\text{equations}),\\ &\mathbf{e}_{\hat{\alpha}}^{\mathrm{T}}\mathbf{q} - \hat{q}_{\hat{\alpha}} = 0 && (1\ \text{equation}). \end{aligned}$$

The resulting system of $N+1$ equations can be written in the form

$$\begin{bmatrix} \mathbf{K} & \mathbf{e}_{\hat{\alpha}} \\ \mathbf{e}_{\hat{\alpha}}^{\mathrm{T}} & \mathbf{0} \end{bmatrix} \begin{bmatrix} \mathbf{q} \\ \lambda \end{bmatrix} = \begin{bmatrix} \mathbf{Q} \\ q_{\hat{\alpha}} \end{bmatrix}. \tag{13.11}$$

Advantages of the above formulation are the preservation of the system matrix symmetry and possibility of direct computation of the multiplier λ. Note that λ has a clear physical sense—if q_α are nodal displacements then λ is the reaction force corresponding to the imposed displacement; if q_α are nodal temperatures then λ is the corresponding nodal heat flux. Disadvantages are the increased system size, increased bandwidth and computationally undesired presence of diagonal zeros in the matrix.

Employing the so-called penalty method we define

$$J^{**}[\mathbf{q}] = \frac{1}{2}\mathbf{q}^{\mathrm{T}}\mathbf{K}\mathbf{q} - \mathbf{q}^{\mathrm{T}}\mathbf{Q} + \frac{\kappa}{2}(q_{\hat{\alpha}} - \hat{q}_{\hat{\alpha}})^2, \tag{13.12}$$

where κ is a large constant such that $\kappa \gg \max_{\alpha,\beta}|K_{\alpha\beta}|$, $\alpha = 1, 2, \ldots, N$. The variation is computed as

$$\delta J^{**} = \frac{\partial J^{**}}{\partial q_\alpha}\,\delta q_\alpha = 0,$$

i.e.

$$\delta\mathbf{q}^{\mathrm{T}}\mathbf{K}\mathbf{q} - \delta\mathbf{q}^{\mathrm{T}}\mathbf{Q} + \kappa\delta\mathbf{q}^{\mathrm{T}}\mathbf{e}_{\hat{\alpha}}(\mathbf{e}_{\hat{\alpha}}^{\mathrm{T}}\mathbf{q} - \hat{q}_{\hat{\alpha}}) = 0.$$

Since the variation $\delta\mathbf{q}$ is arbitrary, the following system of N equations is obtained,

$$\left[\mathbf{K} + \kappa\mathbf{e}_{\hat{\alpha}}\mathbf{e}_{\hat{\alpha}}^{\mathrm{T}}\right]\mathbf{q} = \mathbf{Q} + \kappa\hat{q}_{\hat{\alpha}}\mathbf{e}_{\hat{\alpha}}\,. \tag{13.13}$$

The expression $\mathbf{e}_{\hat{\alpha}}\mathbf{e}_{\hat{\alpha}}^{\mathrm{T}}$ denotes a $N \times N$ matrix in which the only non-zero components is a diagonal unit located at the position $\hat{\alpha}$. Advantages of the formulation are the preservation of the matrix symmetry, size and bandwidth. A disadvantage is the difficulty in making a correct choice of the penalty parameter κ. If it is too small, the boundary condition is fulfilled inaccurately. If it is too large, the system matrix may be ill-conditioned. Typically, it is assumed that

$$\kappa = 10^3 \max_\alpha |K_{\alpha\alpha}| \qquad \text{(no summation over } \alpha).$$

Note that the two above methods may be easily generalized to the case of more than one additional condition. Let M linearly independent conditions be defined as

$$\mathbf{B}_{M\times N}\mathbf{q}_{N\times 1} = \hat{\mathbf{q}}_{M\times 1}\,, \tag{13.14}$$

where $M < N$. Making use of the Lagrange multipliers method, one may write the functional J^* in the form

$$J^{*}[\mathbf{q},\boldsymbol{\lambda}]=\frac{1}{2}\mathbf{q}^{\mathrm{T}}\mathbf{K}\,\mathbf{q}-\mathbf{q}^{\mathrm{T}}\mathbf{Q}+\boldsymbol{\lambda}^{\mathrm{T}}(\mathbf{B}\mathbf{q}-\hat{\mathbf{q}}) \tag{13.15}$$

in which $\boldsymbol{\lambda}$ is now a $M \times 1$ vector. The stationarity condition $\delta J^{*} = 0$ requires satisfaction of the following system of $N+M$ equations:

$$\begin{bmatrix}\mathbf{K} & \mathbf{B}^{\mathrm{T}}\\ \mathbf{B} & \mathbf{0}\end{bmatrix}\begin{bmatrix}\mathbf{q}\\ \boldsymbol{\lambda}\end{bmatrix}=\begin{bmatrix}\mathbf{Q}\\ \hat{\mathbf{q}}\end{bmatrix}. \tag{13.16}$$

In the penalty method we obtain

$$J^{**}[\mathbf{q}]=\frac{1}{2}\mathbf{q}^{\mathrm{T}}\mathbf{K}\mathbf{q}-\mathbf{q}^{\mathrm{T}}\mathbf{Q}+\frac{\kappa}{2}(\mathbf{B}\mathbf{q}-\hat{\mathbf{q}})^{\mathrm{T}}(\mathbf{B}\mathbf{q}-\hat{\mathbf{q}}) \tag{13.17}$$

leading to the system of N equations

$$[\mathbf{K}+\kappa\mathbf{B}^{\mathrm{T}}\mathbf{B}]\mathbf{q}=\mathbf{Q}+\kappa\mathbf{B}^{\mathrm{T}}\hat{\mathbf{q}}. \tag{13.18}$$

Let us finally note that sometimes (especially if iterative solution methods are used) the so-called augmented Lagrange function method is alternatively employed. It comprises features of the two mentioned methods. Assume that

$$\bar{J}[\mathbf{q},\boldsymbol{\lambda}]=\frac{1}{2}\mathbf{q}^{\mathrm{T}}\mathbf{K}\mathbf{q}-\mathbf{q}^{\mathrm{T}}\mathbf{Q}+\frac{\kappa}{2}(\mathbf{B}\mathbf{q}-\hat{\mathbf{q}})^{\mathrm{T}}(\mathbf{B}\mathbf{q}-\hat{\mathbf{q}})+\boldsymbol{\lambda}^{\mathrm{T}}(\mathbf{B}\mathbf{q}-\hat{\mathbf{q}}), \tag{13.19}$$

so that the system of equations assumes the form

$$\begin{bmatrix}\mathbf{K}+\kappa\mathbf{B}^{\mathrm{T}}\mathbf{B} & \mathbf{B}^{\mathrm{T}}\\ \mathbf{B} & \mathbf{0}\end{bmatrix}\begin{bmatrix}\mathbf{q}\\ \boldsymbol{\lambda}\end{bmatrix}=\begin{bmatrix}\mathbf{Q}+\kappa\mathbf{B}^{\mathrm{T}}\hat{\mathbf{q}}\\ \hat{\mathbf{q}}\end{bmatrix}. \tag{13.20}$$

13.2 Solution Methods for Linear Algebraic Equation Systems

Consider a system of N linear equations with N unknowns q_α, $\alpha = 1, 2, \ldots, N$, in the form

$$\mathbf{K}_{N\times N}\mathbf{q}_{N\times 1}=\mathbf{Q}_{N\times 1}\,. \tag{13.21}$$

Symmetry of the matrix $\mathbf{K}$ is not yet assumed. Denote by $\mathbf{K}_Q$ the rectangular $N \times (N+1)$ matrix obtained by appending the column vector $\mathbf{Q}$ to the matrix $\mathbf{K}$ as its $(N+1)$st column:

$$\begin{bmatrix}\mathbf{K}_{Q\;N\times(N+1)}\end{bmatrix}=\begin{bmatrix}\mathbf{K}_{N\times N} & \mathbf{Q}_{N\times 1}\end{bmatrix}. \tag{13.22}$$

Let us denote by $\mathrm{rank}(\mathbf{K})$, $\mathrm{rank}(\mathbf{K}_Q)$ the ranks[1] of $\mathbf{K}$ and $\mathbf{K}_Q$, respectively. According to the known theorem:

1. The system (13.21) has a solution if and only if $\mathrm{rank}(\mathbf{K}) = \mathrm{rank}(\mathbf{K}_Q)$.
2. If $\mathrm{rank}(\mathbf{K}) = \mathrm{rank}(\mathbf{K}_Q) = N$ then the solution is unique while if $\mathrm{rank}(\mathbf{K}) = \mathrm{rank}(\mathbf{K}_Q) = M < N$ then the solution set is a $(N-M)$-parameter family.

The second part of the theorem implies that in the homogeneous case, i.e. for $\mathbf{Q} = \mathbf{0}$, a nontrivial solution may only exist if $\mathrm{rank}(\mathbf{K}) < N$.

In this section, numerical methods of solving systems of equations in the form (13.21) will be presented for cases when unique solutions exist ($\det \mathbf{K} \neq 0$, i.e. $\mathrm{rank}(\mathbf{K}) = N$). They are divided into the following groups:

- direct methods,
- elimination methods,
- iterative methods.

The best known direct method is the application of the so-called Cramer formula. Let $\mathbf{K}^{\alpha}_{N\times N}$ denote a matrix constructed by replacing the αth column in the matrix $\mathbf{K}$ by the column vector $\mathbf{Q}$. The solution of the system (13.21) can be then written in the form

$$q_\alpha = \frac{\det \mathbf{K}^{\alpha}}{\det \mathbf{K}}, \qquad \alpha = 1, 2, \ldots, N. \tag{13.23}$$

This simple formula is unfortunately completely useless in computational practice because the number of elementary arithmetic operations necessary to compute each of the matrix determinants is of the order of $N!$, i.e. greatly higher than in other methods to be discussed.

13.2.1 Elimination Methods

Elimination methods are much more frequently used for the numerical solution of large systems of equations. Let us denote by $\mathbf{L}_{N\times N}$ and $\mathbf{U}_{N\times N}$ an upper and lower triangular matrix, respectively, i.e. $L_{\alpha\beta} = 0$ for $\alpha < \beta$ and $U_{\alpha\beta} = 0$ for $\alpha > \beta$. Particular cases of such matrices are $\overset{*}{\mathbf{L}}_{N\times N}$ and $\overset{*}{\mathbf{U}}_{N\times N}$, for which it is additionally assumed that their main diagonals contain only elements equal to 1, i.e. $L_{\alpha\beta} = U_{\alpha\beta} = 1$ for $\alpha = \beta$. Let us additionally denote by $\mathbf{D}_{N\times N}$ a diagonal matrix, i.e. $D_{\alpha\beta} = 0$ for $\alpha \neq \beta$.

Elimination methods employ the concept of a decomposition of the matrix $\mathbf{K}$ into a product of triangular (and, possibly, also diagonal) matrices. It can be shown that for an arbitrary square matrix $\mathbf{K}_{N\times N}$ the following unique decomposition exists,

$$\mathbf{K} = \overset{*}{\mathbf{L}}\mathbf{D}\overset{*}{\mathbf{U}}. \tag{13.24}$$

[1] The rank of a matrix $\mathbf{A}$ is the maximum number of linearly independent rows or columns it contains.

This can be alternatively written as

$$\mathbf{K} = \mathbf{L}\mathbf{U}, \tag{13.25}$$

where the matrices $\mathbf{L}$ and $\mathbf{U}$ are formed from the matrices $\overset{*}{\mathbf{L}}$ and $\overset{*}{\mathbf{U}}$ by inserting the elements of $\mathbf{D}$ into one of them (or partially into each of them). The decomposition (13.25) is more simple but obviously no longer unique.

Before we discuss methods of determining the matrices $\mathbf{L}$, $\mathbf{U}$ and $\mathbf{D}$, let us first point out advantages of their use. Introduce in the system of equations

$$\mathbf{K}\mathbf{q} \equiv \mathbf{L}\mathbf{U}\mathbf{q} = \mathbf{Q} \tag{13.26}$$

the new unknown

$$\mathbf{s} = \mathbf{U}\mathbf{q} \tag{13.27}$$

which leads to the system

$$\mathbf{L}\mathbf{s} = \mathbf{Q}. \tag{13.28}$$

Since $\mathbf{L}$ is a lower triangular matrix, the system (13.28) can be solved in a trivial way with respect to $\mathbf{s}$ by employing the forward elimination method starting from the first equation. One may symbolically express this step of the algorithm as

$$\mathbf{s} = \mathbf{L}^{-1}\mathbf{Q},$$

remembering that the explicit inversion of the matrix $\mathbf{L}$ is not necessary in this case. Unknown components of the vector $\mathbf{s}$ can be computed for subsequent values of the index α from the formula[2]

$$s_1 = \frac{Q_1}{L_{11}}, \qquad s_\alpha = \frac{1}{L_{\alpha\alpha}}\left(Q_\alpha - \sum_{\beta=1}^{\alpha-1} L_{\alpha\beta}s_\beta\right), \quad \alpha = 2, \ldots, N. \tag{13.29a}$$

Having computed the components of the vector $\mathbf{s}$, we pass to Eq. (13.27). This is a system of equations with respect to $\mathbf{q}$ with an upper triangular matrix. It can also be easily solved by employing the backward elimination method, starting from the last equation, according to the formula

$$q_N = \frac{s_N}{U_{NN}}, \quad q_\alpha = \frac{1}{U_{\alpha\alpha}}\left(s_\alpha - \sum_{\beta=\alpha+1}^{N} U_{\alpha\beta}q_\beta\right), \quad \alpha = N-1, \ldots, 1. \tag{13.29b}$$

[2]In this subsection equations, the summation convention is not used, i.e. there is no summation over α in Eq. (13.29), for instance.

Using the decomposition with the diagonal matrix (13.24) does not make the algorithm more complex—Eq. (13.29) should only be replaced by the formulae

$$s_1 = Q_1, \qquad s_\alpha = Q_\alpha - \sum_{\beta=1}^{\alpha-1} \overset{*}{L}_{\alpha\beta} s_\beta, \quad \alpha = 2, \ldots, N, \tag{13.30a}$$

$$q_N = s_N, \qquad q_\alpha = \frac{s_\alpha}{D_{\alpha\alpha}} - \sum_{\beta=\alpha+1}^{N} \overset{*}{U}_{\alpha\beta}\, q_\beta, \quad \alpha = N-1, \ldots, 1, \tag{13.30b}$$

resulting from the substitution in Eq. (13.24) of

$$\mathbf{s} = \mathbf{D}\overset{*}{\mathbf{U}}\mathbf{q}.$$

This is equivalent to solving the following sequence of equation systems with triangular matrices,

$$\overset{*}{\mathbf{L}}\mathbf{s} = \mathbf{Q}, \qquad (\mathbf{D}\overset{*}{\mathbf{U}})\mathbf{q} = \mathbf{s}. \tag{13.31}$$

It is obvious that the above methods can only succeed if all diagonal components of the matrices **L**, **U** and **D** appearing in denominators in Eqs. (13.29) and (13.30) are non-zero. Otherwise, it is easy to show that $\text{rank}(\mathbf{K}) < N$ and the solution does not exist.

Matrix decomposition methods will be presented here on an example of a symmetric matrix **K**. In such a case, Eq. (13.24) has the form

$$\mathbf{K} = \overset{*}{\mathbf{L}}\mathbf{D}\overset{*}{\mathbf{L}}^{\mathrm{T}}. \tag{13.32}$$

Components of the matrices $\overset{*}{\mathbf{L}}$ and **D** are computed in a recursive way from Eq. (13.33) employed for subsequent increasing values of α and—for a given α—for subsequent increasing values of $\beta = 1, \ldots, \alpha-1$. For $\alpha = 1$ we have

$$D_{11} = K_{11}, \qquad \overset{*}{L}_{11} = 1, \tag{13.33a}$$

while for subsequent $\alpha = 2, \ldots, N$,

$$\begin{aligned} D_{\alpha\alpha} &= K_{\alpha\alpha} - \sum_{\gamma=1}^{\alpha-1} \overset{*2}{L}_{\alpha\gamma} D_{\gamma\gamma}, \\ L_{\alpha\alpha} &= 1, \\ L_{\alpha\beta} &= \frac{1}{D_{\alpha\alpha}} \left(K_{\alpha\beta} - \sum_{\gamma=1}^{\beta-1} \overset{*}{L}_{\alpha\gamma} \overset{*}{L}_{\beta\gamma} D_{\gamma\gamma} \right), \qquad \beta = 1, 2, \ldots, \alpha-1. \end{aligned} \tag{13.33b}$$

Elimination methods are generally much more efficient than direct methods. The above presented algorithm requires $\frac{1}{6}N^3+N^2$ arithmetic operations of multiplication and division (since addition/subtraction operations are significantly faster for a computer than multiplication/division operations, only the latter are used to compare efficiency of specific methods). It is noteworthy that in most problems of mechanics the matrix $\mathbf{K}$ is a banded matrix whose most components (and, consequently, also components of $\overset{*}{L}_{\alpha\beta}$) are zero. This fact significantly decreases the effective number of arithmetic operations necessary to determine the solution.

If all diagonal elements of the matrix $\mathbf{D}$ are positive then one may denote $\mathbf{L} = \overset{*}{\mathbf{L}}\sqrt{\mathbf{D}}$ and come at the yet simpler—and unique in this case—decomposition of $\mathbf{K}$ (the so-called Cholesky decomposition):

$$\mathbf{K} = \mathbf{L}\mathbf{L}^{\mathrm{T}}. \tag{13.34}$$

The lower triangular matrix components $L_{\alpha\beta}$ can be recursively computed as follows. For $\alpha = 1$ we have

$$L_{11} = K_{11}\,, \tag{13.35a}$$

while for subsequent $\alpha = 2, \ldots, N$,

$$\overset{*}{L}_{\alpha\alpha} = \left(K_{\alpha\alpha} - \sum_{\beta=1}^{\alpha-1} L_{\alpha\beta}^2\right)^{\frac{1}{2}}, \tag{13.35b}$$

$$\overset{*}{L}_{\alpha\beta} = \frac{1}{L_{\beta\beta}}\left(K_{\alpha\beta} - \sum_{\gamma=1}^{\beta-1} L_{\alpha\gamma}L_{\beta\gamma}\right), \qquad \beta = 2, \ldots, \alpha-1.$$

Application of the Cholesky method to a full matrix $\mathbf{K}$ requires $\frac{1}{6}N^3 + \frac{3}{2}N^2 + \frac{1}{2}N$ multiplication and division operations and N square root operations—slightly more than in the previous case. Another elimination method, commonly known as the Gauss method, requires $\frac{1}{3}N^3+N^2+\frac{1}{3}N$ operations, which is still less favourable.

13.2.2 Iterative Methods

Yet another group of algorithms for solving linear systems of algebraic equations is formed by the iterative methods. They consist in seeking—given an approximate solution value—another, better approximation, with the use of a much less time consuming algorithms than the above presented elimination methods.

Let us express the considered system of equations in the following residual form,

$$\mathbf{r} = \mathbf{K}\mathbf{q} - \mathbf{Q} = \mathbf{0}. \tag{13.36}$$

Consider an iterative solution method based on the formula

$$\mathbf{q}^{i+1} = \mathbf{q}^{i} + \eta^{i}\boldsymbol{\delta}^{i}, \tag{13.37}$$

in which i is the iteration counter, η^i is a real number and $\boldsymbol{\delta}^i$ is the ith corrector to the approximate solution. Defining the ith residual vector as

$$\mathbf{r}^{i} = \mathbf{K}\mathbf{q}^{i} - \mathbf{Q}, \tag{13.38}$$

one may express $\boldsymbol{\delta}^i$ as

$$\boldsymbol{\delta}^{i} = -\alpha^{i}\tilde{\mathbf{K}}^{-1}\mathbf{r}^{i} + \beta^{i}\boldsymbol{\delta}^{i-1}, \tag{13.39}$$

in which α, β are parameters of the algorithm while $\tilde{\mathbf{K}}$ is an "approximate" coefficient matrix of the equation system. Note that the substitution

$$\tilde{\mathbf{K}} = \mathbf{K}, \qquad \alpha^{i} = \eta^{i} = 1, \qquad \beta^{i} = 0, \tag{13.40}$$

leads for $\mathbf{q}^i = \mathbf{0}$ to the relation

$$\mathbf{q} = -\mathbf{K}^{-1}(-\mathbf{Q}) = \mathbf{K}^{-1}\mathbf{Q}, \tag{13.41}$$

i.e. to the direct solution of the system (13.29b).

Employing iterative methods based on the above relations make only sense if the matrix $\tilde{\mathbf{K}}$ is much easier to invert or decompose into triangular matrices than the matrix $\mathbf{K}$.

The simplest iterative method is the Jacobi method in which

$$\alpha^{i} = \eta^{i} = 1, \qquad \beta^{i} = 0, \qquad \tilde{\mathbf{K}} = \mathbf{I} \tag{13.42}$$

($\mathbf{I}$ is a unit matrix), i.e.

$$\boldsymbol{\delta}^{i} = -\mathbf{r}^{i}, \tag{13.43}$$

and the subsequent, $(i+1)$st approximate solution is given by Eq. (13.37) as

$$\mathbf{q}^{i+1} = \mathbf{q}^{i} - \mathbf{r}^{i}. \tag{13.44}$$

The Jacobi iteration algorithm can be also expressed as follows. Let us decompose the matrix $\mathbf{K}$ into a sum of the matrices $\mathbf{K}-\mathbf{I}$ and $\mathbf{I}$. The product $\mathbf{Kq}$ can be thus expressed as $(\mathbf{K}-\mathbf{I})\mathbf{q} + \mathbf{q}$. We have then

$$\mathbf{q} = \mathbf{Q} - (\mathbf{K}-\mathbf{I})\mathbf{q},$$

which implies the following recursive formula

$$\mathbf{q}^{i+1} = \mathbf{Q} - (\mathbf{K}-\mathbf{I})\mathbf{q}^{i} = \underbrace{\mathbf{Q} - \mathbf{K}\mathbf{q}^{i}}_{-\mathbf{r}^{i}} + \mathbf{q}^{i}. \tag{13.45}$$

Unfortunately, this algorithm is frequently not convergent. Somewhat better properties can be achieved by assuming, cf. Eq. (13.42),

$$\alpha^{i} = \eta^{i} = 1, \qquad \beta^{i} = 0, \qquad \tilde{\mathbf{K}} = \mathbf{D}, \tag{13.46}$$

where $\mathbf{D}$ is a diagonal matrix whose components equal diagonal components of $\mathbf{K}$, i.e. $D_{\alpha\alpha} = K_{\alpha\alpha}$. We have then

$$\boldsymbol{\delta}^{i} = -\mathbf{D}^{-1}\mathbf{r}^{i}, \tag{13.47}$$

and the subsequent, $(i+1)$st approximate solution is given as

$$\mathbf{q}^{i+1} = \mathbf{q}^{i} - \mathbf{D}^{-1}\mathbf{r}^{i}. \tag{13.48}$$

In other words, the matrix $\mathbf{K}$ is decomposed into a sum of matrices $\mathbf{K}-\mathbf{D}$ and $\mathbf{D}$ and the system of equations is then expressed in the form

$$\mathbf{D}\mathbf{q} = \mathbf{Q} - (\mathbf{K}-\mathbf{D})\mathbf{q},$$

i.e.

$$\mathbf{q} = \mathbf{D}^{-1}[\mathbf{Q} - (\mathbf{K}-\mathbf{D})\mathbf{q}].$$

This yields the following recursive formula

$$\mathbf{q}^{i+1} = \mathbf{D}^{-1}[\mathbf{Q} - (\mathbf{K}-\mathbf{D})\mathbf{q}^{i}] = \mathbf{q}^{i+1} = \mathbf{D}^{-1}\underbrace{[\mathbf{Q} - \mathbf{K}\mathbf{q}^{i}]}_{-\mathbf{r}^{i}} + \mathbf{q}^{i}. \tag{13.49}$$

Further improvement of the Jacobi iteration algorithm can be achieved by introducing the so-called *line search* routine in order to determine in each iteration the optimum step length η^{i}, cf. Eq. (13.37). In other words, we assume

$$\alpha^{i} = 1, \qquad \beta^{i} = 0, \qquad \tilde{\mathbf{K}} = \mathbf{I}, \qquad \eta^{i} = ?\,, \tag{13.50}$$

i.e.

$$\mathbf{q}^{i+1} = \mathbf{q}^{i} + \eta^{i}\boldsymbol{\delta}^{i} = \mathbf{q}^{i} - \eta^{i}\mathbf{r}^{i}. \tag{13.51}$$

Let us recall the fact that the considered system of equations is a minimum condition for the functional J in the form

$$J[\mathbf{q}] = \frac{1}{2}\mathbf{q}^{\mathrm{T}}\mathbf{K}\mathbf{q} - \mathbf{q}^{\mathrm{T}}\mathbf{Q}. \tag{13.52}$$

Note that the residual vector $\mathbf{r}$ defined in Eq. (13.36) is the gradient of this functional:

$$\mathbf{r} = \mathbf{K}\mathbf{q} - \mathbf{Q} = \frac{\mathrm{d}J}{\mathrm{d}\mathbf{q}^{\mathrm{T}}}\,. \tag{13.53}$$

Let us write down the value of J in Eq. (13.52) after the $(i+1)$st iteration i.e., cf. Eq. (13.51),

$$\begin{aligned}
J[\mathbf{q}^{i+1}] &= \frac{1}{2}(\mathbf{q}^i + \eta^i\boldsymbol{\delta}^i)^{\mathrm{T}}\mathbf{K}(\mathbf{q}^i + \eta^i\boldsymbol{\delta}^i) - (\mathbf{q}^i + \eta^i\boldsymbol{\delta}^i)^{\mathrm{T}}\mathbf{Q}\\
&= \frac{1}{2}\mathbf{q}^{i\mathrm{T}}\mathbf{K}\mathbf{q}^i - \mathbf{q}^{i\mathrm{T}}\mathbf{Q} + \frac{1}{2}\eta^i\left[\mathbf{q}^{i\mathrm{T}}\mathbf{K}\boldsymbol{\delta}^i + \boldsymbol{\delta}^{i\mathrm{T}}\mathbf{K}\mathbf{q}^i\right]\\
&\quad - \eta^i\boldsymbol{\delta}^{i\mathrm{T}}\mathbf{Q} + \frac{1}{2}\eta^{i2}\boldsymbol{\delta}^i\mathbf{K}\boldsymbol{\delta}^i\\
&= \underbrace{\frac{1}{2}\mathbf{q}^{i\mathrm{T}}\mathbf{K}\mathbf{q}^i - \mathbf{q}^{i\mathrm{T}}\mathbf{Q}}_{J[\mathbf{q}^i]} + \eta^i\boldsymbol{\delta}^{i\mathrm{T}}\underbrace{(\mathbf{K}\mathbf{q}^i - \mathbf{Q})}_{\mathbf{r}^i} + \frac{1}{2}\eta^{i2}\boldsymbol{\delta}^{i\mathrm{T}}\mathbf{K}\boldsymbol{\delta}^i\\
&= J[\mathbf{q}^i] + \eta^i\boldsymbol{\delta}^{i\mathrm{T}}\mathbf{r}^i + \frac{1}{2}\eta^{i2}\boldsymbol{\delta}^{i\mathrm{T}}\mathbf{K}\boldsymbol{\delta}^i.
\end{aligned} \tag{13.54}$$

Since the vector $\mathbf{q}^i$ is known from the previous iteration and the direction $\boldsymbol{\delta}^i$ is known from Eq. (13.51), i.e.

$$\boldsymbol{\delta}^i = -\mathbf{r}^i, \tag{13.55}$$

the minimum condition for the potential energy $J[\mathbf{q}^{i+1}]$ in the $(i+1)$th iteration can be written in the form

$$\frac{\partial J}{\partial \eta^i} = -\mathbf{r}^{i\mathrm{T}}\mathbf{r}^i + \eta^i\mathbf{r}^{i\mathrm{T}}\mathbf{K}\mathbf{r}^i = 0, \tag{13.56}$$

as η^i is the only unknown here. In other words, the scalar quantity η^i

$$\eta^i = \frac{\mathbf{r}^{i\mathrm{T}}\mathbf{r}^i}{\mathbf{r}^{i\mathrm{T}}\mathbf{K}\mathbf{r}^i} \tag{13.57}$$

minimizes the potential energy $J[\mathbf{q}^{i+1}]$.[3]

---

[3]Note: the term "line search" refers in fact to nonlinear problems—here, instead of performing the search, one needs to only determine the value of η^i from Eq. (13.57).

Note that the method is based on Eq. (13.54) in which choice of the direction $\boldsymbol{\delta}$ has not been defined—Eq. (13.55) is just one of many other choices. Another possible assumption might be the one given in Eq. (13.47).

The Jacobi iterative method including the line search routine is called in the literature the *steepest descent method.*

It is convenient to transform Eq. (13.57) to another form. Let us assume that the vector $\mathbf{r}^{i+1}$ has been computed for $\eta^i = 1$ as

$$\mathbf{r}^{i+1}|_{\eta^i=1} = \mathbf{K}\underbrace{(\mathbf{q}^i + \boldsymbol{\delta}^i)}_{\mathbf{q}^{i+1}|_{\eta^i=1}} - \mathbf{Q}. \tag{13.58}$$

Subtracting Eq. (13.38) from Eq. (13.58), one obtains

$$\mathbf{r}^{i+1}|_{\eta^i=1} - \mathbf{r}^i = \mathbf{K}\boldsymbol{\delta}^i, \tag{13.59}$$

which allows to write Eq. (13.57) (without making use of Eq. (13.55)) in the form

$$\eta^i = -\frac{\boldsymbol{\delta}^{i\mathrm{T}}\mathbf{r}^i}{\boldsymbol{\delta}^{i\mathrm{T}}\mathbf{K}\boldsymbol{\delta}^i}. \tag{13.60}$$

The last iterative method of solving linear systems of algebraic equations discussed in this section is the *conjugate gradient method.* Up to now we used to assume in Eqs. (13.37) and (13.39) that $\beta^i = 0$, limiting ourselves to one-parameter iterative methods (with the parameter α^i). The conjugate gradient method is a two-parameter method in which it is assumed that

$$\begin{cases} \alpha^i = 1, \quad \tilde{\mathbf{K}} = \mathbf{I}, \quad \eta^i = ? \quad \text{(line search)} \\ \beta^i \qquad\qquad\qquad\qquad \text{resulting from } \boldsymbol{\delta}^{i\mathrm{T}}\mathbf{K}\boldsymbol{\delta}^j = 0 \text{ for } i \neq j. \end{cases} \tag{13.61}$$

We have thus this time

$$\mathbf{q}^{i+1} = \mathbf{q}^i + \eta^i\boldsymbol{\delta}^i = \mathbf{q}^i - \eta^i(\mathbf{r}^i - \beta^i\boldsymbol{\delta}^{i-1}), \tag{13.62}$$

where the value η^i is given as before by Eq. (13.60). The parameter β^i can be defined in a variety of ways—in FEM the most frequent choice is the Hestenes–Steifel formula:

$$\beta_i = \frac{\mathbf{r}^{i\mathrm{T}}\,\mathbf{r}^i}{\mathbf{r}^{i-1\mathrm{T}}\,\mathbf{r}^{i-1}}. \tag{13.63}$$

It can be shown that the so-defined value of β^i ensures $\mathbf{K}$-orthogonality of the vector $\boldsymbol{\delta}^i$ and fulfillment of the condition

$$\mathbf{r}^{i+1\mathrm{T}}\,\mathbf{r}^k = 0 \qquad \text{for } k = 0, 1, \ldots, i, \tag{13.64}$$

justifying the name of the method (note that $\mathbf{r}$ is the gradient of the functional J),

13.3 Multigrid Methods

While discussing iterative methods of solving large linear systems of algebraic equations, one should not forget mentioning quite a natural approach in which more than one, sparse and dense discretization meshes are used. The dense mesh may be either a result of enrichment of the sparse one or defined independently. The idea of the *multigrid method* consists in

1. computing the solution corresponding to the sparse mesh (i.e. solving the problem with a relatively low number of unknowns),
2. estimating on this basis the approximate solution corresponding to the dense mesh,
3. apply an iterative solution method to the dense mesh.

Let the system of equations for the sparse mesh has the form

$$\tilde{\mathbf{K}}_{M\times M}\tilde{\mathbf{q}}_{M\times 1} = \tilde{\mathbf{Q}}_{M\times 1}, \tag{13.65}$$

and the analogous system for the dense mesh

$$\mathbf{K}_{N\times N}\mathbf{q}_{N\times 1} = \mathbf{Q}_{N\times 1} \tag{13.66}$$

(it is assumed that M is significantly smaller than N). Let us define a matrix $\mathbf{P}_{N\times M}$ that allows to determine approximate values of the dense mesh solution given the sparse mesh solution:

$$\mathbf{q}_{N\times 1} = \mathbf{P}_{N\times M}\tilde{\mathbf{q}}_{M\times 1}. \tag{13.67}$$

The simplest version of the multigrid method is expressed by the following algorithm:

1. solution of the system $\tilde{\mathbf{q}} = \tilde{\mathbf{K}}^{-1}\tilde{\mathbf{Q}}$,
2. transformation $\mathbf{q}^0 = \mathbf{P}\tilde{\mathbf{q}}$,
3. iterative solution of the system $\mathbf{Kq} = \mathbf{Q}$ with the vector $\mathbf{q}^0$ as the first approximate.

13.4 Solution Methods for Nonlinear Algebraic Equation Systems

While solving a nonlinear time-dependent FEM problem, one usually assumes that at a typical time instant t internal and external forces acting on the system nodal points remain in equilibrium up to an acceptable error level assumed in the iterative method, i.e.

$$\mathbf{F}^t(\mathbf{q}^t) \approx \mathbf{Q}^t, \tag{13.68}$$

which can also be written in a residual form as

$$\mathbf{r}^t = \mathbf{Q}^t - \mathbf{F}^t \approx \mathbf{0}. \tag{13.69}$$

At a subsequent time instant $t+\Delta t$ we have

$$\mathbf{F}^{t+\Delta t}(\mathbf{q}^t, \Delta\mathbf{q}) = \mathbf{Q}^{t+\Delta t} \tag{13.70}$$

i.e.

$$\mathbf{r}^{t+\Delta t} = \mathbf{Q}^{t+\Delta t} - \mathbf{F}^{t+\Delta t} = \mathbf{0}. \tag{13.71}$$

Let us stress that the dependence of the vector $\mathbf{F}^t$ on $\mathbf{q}^t$ is expressed here in only a symbolic way, since for inelastic materials the internal forces may generally depend on the entire history of the deformation process and not on just the final values of generalized coordinates. In an incremental problem, the vector $\mathbf{F}^t$ is evaluated on the basis of the current (at the time instant t) stress state $\boldsymbol{\sigma}$ as

$$\mathbf{F}^t = \int_\Omega \mathbf{B}^{\mathrm{T}} \boldsymbol{\sigma}^t \mathrm{d}V, \tag{13.72}$$

while at $t+\Delta t$ this vector is expressed as

$$\mathbf{F}^{t+\Delta t}(\mathbf{q}^t, \Delta\mathbf{q}) = \mathbf{F}^t(\mathbf{q}^t) + \Delta\mathbf{F}(\mathbf{q}^t, \Delta\mathbf{q}), \tag{13.73}$$

$$\Delta\mathbf{F}(\mathbf{q}^t, \Delta\mathbf{q}) \quad = \hat{\mathbf{K}}(\mathbf{q}^t, \Delta\mathbf{q})\, \Delta\mathbf{q}, \tag{13.74}$$

where $\hat{\mathbf{K}}$ is an appropriate, unknown at t secant matrix in the considered time interval (the secant stiffness matrix in mechanical problems, for instance).

The system (13.70) can be solved with the use of the iterative Newton–Raphson procedure, presented in Fig. 13.1 for a system with one degree of freedom (one unknown). The key role in this method is played by the algorithmic tangent stiffness matrix $\mathbf{K}$ defined as

$$\bar{\mathbf{K}}^{t+\Delta t} = -\frac{\mathrm{d}\mathbf{r}^{t+\Delta t}}{\mathrm{d}\mathbf{q}^{t+\Delta t}} = \frac{\mathrm{d}\mathbf{F}^{t+\Delta t}}{\mathrm{d}\mathbf{q}^{t+\Delta t}} = \frac{\mathrm{d}\mathbf{F}^{t+\Delta t}}{\mathrm{d}\Delta\mathbf{q}} = \frac{\mathrm{d}\Delta\mathbf{F}}{\mathrm{d}\Delta\mathbf{q}}. \tag{13.75}$$

Given an approximate solution $\Delta\mathbf{q}^{(i)}$ computed at the ith iteration, one can determine the ith approximate of the vectors $\mathbf{F}^{t+\Delta t}$, $\mathbf{r}^{t+\Delta t}$ and the matrix $\mathbf{K}^{t+\Delta t}$ as

$$\mathbf{F}^{(i)} = \mathbf{F}^{t+\Delta t}(\mathbf{q}^t, \Delta\mathbf{q}^{(i)}), \qquad \mathbf{r}^{(i)} = \mathbf{Q}^{t+\Delta t} - \mathbf{F}^{(i)}, \qquad \bar{\mathbf{K}}^{(i)} = \left.\frac{\mathrm{d}\Delta\mathbf{F}}{\mathrm{d}\Delta\mathbf{q}}\right|_{\Delta\mathbf{q}=\Delta\mathbf{q}^{(i)}}$$

and then compute the solution corrector $\delta\mathbf{q}^{(i+1)}$ and the subsequent approximate $\Delta\mathbf{q}^{(i+1)}$ from the following linear system of equations:

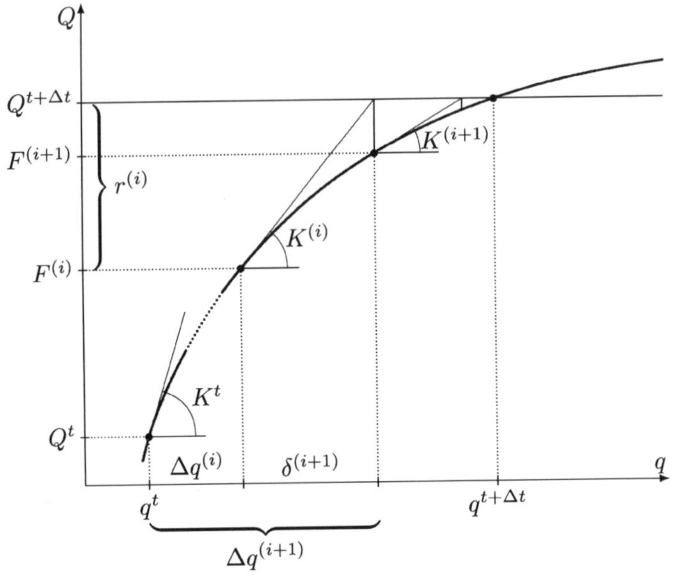

Fig. 13.1 Newton–Raphson method

$$\bar{\mathbf{K}}^{(i)} \delta \mathbf{q}^{(i+1)} = \mathbf{r}^{(i)} , \tag{13.76}$$

$$\Delta \mathbf{q}^{(i+1)} = \Delta \mathbf{q}^{(i)} + \delta \mathbf{q}^{(i+1)} . \tag{13.77}$$

The initial approximates can be assumed as, for instance,

$$\delta \mathbf{q}^{(0)} = \mathbf{0}, \qquad \mathbf{F}^{(0)} = \mathbf{F}^t , \qquad \bar{\mathbf{K}}^{(0)} = \bar{\mathbf{K}}^t , \tag{13.78}$$

where $\bar{\mathbf{K}}^t$ is the stiffness matrix explicitly computed at the time instant t. Iterations are repeated until

$$||\delta \mathbf{q}^{(i)}|| < \varepsilon_1 , \tag{13.79}$$

or

$$||{}^{t+\Delta t}\mathbf{r}^{(i)}|| < \varepsilon_2 , \tag{13.80}$$

where ε_1 and ε_2 are predefined parameters of the routine.

The Newton–Raphson method in the above formulation requires solving the linear system of equations (13.76) (i.e. performing triangularization of the stiffness matrix or applying iterative solution algorithms) at each iteration during each time step of the analysis. In order to decrease the computational cost, a modified Newton–Raphson method is frequently used instead, in which triangularization of the stiffness matrix is necessary only once, i.e. at the first iteration. Equation (13.76) is replaced by

$$\bar{\mathbf{K}}^{(0)} \delta \mathbf{q}^{(i+1)} = \mathbf{r}^{(i)} . \tag{13.81}$$

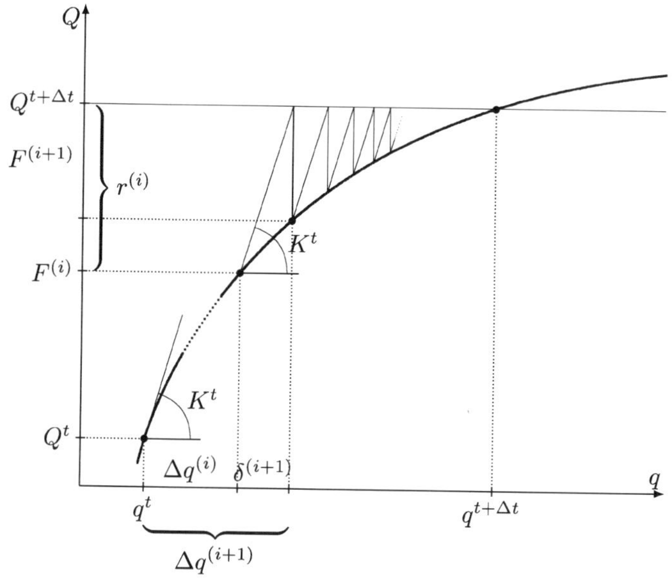

Fig. 13.2 Modified Newton–Raphson method

The modified procedure is illustrated in Fig. 13.2 for a system with one degree of freedom. The price to pay for a lower cost of the method are its much worse convergence properties. It is worth mentioning that the modified method can have several variants. For instance, one may make use of the same matrix $\bar{\mathbf{K}}$ in a number of time steps or—to make convergence faster—update the matrix after a number of iterations during the same time step.

Currently, much attention is also paid to the so-called quasi-Newton methods, which form an approach comprising features of both the full and the modified Newton–Raphson method. In this approach, instead of the tangent matrix, a secant matrix (more strictly—its certain approximate) is used in the iterative procedure. The matrix is updated at each iteration with a very convenient method that makes it unnecessary to inverse the modified matrix while solving the system of equations. This is due to the following relation,

$$\left(\bar{\mathbf{K}}^{(i)}\right)^{-1} = \mathbf{A}^{(i)\mathrm{T}} \left(\bar{\mathbf{K}}^{(i-1)}\right)^{-1} \mathbf{A}^{(i)}, \tag{13.82}$$

which allows to compute the inverse matrix at the ith iteration given the inverse matrix at the previous iteration, with the help of the $N \times N$ matrix $\mathbf{A}^{(i)}$ defined as

$$\mathbf{A}^{(i)} = \mathbf{I} + \mathbf{v}^{(i)}\mathbf{w}^{(i)\mathrm{T}}, \tag{13.83}$$

where

$$\mathbf{v}^{(i)} = \left(\frac{\delta\mathbf{q}^{(i)\mathrm{T}}\boldsymbol{\gamma}^{(i)}}{\delta\mathbf{q}^{(i)\mathrm{T}}\,\bar{\mathbf{K}}^{(i-1)}\delta\mathbf{q}^{(i)}} \right)^{\frac{1}{2}} \bar{\mathbf{K}}^{(i-1)}\delta\mathbf{q}^{(i)} - \boldsymbol{\gamma}^{(i)},$$

$$\mathbf{w}^{(i)} = \frac{\delta\mathbf{q}^{(i)}}{\delta\mathbf{q}^{(i)\mathrm{T}}\boldsymbol{\gamma}^{(i)}}$$

and

$$\delta\mathbf{q}^{(i)} = \Delta\mathbf{q}^{(i)} - \Delta\mathbf{q}^{(i-1)}, \qquad \boldsymbol{\gamma}^{(i)} = \mathbf{F}^{(i)} - \mathbf{F}^{(i-1)} = \mathbf{r}^{(i-1)} - \mathbf{r}^{(i)}.$$

All quantities necessary for efficient application of Eqs. (13.82) and (13.83) can be determined from the results of computations at the iteration $(i-1)$.

13.5 Solution Methods for Linear and Nonlinear Systems of First Order Ordinary Differential Equations

Equations of this type usually appear in heat conduction problems. Consider first the linear case, i.e. the system of N linear equations in the form, cf. Eq. (11.7),

$$\mathbf{C}\dot{\mathbf{q}} + \mathbf{K}\mathbf{q} = \mathbf{Q}, \tag{13.84}$$

where $\mathbf{q}(\tau)$ is a N-dimensional vector of time-dependent generalized coordinates of the discrete system. The initial condition has the form

$$\mathbf{q}(0) = \hat{\mathbf{q}}^0. \tag{13.85}$$

There exists an analytical solution of the initial value problem (13.84) and (13.85) of the form

$$\mathbf{q}(\tau) = e^{-\tau}\mathbf{C}^{-1}\mathbf{K}\hat{\mathbf{q}}_0 + \int_0^\tau e^{-(\tau-\tau')\mathbf{C}^{-1}\mathbf{K}}\,\mathbf{C}^{-1}\mathbf{Q}(\tau')\,\mathrm{d}\tau'. \tag{13.86}$$

Since

$$e^{\mathbf{A}} = I + \mathbf{A} + \cdots + \frac{\mathbf{A}^n}{n!} + \cdots \tag{13.87}$$

while for

$$f(\tau) = \int_0^\tau F(\tau,\tau')\,\mathrm{d}\tau' \tag{13.88}$$

the following equation holds true,

$$\frac{\mathrm{d}f(\tau)}{\mathrm{d}\tau} = F(\tau,\tau) + \int_0^{\tau} \frac{\partial F}{\partial \tau}(\tau,\tau')\,\mathrm{d}\tau', \tag{13.89}$$

it is straightforward to show by direct substitution that (13.86) is indeed the solution of Eqs. (13.84) and (13.85).

Since the formula (13.87) is very troublesome in practical use, the solution of Eq. (13.84) is usually obtained in a different way. The most frequent one-step scheme of time discretization of Eq. (13.84) is the generalized trapezoidal rule defined as

$$\mathbf{C}\dot{\mathbf{q}}^{t+\Delta t} + \mathbf{K}\mathbf{q}^{t+\Delta t} = \mathbf{Q}^{t+\Delta t}, \tag{13.90}$$

$$\mathbf{q}^{t+\Delta t} = \mathbf{q}^{t} + \Delta t[(1-\theta)\dot{\mathbf{q}}^{t} + \theta\dot{\mathbf{q}}^{t+\Delta t}]. \tag{13.91}$$

The parameter θ assumes values from the range $0 \le \theta \le 1$. Its various characteristic values correspond to different known finite difference schemes such as

- $\theta = 0$—forward difference method,
- $\theta = \frac{1}{2}$—central difference (Crank–Nicholson) method,
- $\theta = 1$—backward difference method.

For $\theta \neq 0$ the considered method is called implicit. Substituting Eq. (13.91) into Eq. (13.90), one arrives at the system of equations

$$(\mathbf{C} + \theta\Delta t\mathbf{K})\,\dot{\mathbf{q}}^{t+\Delta t} = \mathbf{Q}^{t+\Delta t} - \mathbf{K}\left[\mathbf{q}^{t} + (1-\theta)\Delta t\dot{\mathbf{q}}^{t}\right], \tag{13.92}$$

from which the vector $\dot{\mathbf{q}}^{t+\Delta t}$ can be computed with the help of any of the solution methods discussed in Sect. 13.2 for linear systems of algebraic equations. Note that the right-hand side of the system (13.92) is known at the beginning of the time step, i.e. at the time instant t. Next, the vector $\mathbf{q}^{t+\Delta t}$ is computed directly from Eq. (13.91).

For $\theta = 0$, i.e. in the explicit method, the procedure is more simple and consists in subsequent determining the quantities

$$\mathbf{q}^{t+\Delta t} = \mathbf{q}^{t} + \Delta t\dot{\mathbf{q}}^{t}, \tag{13.93}$$

$$\mathbf{C}\dot{\mathbf{q}}^{t+\Delta t} = \mathbf{Q}^{t+\Delta t} - \mathbf{K}\mathbf{q}^{t+\Delta t}. \tag{13.94}$$

Computation of $\dot{\mathbf{q}}^{t+\Delta t}$ from Eq. (13.94) requires again solution of a linear system of equations with $\mathbf{C}$ being its coefficient matrix. In many problems this matrix appears to have approximately a diagonal form, in which case the solution is a trivial task.

Another approach to solution of the system (13.84) and (13.85) is its modal analysis. Assume $\mathbf{Q}(\tau) = \mathbf{0}$ and the solution in the form

$$\mathbf{q}(\tau) = \boldsymbol{\phi}e^{-\lambda\tau} \tag{13.95}$$

where $\boldsymbol{\phi}$ denotes a vector of coefficients. Substituting Eq. (13.95) into Eq. (13.84) one obtains

$$\mathbf{C}(-\lambda)\boldsymbol{\phi}\, e^{-\lambda\tau} + \mathbf{K}\boldsymbol{\phi}\, e^{-\lambda\tau} = \mathbf{0}, \tag{13.96}$$

i.e., since $e^{-\lambda\tau} \neq 0$,

$$\mathbf{K}\boldsymbol{\phi} = \lambda\mathbf{C}\boldsymbol{\phi}. \tag{13.97}$$

This equation defines the so-called generalized eigenproblem whose solution are N pairs $(\lambda_\alpha,\ \boldsymbol{\phi}_\alpha)$, $\alpha = 1, 2, \ldots, N$.

Since the matrices $\mathbf{C}$ and $\mathbf{K}$ are symmetric and positive definite, the eigenvalues $\lambda_1, \ \ldots, \ \lambda_N$ are real and positive and the eigenvectors $\boldsymbol{\phi}_1, \ \ldots, \ \boldsymbol{\phi}_N$ are $\mathbf{C}$-orthonormal, i.e. they fulfill the conditions

$$\boldsymbol{\phi}_\alpha^{\mathrm{T}}\mathbf{C}\boldsymbol{\phi}_\beta = \delta_{\alpha\beta}\,, \tag{13.98}$$

which imply

$$\boldsymbol{\phi}_\alpha^{\mathrm{T}}\mathbf{K}\boldsymbol{\phi}_\beta = \lambda_\alpha\delta_{\alpha\beta} \qquad \text{(no summation over } \alpha). \tag{13.99}$$

It will be assumed below that

$$\lambda_1 < \lambda_2 < \cdots < \lambda_N\,, \tag{13.100}$$

i.e., for simplicity, the case of multiple eigenvalues will be neglected. Let us denote

$$\boldsymbol{\Lambda}_{N\times N} = \begin{bmatrix} \lambda_1 & & & \\ & \lambda_2 & & \\ & & \ddots & \\ & & & \lambda_N \end{bmatrix},$$

$$\tilde{\boldsymbol{\Phi}}_{N\times N} = \begin{bmatrix} \boldsymbol{\phi}_1 & \boldsymbol{\phi}_2 & \ldots & \boldsymbol{\phi}_N \end{bmatrix} = \begin{bmatrix} \phi_{11} & \phi_{21} & \ldots & \phi_{N1} \\ \phi_{12} & \phi_{22} & \ldots & \phi_{N2} \\ \vdots & \vdots & \ddots & \vdots \\ \phi_{1N} & \phi_{2N} & \ldots & \phi_{NN} \end{bmatrix},$$
$$\qquad\qquad\qquad\qquad\underbrace{\phantom{\phi_{11}}}_{\boldsymbol{\phi}_1}\ \underbrace{\phantom{\phi_{21}}}_{\boldsymbol{\phi}_2}\ \ \underbrace{\phantom{\phi_{N1}}}_{\boldsymbol{\phi}_N}$$

and consequently write the conditions (13.98) and (13.99) in the form

$$\tilde{\boldsymbol{\Phi}}^{\mathrm{T}}\mathbf{C}\tilde{\boldsymbol{\Phi}} = \mathbf{I}, \qquad \tilde{\boldsymbol{\Phi}}^{\mathrm{T}}\mathbf{K}\tilde{\boldsymbol{\Phi}} = \boldsymbol{\Lambda}. \tag{13.101}$$

Let us now transform the system (13.84) and (13.85) assuming

$$\mathbf{q}(\tau) = \tilde{\boldsymbol{\Phi}}\mathbf{x}(\tau), \qquad \mathbf{X}(\tau) = \tilde{\boldsymbol{\Phi}}^{\mathrm{T}}\mathbf{Q}(\tau), \qquad \hat{\mathbf{x}}^0 = \tilde{\boldsymbol{\Phi}}^{\mathrm{T}}\mathbf{C}\hat{\mathbf{q}}^0. \tag{13.102}$$

Left-multiplication of Eq. (13.84) by $\tilde{\mathbf{\Phi}}^{\mathrm{T}}$ and Eq. (13.85) by $\tilde{\mathbf{\Phi}}^{\mathrm{T}}\mathbf{C}$ leads to

$$\dot{\mathbf{x}}(\tau) + \mathbf{\Lambda}\mathbf{x}(\tau) = \mathbf{X}(\tau), \tag{13.103}$$

$$\mathbf{x}(0) = \hat{\mathbf{x}}^0. \tag{13.104}$$

Since the matrix $\mathbf{\Lambda}$ is diagonal, the system of equations (13.103) and (13.104) consists of N independent scalar equations

$$\dot{x}_\alpha(\tau) + \lambda_\alpha x_\alpha(\tau) = X_\alpha(\tau) \qquad \text{(no summation over } \alpha) \tag{13.105}$$

with the initial conditions

$$x_\alpha(0) = \hat{X}_\alpha\,. \tag{13.106}$$

Having solved each of the N equations (13.105), we can compute $\mathbf{q}(\tau)$ from Eq. (13.102)—note, however, that the generalized eigenproblem (13.97) has to be solved first. The scalar equations (13.105) can be solved with one of known methods, either analytical, cf. Eq. (13.86), or numerical, by integrating step-by-step.

Modal analysis has a much wider applicability and higher importance than it could be concluded from the solution scheme sketched above. It can, for instance, be employed to analyse such essential properties of numerical algorithms as stability or consistency.

In the case of a nonlinear nonstationary heat conduction problem we have, cf. Eq. (11.11),

$$\mathbf{C}(\mathbf{q})\dot{\mathbf{q}} + \mathbf{K}(\mathbf{q})\mathbf{q} = \mathbf{Q}(\tau, \mathbf{q}), \tag{13.107}$$

$$\mathbf{q}(0) = \hat{\mathbf{q}}^0\,. \tag{13.108}$$

Assuming that, at a typical time instant $\tau = t$, the solution $\mathbf{q}^t$ is known and so are $\mathbf{C}^t$, $\mathbf{K}^t$ and $\mathbf{Q}^t$, we can write down the following system of equations at $t{+}\Delta t$:

$$\mathbf{C}^{t+\Delta t}\,\dot{\mathbf{q}}^{t+\Delta t} + \mathbf{K}^{t+\Delta t}\,\mathbf{q}^{t+\Delta t} = \mathbf{Q}^{t+\Delta t}. \tag{13.109}$$

The system is obviously nonlinear and iterative techniques are necessary to solve it. The simplest of them can be described as follows.

Denote by $\mathbf{q}^{(i)}$ and $\dot{\mathbf{q}}^{(i)}$ approximate values of the unknown vector and its rate, respectively, computed at the ith iteration. Besides, denote by

$$\mathbf{C}^{(i)} = \mathbf{C}(\mathbf{q}^{(i)}), \qquad \mathbf{K}^{(i)} = \mathbf{K}(\mathbf{q}^{(i)}), \qquad \mathbf{Q}^{(i)} = \mathbf{Q}(t{+}\Delta t, \mathbf{q}^{(i)}), \tag{13.110}$$

the approximate matrices and the right-hand side vector at this iteration. With this notation, let us modify the scheme (13.92) known from our discussion of the linear problem. The following system of equations is solved at the next, $(i{+}1)$st iteration:

$$\left(\mathbf{C}^{(i)} + \theta\Delta t\mathbf{K}^{(i)}\right)\dot{\mathbf{q}}^{(i+1)} = \mathbf{Q}^{(i)} - \mathbf{K}^{(i)}\left[\mathbf{q}^t + (1-\theta)\Delta t\dot{\mathbf{q}}^t\right]. \tag{13.111}$$

We then compute

$$\mathbf{q}^{(i+1)} = \mathbf{q}^t + \Delta t \left[(1-\theta)\dot{\mathbf{q}}^t + \theta \dot{\mathbf{q}}^{(i+1)} \right]. \tag{13.112}$$

At the first iteration we can assume $\mathbf{q}^{(0)} = \mathbf{q}^t$, for instance. The condition to halt iterations may have the form

$$|\mathbf{q}^{(i+1)} - \mathbf{q}^{(i)}| < \varepsilon,$$

where ε is the required accuracy of the scheme.

The above simple iteration scheme is usually inefficient and rarely used in practice. Instead, various versions of the Newton–Raphson scheme discussed in Sect. 13.4 are rather used.

13.6 Solution Methods for Linear and Nonlinear Systems of Second Order Ordinary Differential Equations

Let us start our considerations from the linear equation of motion in the form[4]

$$\mathbf{M}\ddot{\mathbf{q}} + \mathbf{C}\dot{\mathbf{q}} + \mathbf{K}\mathbf{q} = \mathbf{Q} \tag{13.113}$$

with the initial conditions

$$\mathbf{q}(0) = \hat{\mathbf{q}}^0, \qquad \dot{\mathbf{q}}(0) = \hat{\dot{\mathbf{q}}}^0, \tag{13.114}$$

where $\mathbf{Q}$ is a known vector function of time. Defining a new variable in the form of a $2N$-dimensional vector

$$\mathbf{g} = \begin{bmatrix} \mathbf{q} \\ \dot{\mathbf{q}} \end{bmatrix}, \tag{13.115}$$

one can write down the initial value problem (13.113) and (13.114) in the following equivalent form,

$$\dot{\mathbf{g}} + \mathbf{A}\mathbf{g} = \mathbf{B}, \tag{13.116}$$

$$\mathbf{g}(0) = \begin{bmatrix} \hat{\mathbf{q}} \\ \hat{\dot{\mathbf{q}}} \end{bmatrix}, \tag{13.117}$$

[4]Discrete equations of dynamics derived in Chap. 11 had a simple form which did not include the term with $\dot{\mathbf{q}}$. In computational practice, this term is frequently added for physical reasons (the matrix $\mathbf{C}$ describes damping in the system) as well as for numerical reasons—to assure a better stability of solution algorithms.

where

$$\mathbf{A}_{2N\times 2N} = \begin{bmatrix} \mathbf{0} & -\mathbf{I} \\ \mathbf{M}^{-1} & \mathbf{M}^{-1}\mathbf{C} \end{bmatrix}, \qquad \mathbf{B} = \begin{bmatrix} \mathbf{0} \\ \mathbf{M}^{-1}\mathbf{Q} \end{bmatrix}. \tag{13.118}$$

Equation (13.116) is a linear system of $2N$ ordinary differential equations which may be solved with algorithms discussed in Sect. 13.5. For some reasons, however, it has become a common practice to solve problems of dynamics by directly using Eq. (13.113) and such an approach will be presented below.

Let us start from the modal analysis. The problem of free vibration in a system with no damping is described by the system of equations

$$\mathbf{M}\ddot{\mathbf{q}} + \mathbf{K}\mathbf{q} = \mathbf{Q}. \tag{13.119}$$

Assuming the solution of the system (13.113) in the form

$$\mathbf{q}(\tau) = \boldsymbol{\phi} \sin \omega(\tau), \tag{13.120}$$

where $\boldsymbol{\phi} = \{\phi_\alpha\}$ is a N-dimensional vector and ω is a constant called the angular frequency of vibrations, one obtains

$$-\omega^2 \mathbf{M}\boldsymbol{\phi} + \mathbf{K}\boldsymbol{\phi} = \mathbf{0} \tag{13.121}$$

or

$$(\mathbf{K} - \omega^2 \mathbf{M})\boldsymbol{\phi} = 0. \tag{13.122}$$

Equation (13.122) is a generalized eigenproblem. In the considered case it has N real solutions in the form of pairs eigenvalue—eigenvector, $(\omega_1^2, \boldsymbol{\phi}_1)$, $(\omega_2^2, \boldsymbol{\phi}_2)$, …, $(\omega_N^2, \boldsymbol{\phi}_N)$, fulfilling the **M**-orthonormality conditions:

$$\boldsymbol{\phi}_\alpha^{\mathrm{T}} \mathbf{M} \boldsymbol{\phi}_\beta = \delta_{\alpha\beta}, \quad \alpha, \beta = 1, 2, \ldots, N. \tag{13.123}$$

The vector $\boldsymbol{\phi}_\alpha = \{\phi_{\alpha 1}, \phi_{\alpha 2}, \ldots \phi_{\alpha N}\}$ is called the αth free vibration form while ω_α its corresponding angular frequency.

Without limiting generality of the discussion, one can assume $0 \le \omega_1^2 < \omega_2^2 \le \cdots \le \omega_N^2$. Defining the $N \times N$ matrices

$$\boldsymbol{\Omega}^2_{M\times N} = \begin{bmatrix} \omega_1^2 & & & \\ & \omega_2^2 & & \\ & & \ddots & \\ & & & \omega_N^2 \end{bmatrix},$$

$$\tilde{\boldsymbol{\Phi}}_{N\times N} = \begin{bmatrix} \boldsymbol{\phi}_1 & \boldsymbol{\phi}_2 & \dots & \boldsymbol{\phi}_N \end{bmatrix} = \begin{bmatrix} \phi_{11} & \phi_{21} & \dots & \phi_{N1} \\ \phi_{12} & \phi_{22} & \dots & \phi_{N2} \\ \vdots & \vdots & \ddots & \vdots \\ \phi_{1N} & \phi_{2N} & \dots & \phi_{NN} \end{bmatrix},$$

one may write down Eq. (13.122) as

$$\mathbf{K}\tilde{\boldsymbol{\Phi}} = \boldsymbol{\Omega}^2\mathbf{M}\tilde{\boldsymbol{\Phi}}, \tag{13.124}$$

while Eq. (13.123) as

$$\tilde{\boldsymbol{\Phi}}^{\mathrm{T}}\mathbf{M}\tilde{\boldsymbol{\Phi}} = \mathbf{I}. \tag{13.125}$$

The latter equation implies the relation

$$\tilde{\boldsymbol{\Phi}}^{\mathrm{T}}\mathbf{K}\tilde{\boldsymbol{\Phi}} = \boldsymbol{\Omega}^2. \tag{13.126}$$

Denoting by $\mathbf{x}_{N\times 1}(\tau)$ and $\mathbf{X}_{N\times 1}(\tau)$ vectors fulfilling the conditions

$$\mathbf{q}(\tau) = \tilde{\boldsymbol{\Phi}}\mathbf{x}(\tau), \qquad \mathbf{X}(\tau) = \tilde{\boldsymbol{\Phi}}^{\mathrm{T}}\mathbf{Q}(\tau), \tag{13.127}$$

respectively, substituting them into Eq. (13.119) and finally left-multiplying it by $\tilde{\boldsymbol{\Phi}}^{\mathrm{T}}$, we arrive at

$$\ddot{\mathbf{x}}(\tau) + \boldsymbol{\Omega}^2\mathbf{x}(\tau) = \mathbf{X}(\tau). \tag{13.128}$$

The initial conditions assume the form

$$\mathbf{x}(0) = \tilde{\boldsymbol{\Phi}}^{\mathrm{T}}\mathbf{M}\hat{\mathbf{q}}, \qquad \dot{\mathbf{x}}(0) = \tilde{\boldsymbol{\Phi}}^{\mathrm{T}}\mathbf{M}\hat{\dot{\mathbf{q}}}. \tag{13.129}$$

Analysis of the system (13.128) shows that in the absence of damping the resulting N differential equations are decoupled, i.e. they may be solved independently of each other. The unknown functions $\mathbf{q}(\tau)$ can be then found from Eq. (13.127). We have thus N equations in the form

$$\ddot{\mathbf{x}}_\alpha(\tau) + \omega_\alpha^2 x_\alpha(\tau) = X_\alpha(\tau) \qquad \text{(no summation over } \alpha\text{)}. \tag{13.130}$$

with the initial conditions (13.129). To solve each of the equations, any of known integration methods for second order ordinary differential equations may be employed. Alternatively, one may make use of an analytical solution in the form of the so-called Duhamel integral, i.e.

$$x_\alpha(\tau) = \frac{1}{\omega_\alpha}\int_0^\tau X_\alpha(\tau')\,\sin(\tau - \tau')\mathrm{d}\tau' + A_\alpha \sin\omega_\alpha\tau + B_\alpha \cos\omega_\alpha\tau\,, \tag{13.131}$$

where the constants A_α and B_α should be determined from the initial conditions for each of the vibration form $\alpha = 1, 2 \ldots, N$.

If damping is non-zero, one may benefit from the above transformation, too. The system of equations (13.128) takes then the form

$$\ddot{\mathbf{x}}(\tau) + \tilde{\boldsymbol{\Phi}}^{\mathrm{T}}\mathbf{C}\tilde{\boldsymbol{\Phi}}\dot{\mathbf{x}}(\tau) + \boldsymbol{\Omega}^2\mathbf{x}(\tau) = \mathbf{X}(\tau) \tag{13.132}$$

which, however, cannot be decoupled.

The solution procedure for linear dynamics problems described above is called the *modal superposition method.*

Let us now pass to numerical integration methods for nonlinear second order differential equations. Consider once again the system of differential equations in the form, cf. Eq. (11.31),

$$\mathbf{M}\ddot{\mathbf{q}}^{t+\Delta t} + \mathbf{C}\dot{\mathbf{q}}^{t+\Delta t} + \mathbf{F}^{t+\Delta t}(\mathbf{q}^t, \Delta\mathbf{q}) = \mathbf{Q}^{t+\Delta t}, \tag{13.133}$$

where the damping term with the first derivative $\dot{\mathbf{q}}^{t+\Delta t}$ has been again included.

Due to complexity of various physical processes described by equations of non-linear dynamics, one cannot indicate a single time integration algorithm that might be called optimum. To the contrary, there are several different algorithms that appear appropriate for different particular cases, i.e. for specific forms of Eq. (13.133). The algorithms are in fact associated with methods of direct integration of first order differential equations discussed in the previous section. They can be divided into explicit and implicit ones. Two examples of such algorithms are presented below.

An implicit algorithm, called the Newmark algorithm, is based on the following difference formulae:

$$\begin{aligned} \ddot{\mathbf{q}}^{t+\Delta t} &= \frac{1}{\alpha(\Delta t)^2}\left[\Delta\mathbf{q} - \Delta t\dot{\mathbf{q}}^t - (\Delta t)^2\left(\frac{1}{2} - \alpha\right)\ddot{\mathbf{q}}^t\right], \\ \dot{\mathbf{q}}^{t+\Delta t} &= \frac{\delta}{\alpha\Delta t}\Delta\mathbf{q} + \left(1 - \frac{\delta}{\alpha}\right)\dot{\mathbf{q}}^t + \Delta t\left(1 - \frac{\delta}{2\alpha}\right)\ddot{\mathbf{q}}^t. \end{aligned} \tag{13.134}$$

The parameters α and δ are selected so as to ensure the best properties of the algorithm in terms of its accuracy and stability. Initially, the algorithm has been proposed for linear problems with the parameter values $\delta = 0.5$ and $\alpha = 0.25$.

The key feature of the expressions (13.134) is that both the acceleration $\ddot{\mathbf{q}}$ and rate $\dot{\mathbf{q}}$ at the time instant $t+\Delta t$ depend only on the value of $\Delta\mathbf{q}$ and known values of $\mathbf{q}$, $\dot{\mathbf{q}}$ and $\ddot{\mathbf{q}}$ at the time instant t. Substituting Eq. (13.134) into Eq. (13.133) one obtains the system of equations

$$\left[\frac{1}{(\Delta t)^2\alpha}\mathbf{M}+\frac{\delta}{\alpha\Delta t}\mathbf{C}\right]\Delta\mathbf{q}+\mathbf{F}^{t+\Delta t}(\mathbf{q}^t,\Delta\mathbf{q})$$
$$=\mathbf{Q}^{t+\Delta t}+\left[\frac{1}{\alpha\Delta t}\dot{\mathbf{q}}^t+\left(\frac{1}{2\alpha}-1\right)\ddot{\mathbf{q}}^t\right]\mathbf{M}$$
$$+\left[\left(\frac{\delta}{\alpha}-1\right)\dot{\mathbf{q}}^t+\Delta t\left(\frac{\delta}{2\alpha}-1\right)\ddot{\mathbf{q}}^t\right]\mathbf{C}, \tag{13.135}$$

whose left-hand side depends nonlinearly on $\Delta\mathbf{q}$ while its right-hand side is known. This can be written briefly as

$$\bar{\mathbf{F}}(\mathbf{q}^t,\Delta\mathbf{q})=\bar{\mathbf{Q}}, \tag{13.136}$$

where $\bar{\mathbf{F}}$ and $\bar{\mathbf{Q}}$ denote the left- and right-hand side terms of Eq. (13.135), respectively. The system is nonlinear and thus the application of an iterative approach is necessary. Since the form of Eq. (13.136) is analogous to the form of the nonlinear statics equation, cf. Eq. (13.70), one of the previously presented iterative methods may be employed here.

An example of an explicit algorithm is the one based on the following difference formulae:

$$\dot{\mathbf{q}}^{t+\Delta t}=\dot{\mathbf{q}}^t+\Delta t\ddot{\mathbf{q}}^{t+\Delta t}, \qquad \mathbf{q}^{t+\Delta t}=\mathbf{q}^t+\Delta t\dot{\mathbf{q}}^{t+\Delta t}. \tag{13.137}$$

The right-hand sides of the two above equations contain quantities known at the time instant t and the unknown acceleration $\ddot{\mathbf{q}}^{t+\Delta t}$ (the rate $\dot{\mathbf{q}}^{t+\Delta t}$ on the right-hand side of the second equation is determined from the first one). Substituting Eq. (13.137) into Eq. (13.133), one obtains the system of equations

$$\mathbf{M}\ddot{\mathbf{q}}^{t+\Delta t}=\mathbf{Q}^{t+\Delta t}-\mathbf{F}^{t+\Delta t}(\mathbf{q}^t,\Delta\mathbf{q})-\mathbf{C}\dot{\mathbf{q}}^{t+\Delta t} \tag{13.138}$$

from which the unknown vector $\ddot{\mathbf{q}}^{t+\Delta t}$ can be computed.

An important feature of this system is its linearity—note that neither the coefficient matrix $\mathbf{M}$ nor any of the right-hand side terms depend on the solution $\ddot{\mathbf{q}}^{t+\Delta t}$. Moreover, in several cases, the matrix $\mathbf{M}$ can be approximated with sufficient accuracy by a diagonal matrix which makes the solution procedure of the system trivial. There is a price to pay for these advantages, though. Contrary to the implicit Newmark algorithm, the presented explicit scheme is only conditionally stable. The issue of stability of numerical algorithms is not discussed here in detail—let us only mention that stability of this particular algorithm is conditioned by the time step length Δt. In other words, time steps in the explicit scheme must be sufficiently small which may significantly increase the computation time in many cases. These problems may be avoided by using the Newmark algorithm, at least for the values of parameters $\alpha,\ \delta$ given above.

Index

M. Kleiber and P. Kowalczyk, *Introduction to Nonlinear Thermomechanics of Solids*, Lecture Notes on Numerical Methods in Engineering and Sciences,
DOI 10.1007/978-3-319-33455-4

Printed in the United States
By Bookmasters